MOUNTAIN TIMBER

The Comox Logging Company in the Vancouver Island Mountains

To the Curtis family children –
of Jim + Elizabeth Curtis –

RICHARD SOMERSET MACKIE

from Richard Mackie
December 2009
Comox B.C.

sononis
PRESS

Winlaw, British Columbia

Since *Island Timber* was published in 2000, many people in the logging community of the Comox Valley have, unfortunately, passed on. Their invaluable recollections and photographs of woods work—and of the camps and communities that supported it—contributed greatly to our understanding of the logging history of British Columbia. This book is dedicated to those men and women and to their descendants who still call the valley home.

LIBRARY AND ARCHIVES CANADA CATALOGUING IN PUBLICATION
Mackie, Richard Somerset
Mountain timber : the Comox Logging Company in the Vancouver Island mountains / Richard Somerset Mackie.

Includes bibliographical references and index.
ISBN 978-1-55039-171-8

1. Comox Logging Company—History. 2. Logging--British Columbia—Vancouver Island—History. 3. Forests and forestry—British Columbia—Vancouver Island—History. 4. Vancouver Island (B.C.)—History. I. Title.

SD538.3.C2M332 2009 338.7'63498097112 C2009-904073-5

Sono Nis Press most gratefully acknowledges support for our publishing program provided by the Government of Canada through the Book Publishing Industry Development Program (BPIDP) and the Canada Council for the Arts, and by the Province of British Columbia through the British Columbia Arts Council and the Book Publishing Tax Credit, Ministry of Provincial Revenue.

Cover and interior design: Frances Hunter
Photo reproduction: Len Todd
Cartography: Ole Heggen
Editing: John Eerkes-Medrano and Dawn Loewen
Proofreading: Audrey McClellan

Cover photo: Wilmer Gold photograph of part of the crew of Skidder 3, Comox Lake, 1934. Back row, left to right: Wattie (Walter) Whyte, Bob Phye, Roy Morrison, Ron Beaumont, George Colwell, Bill Stant, Oswald Harmston, Herb Webber. Front row, left to right: Bob McQuillan, Walter Widen (Viden), Oscar Berkenstock. *Wilmer Gold Collection, United Steelworkers Local 1-80, Duncan, BC*

Published by
Sono Nis Press
PO Box 160
Winlaw, BC V0G 2J0

1-800-370-5228 (North America)
1-250-226-0077 (International)
www.sononis.com
books@sononis.com

Printed and bound in Canada by Friesens Printing.

contents

illustrations

Those trees weren't put on that mountain by God to be praised, they were put there to be cut down.

—PHIL GAGLARDI, MINISTER OF PUBLIC WORKS AND TRANSPORT, BRITISH COLUMBIA, 1952–68

preface

When I was growing up in North Saanich, Vancouver Island, my mother, Gillian, an art historian, had a poster on the inside of her kitchen door. It showed an eroding mountainside of west coast logging slash. The caption read, "O pardon me thou bleeding piece of earth that I am meek and gentle with these butchers." The source was Mark Antony's funeral oration in Shakespeare's *Julius Caesar*, act 3, scene 1.

Images like this fuelled my worst suspicions of people who lived north of the Malahat—the high mountain road that serves as the natural divide between Greater Victoria and the northern parts of the island. North of the Malahat, people lived in logging camps and pulp-mill towns, drove pickups and logging trucks, wore checked jackets, smoked Export A Regular, and demolished old forests for a living. All this seemed distinctly remote and otherworldly. Accordingly, I embarked on my history of the Comox Logging & Railway Company with some misgivings, but I had always had good luck writing in the Comox Valley and I welcomed the work.

Soon I was overwhelmed by the quality of the people, their stories, and their photos. I realized that the story of logging was more typical of 20th-century Vancouver Island than was my own experience. I had grown up south of the Malahat in an academic family, while to the north, most communities owed their existence to the removal of the old-growth forests that surrounded them, and the forests provided a natural frontier of life and work that lasted for much of the 20th century.

As my first book unfolded, I saw that the destruction of millions of acres of old-growth on Vancouver Island, which was largely complete by the 1970s, could best be told without impassioned editorializing, though Mark Antony's oration always remained in the back of my mind. Perhaps a meek and gentle response was appropriate after all; the story seemed to suit my descriptive skills as a historian, biographer, and geographer, and the photographs would say much that I could not.

This has been a challenging history to compile. The company was large, the region was vast, and the people were legion. I had intended to write a one-volume history, but with the sheer volume and quality of reminiscences and surviving photographs, it has ballooned to three, of which this book is the second. Big trees, at least for commercial photographers, mean big photographs, and big photographs best suit big books. I could see why Jack Hodgins, whose father and uncles worked for Comox Logging, has written so prodigiously about the Comox Valley. Like the paintings of E.J. Hughes, the history of Vancouver Island combines equal strands of industry, agriculture, forested landscape, and working people, and a large canvas is required to portray it all.

MOUNTAIN TIMBER OWES ITS ORIGIN TO a grant received in December 2001 from the Filberg Heritage Lodge and Park Association of Comox, via the Vancouver Foundation. I must thank the Filberg chairs and directors, especially Alice Bullen and Pauline Thompson, for their initial support and subsequent patience. I am grateful to Mauro Vescera and John Binsted at the Vancouver Foundation. This book is an example of the kind of tree that can, in the right conditions, take root and flourish from a modest outlay of seed money.

From the outside, books, like houses, should be accessible and inviting. The facade shows what the architect or author wants us to see. We do not see the wiring, plumbing, foundations, structural timbers, joists, or rafters. We take these elements for granted and assume their presence. These structural components should be inconspicuous, but the house would crumble without them.

Similarly with a book. My name is on the cover but it is a collaboration by a team of experts who provided vital and sometimes invisible contributions and support. Diane Morriss, the publisher of Sono Nis Press, assembled a fine editorial and design team consisting of John Eerkes-Medrano, Dawn Loewen, Frances Hunter, and Audrey McClellan. The maps and diagrams are the work of Ole Heggen of the Geography Department at the University of Victoria. I am grateful to all of them for their painstaking work and expertise.

For all images supplied by individuals (loggers or their families), photo reproduction is by Len Todd. In his simple darkroom at Qualicum Beach, Len has transformed thousands of forgotten and faded photos into evocative statements of character and interest. Typically, he refuses to take much credit for his peculiar magic by saying, modestly,"Well, it helps when they had a good lens—and a clean one."

Original photographs were the choicest fruit of the many interviews I conducted. They triggered memories of names and places and offered fresh, first-hand reflections on the past. I have embraced Walter Benjamin's observation that "the past does not break down into words, but into images." Indeed, the book is built around the photographs I was shown in interviews and subsequently borrowed. Most are unassuming private snapshots taken by company employees. The 20th century was the era of the cheap box camera with basic lens, shutter, and viewfinder, such as the Kodak Brownie, introduced in 1900, and similar models. Some logging families kept their photos piled in old chocolate boxes, but fortunately many preserved them in photo albums, and some even included original captions.

Conceptually this book was sparked by the creative forays of Harold Innis and Cole Harris into the economic patterns that unite the Canadian past. Their work is referenced below. Local historians have also provided much inspiration, especially Rene (Irene) Harding, who carried Comox Valley history almost single-handedly in the 1960s and 1970s, and also Betty Annand, Harper and Wallace Baikie, Alex Buckham, Helena Cartwright, Eric Duncan, Janette Glover-Geidt, Judy Hagen, Lyn Henderson, Arthur Hilton, Reta Hodgins, Chris Holmes, Ben Hughes, Dick Isenor, Rick James, Ruth Masters, Allan Pritchard, Henry Schulz, Eustace Smith, (Dorothy) Isabelle Stubbs, Jeanette Taylor, and Vida Wellwood. I have relied on their work, much of which appeared in the form of newspaper articles, and I have quietly thanked them many times.

I am grateful to Betty Annand for permission to use extracts from *Voices from Bevan* and to Bruce Baikie for permission to use extracts from Wallace Baikie's *Rolling with the Times*. I have fallen back often on Bob Turner's transport classics, especially *Logging by Rail*. Bob also identified some of the locomotives in the photographs, and Bob Spearing identified some of the photographs of tugboats.

Land of plenty: view west from the Back Road, Comox, *c.* 1960. In the early 20th century Comox Logging moved up from the valley bottom and into the tempting but remote mountain timber around it. For many local men, logging work was a natural response to the arrival of industrial capital on the edges of a farming settlement. This view shows the rich farmland of the lower Tsolum River, cultivated since the 1860s; the city and growing suburbs of Courtenay; Comox Glacier at top left; and the mountainous ridge of Forbidden Plateau at right. The high peak is Mt. Becher (4,540 feet), and the wide valley in front of the glacier holds Comox Lake, where logging started in 1927.—British Columbia Archives, I-21458

Specific research was provided by the staff of many museums and archives, especially Catherine Siba of the Courtenay Museum. It has been a pleasure to work with her over the years. Rosemary Cookson of the Vancouver Public Library and David Mattison of the British Columbia Archives helped with the career and work of photographer Walter Montgomery. Allan Lundgren and Tim McGonigle of United Steelworkers Local 1-80 (previously IWA Canada Local 1-80), of Duncan, B.C., gave complete access to Wilmer Gold's remarkable collection of logging phtographs. John Parminter and Ken Bowen of the Ministry of Forests and Range provided maps, as did Lori Sugden of the map library at the University of Victoria. Gord Fish and Mark Poiré at the British Columbia government's Air Photo Warehouse provided historical aerial photos.

Arlene Halme, Marguerite Holgate, Sharon Keen, Allan Pritchard, Jeanette Taylor, and Leona Taylor sent specific archival references. Frank Davison, Bob Muir, and Jim Rainer lent me their original copies of the *Circle F Newsletter* and *Crown Zellerbach News*. Jenny Clayton, Sherryll Harris, and Wendy Fried helped with newspaper research. Jackie Morris assisted with interviews, and Arlene Halme transcribed Comox Logging's personnel records. Rolf Christensen, Adam Holmgren, Christina Löwenborg, and Cathy Richardson provided Swedish translations. I appreciated the informative websites of the Comox Valley Family Research Group and Piercy's Funeral Home.

Former employees of the Comox Logging Company showed an extraordinary willingness to help resurrect this company in retrospective book form. They knew what it looked like as a living and three-dimensional entity. I have borrowed many collections of books or private papers but must thank especially Rolanda Buck, Marguerite Holgate, Bob Muir, and Les Pidcock. Many people helped with references and it is hard to single them out, but Chris Michel introduced me to the carpenters and experts in demolition timber in the introduction. Steve Barber, Jennifer Barr, Kathryn Bridge, and Nick Russell helped find the plans of the Hoult Horton house, and Gary Chilibeck, the owner, gave me permission to reproduce one of them. Noëlle Clark, Roger Emery, and Madge Herbert gave complete access to Major Harry Steere-Clark's collection of photographs, and Roger scanned them. Gilbert Taillefer of Merville allowed me to photocopy his original album of Steere-Clark photographs.

A complete list of interviews will be found in the bibliography, but the social history chapters (1, 5, 8, 9) benefited from the generous help of Dick Downey, Muriel Holmes, Bob Jennings, Don McIver, Dorothy Tukham, and Jack Ware Jr. Doug Meredith introduced me to Al Knappett, and I must thank those to whom I sent photos repeatedly: Frank Davison, Harry Erickson, Myrtle Heron, Bob Muir, Fred Parkin, and George Yates Jr. The logging chapters (2, 3, 4, 6, 7) also required a lot of help. Bob Muir helped me untangle some rigging, and the assistance of Frank Davison, Buck Pedersen, and Erik Piikkila was invaluable. The Comox Lake chapters (4 through 7) benefited from the comments and assistance of Fred Blackburn, John Holmes, Darrell McQuillan, Laurie Mathers, Fred Parkin, Sam Telosky, Pat Williams, and George Yates Jr.

Since 2000 I have spoken at museums and historical societies at Campbell River, Comox, Courtenay, Cumberland, Duncan, Ladysmith, Nanaimo, Parksville, and Port Alberni. People in the audiences have commented helpfully on my logging slides and sometimes identified people in them. As a result my talks have sometimes resembled exercises in collective research and discovery. I have also spoken on forest history topics at the B.C. Forest Discovery Centre, the Canadian Institute of Forestry, North Island College, the Pacific Forestry Centre, the University of British Columbia, the University of Victoria, and Vancouver Island University. For their invitations I am grateful to Ken Blackburn, Mandy Crocker, Emily Jane Davis, Aimée Greenaway, Deb Griffiths, Linda Hogarth, Krista Kaptein, Barb Lemky, Larry McCann, Shirley McLoughlin, Wendy Masterton, Pam Moughton, Olaf Niemann, Sharon Niscak, Catherine Siba, Brian Titus, Chris Tollefson, Tony Trofymow, Michel Vallée, Kit Willmot, and Graeme Wynn.

In 2003 I presented an early draft of part of the introduction to a writers' group in Victoria at which Katherine Gordon, Sylvia Olsen, and Nikki Tate provided some good suggestions. In 2008 Erik Piikkila kindly read the entire manuscript in draft.

Paul O'Rorke and Doug Watson provided serious computer advice. I also received the assistance of Betty Brooks, Jim and Doreen Rainer, Julian Roberts, John Scull, James Tirrul-Jones, Bill Verchère, and Warrick Whitehead. John Belshaw, Cheryl Fraser, Elizabeth Gorla, Cole Harris, Gretchen Hartley, Jack Hodgins, Mike Kennedy, Tyler McLennan, Dan Marshall, Carol Neufeld, John Salmen, and Julian Wake supported the project in important ways.

Finally, I must thank my wife, Cathy Richardson, for her constant and cheerful support and for making it possible to pursue my work. She and our children—Juliet, Raffie, and Rupert—have acted as a balance wheel to the irregular beat of a freelance writer's life.

Richard Mackie
Cowichan Bay, British Columbia
June 2009

A group poses in the sleet at Anderson's Camp, Union Bay, *c.* 1910. The men in the upper row are standing on the log rollway or loading platform. They are loading the flatcars (behind) with the help of a gin pole (above), donkey engine, and parbuckle slings. Parbuckle loading technology would shortly become obsolete with the advent of highlead skidder logging.—Muriel Holmes

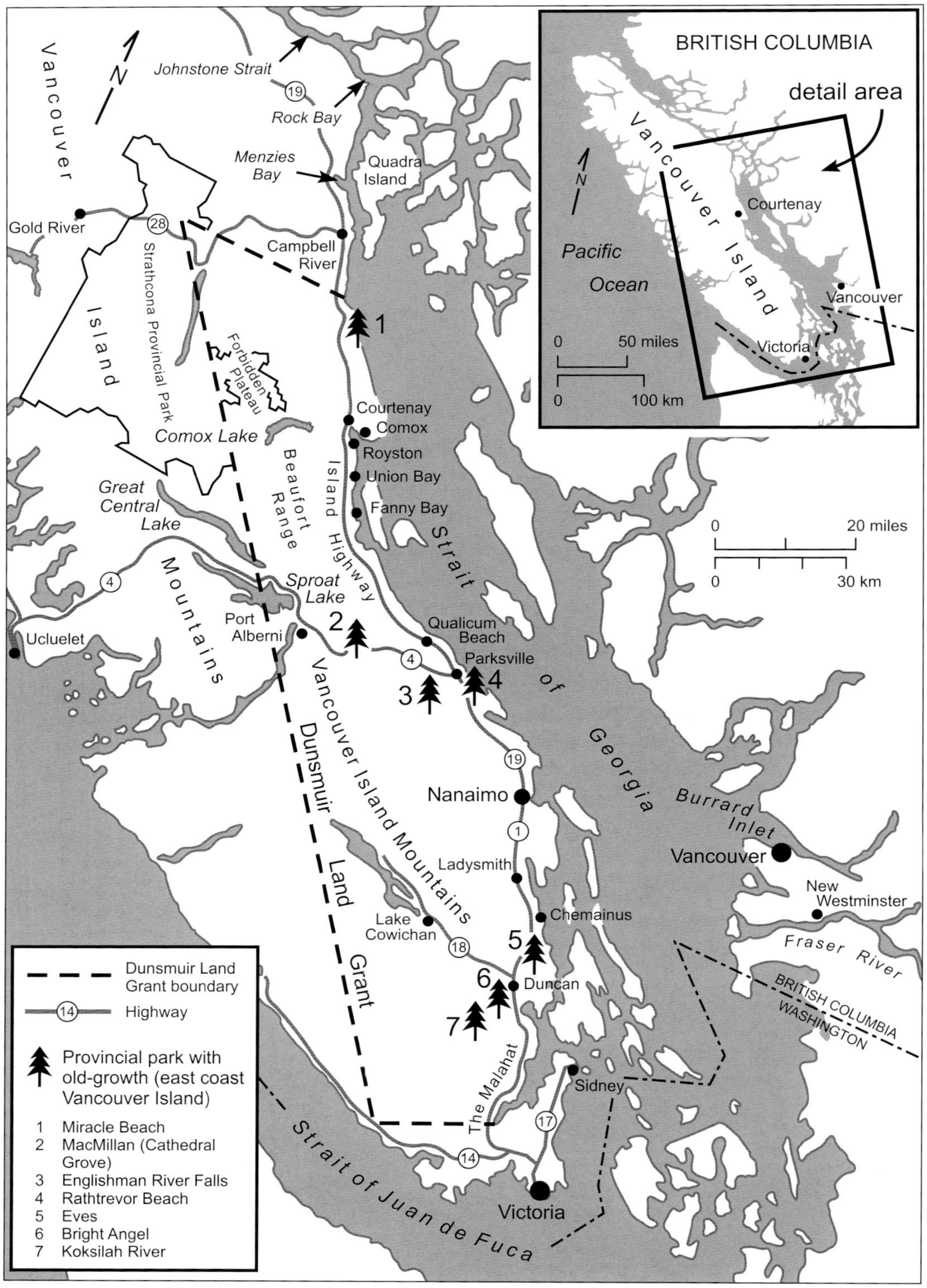

Eastern Vancouver Island, 2009

The first almighty fact about British Columbia is mountains.

—RODERICK HAIG-BROWN, 1959

INTRODUCTION: Vertical Challenges

GO INTO THE BASEMENT OF A HOUSE IN VICTORIA'S FAIRFIELD OR FERNWOOD DISTRICTS, in neighbouring Oak Bay or Vancouver's Kitsilano or Kerrisdale, or in the older parts of Nanaimo, Port Alberni, or Powell River. Many such houses were built in 1912, the peak year of the building boom that marked the premiership of Richard McBride in British Columbia from 1903 to 1915. You will find yourself standing within a dark timber frame of what, in the Victoria building trade, is known as a "1912 house." Above your head you will see floor joists supporting the second floor. Sometimes the original vertical basement timbers are visible. If you measure them, chances are they will be 6 × 8 inches. This timber frame supported the weight of another two storeys as well as the heavy household wares of 1912—the sideboards, tables, cabinets, bookcases, pianos, chests of drawers, wardrobes, and beds.

You will find that all these timber posts, beams, and joists, of all dimensions, are solid and indestructible old-growth Douglas fir, still the tree of choice in the coastal logging and construction industries. The growth rings are close together on this tight-grained old-growth timber, and the posts have hardened with age. You will be told by frustrated homeowners that these posts are so hard you can't pound a nail into them. Holes for anything attached to the timbers—light and electrical fixtures, coat and picture hooks, clotheslines—have to be laboriously predrilled. Tacks and pins will not enter the dense Douglas fir wood, which can have 50 or 60 growth rings per inch.

This Douglas fir timber and lumber has substantial resale value. When a 1912 house is demolished, wreckers sell the fir locally and abroad. Some of it gets "sucked up into the local market," according to Victoria carpenter Chris Michel, but most is shipped to the United States, where it is resawn and resold. According to Bevin and Dale Hodgins, who own a house demolition business in Victoria, the 1912 houses are known for their "rough-cut" structural timbers and lumber. "They'd circular-saw the green wood, and the cut would pull on the sapwood so that the resins moved out. This left a coarse surface known as rough cut," said Dale Hodgins.

According to Bevin Hodgins, the typical length of a basement timber from a 1912 house is 14 or 15 feet. The old Oak Bay Beach Hotel, recently demolished, contained 32-foot timbers; a demolished hotel in Vancouver, according to retired logger Eddy Berkeley, had 80- to 90-foot timbers. Victoria's Mount View High School (built 1930), also recently demolished, contained solid

12 × 12 inch basement timbers. Floor joists are more valuable than the basement timbers because an average house requires far more of them, "and if it's clear wood it's worth twice as much as wood with knots," said builder Peter Murphitt of Cowichan Bay. Joists on the 1912 houses are 2 × 10, 2 × 8, or 2 × 6 inches, usually depending on the floor. The Mount View joists were massive slabs measuring 3 × 15.

Bevin Hodgins estimated that he gets an average of 4,000 board feet of rough-cut two-inch demolition lumber from a 1912 house in Victoria. He sells this, at $1.50 a foot, to a timber merchant in Duluth, Minnesota, named Max Taubert, owner of the Duluth Timber Company. "Anything over two inches brings $1.50 a board foot locally," said Hodgins. Taubert buys over half the rough-cut joists and timbers from the Victoria demolition business, and after importing and resawing it, he sells it for US$7.00 a board foot. He turns it into what is called "boutique wood," as Dale Hodgins explained: "Rough cut is all the rage. They'll take a 2 × 10 and split it in half so that each 1 × 10 retains an aspect of the rough cut, then sand-blast or brusher-wash it to highlight the rough cut. They use these as panelling with the rough cut exposed. In the U.S. market they're lapping the stuff up." The big-dimension rough-cut timbers, the 6 × 8s, go into modified post-and-beam timber-frame houses. These appear to be genuine post-and-beam houses, but a more modern structural framework is hidden behind the walls.

The bigger-dimension timbers also go straight to the U.S. market. "Big timbers salvaged from barns or old sawmills are much superior to the domestic demolition timber," asserted Victoria builder Robin Lindorff. "The timber is bigger and in better condition. You can clean it up easily. It gets sent straight to the U.S. buyers and then shipped all over the U.S." Timber from the Second World War aircraft hangar at Comox and from the Youbou sawmill at Lake Cowichan, both of which were recently dismantled, is now in California.

Clearly, old-growth Douglas fir timber is still in demand long after the original forests were cut.

NOW LEAVE THE BASEMENT and take a drive up Vancouver Island to what remains of the forests in the coastal Douglas fir and coastal western hemlock ecological zones. These zones cover the entire east coast of the island (see map). Protected by the rain shadow of the Olympic and Vancouver Island mountains,[1] the coastal Douglas fir zone is characterized by a dry Mediterranean climate, and Garry oak and arbutus are often found in association with Douglas fir. Although forest of that type once dominated southern Vancouver Island, now only an estimated 2 per cent of it remains. To the west and north of the Douglas fir zone lies the coastal western hemlock zone, which generally extends to wetter regions and higher elevations.

Only a few remnants of original forest survive in either ecological zone between the Malahat Summit, just north of Victoria, and Campbell River. Those remnants can be found in seven provincial parks: Miracle Beach, north of Courtenay; MacMillan Provincial Park (Cathedral Grove), near Coombs; Englishman River Falls, near Errington; Rathtrevor Beach, near Parksville; Eves, near Crofton; Bright Angel, on the Koksilah River near Duncan; and Koksilah River, on the Koksilah west of Shawnigan Lake. Often only a few old trees remain; at Eves Park, for example, they are in a ravine, where they were inaccessible to steam-powered logging equipment. Much of the rest, covering up to two million acres of Vancouver Island, was cut in the heyday of coastal logging, between 1910 and 1950. Most of Vancouver Island consisted of one continuous forest until the 1920s, when 50,000-acre incursions of logging capital, infrastructure, and manpower first interrupted its continuity. Even now, if you choose your course, you can walk from the

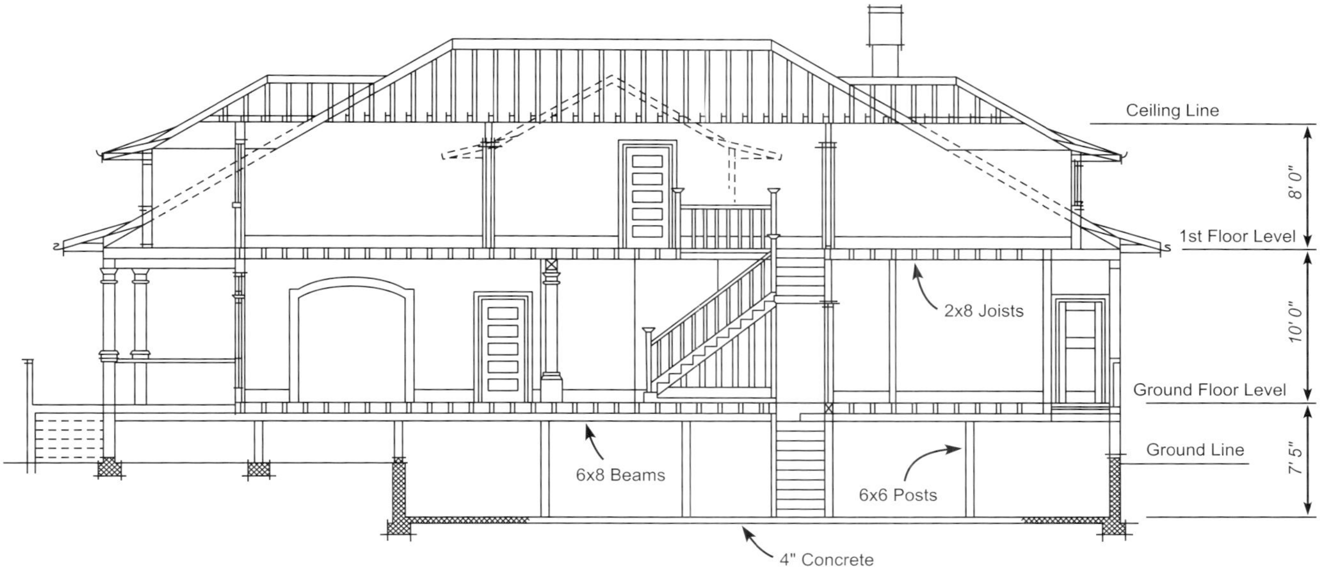

House at 614 Seaforth Street, Victoria West, designed in the colonial bungalow style by Hoult Horton and built by Duncan McBeath in 1911. This cross-section shows the variety of lumber required in house construction in B.C. in the early 20th century, including joists, beams, and 6 x 6 inch timber posts.—REDRAWN FROM ORIGINAL PLANS AT THE BUILDING DEPARTMENT, CITY OF VICTORIA, AND REPRODUCED WITH PERMISSION OF THE OWNER, GARY CHILIBECK

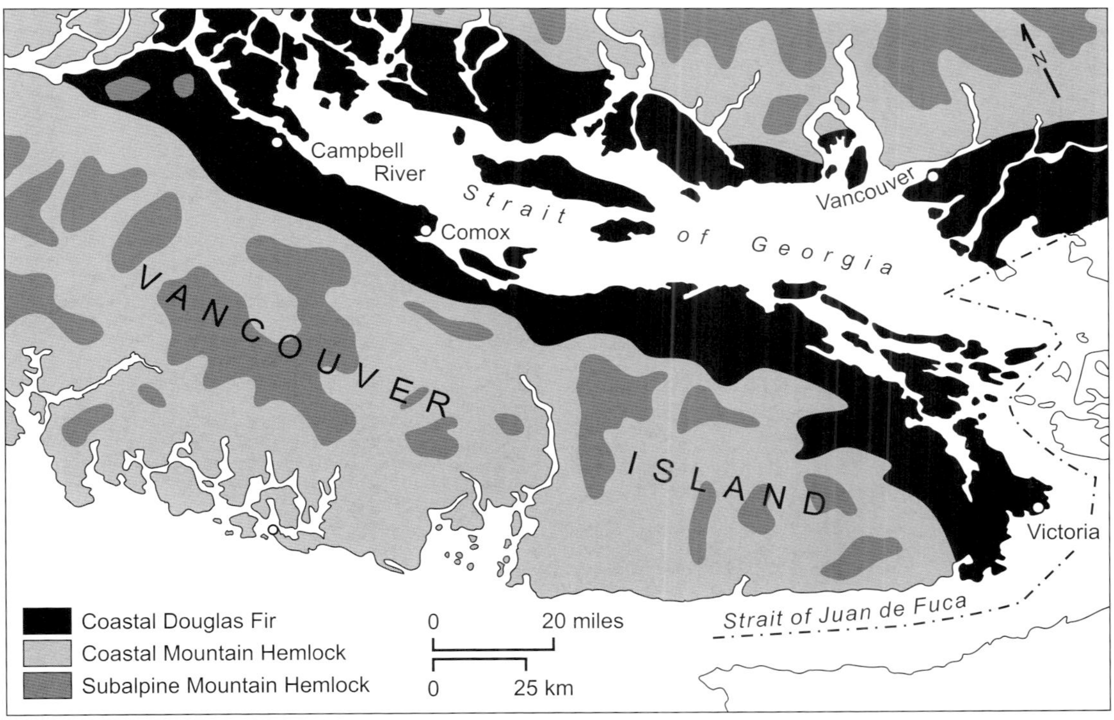

Coastal Douglas fir and coastal western hemlock ecological zones, southern Vancouver Island. Much of the forest in these zones was cut by large companies like Comox Logging in the heyday of coastal logging, c. 1910–50, leaving only scattered remnants in several provincial parks.—REDRAWN FROM FARLEY, *ATLAS OF BRITISH COLUMBIA*

Malahat all the way to Cape Scott, at the Island's northern tip, without seeing many overt signs of human activity apart from 280 miles of logging roads, logging slash, tree farms, and other artifacts of past and future forestry.

The Island Highway (or, south of Nanaimo, the Trans-Canada Highway) generally follows the western edge of the Island's populous eastern coastal strip. From the Malahat to Parksville, the highway skirts suburbs, strip malls, convenience stores, gas stations, motels, and car lots—monuments to the automobile era of the late 20th century. All the big towns—Duncan, Ladysmith, Nanaimo, Parksville, Courtenay, Comox, and Campbell River—are on or near the Strait of Georgia. All are on the low-elevation coastal strip, and most Vancouver Islanders are city dwellers who rarely enter the vast tree farms west of the main highway.

Few public roads connect the east and west coasts of the Island in the 120 miles between Duncan and Campbell River. Even today, the names of the major roads leading westward off the highway are of industrial origin: from the mining era come Mount Sicker Road, Harewood Mines Road, Jingle Pot Road, East Wellington Road; from logging come Bloedel Creek Road, CPFP (Canadian Pacific Forest Products) Road, MacMillan Bloedel Haul Road, Fanny Bay Main, Van West Road, Comox Main Logging Road, Oyster River Main, Duncan Bay Main. These originated as railways or access, tote, or haul roads that led from work camps in the Vancouver Island Mountains to the Strait of Georgia. The names of many defunct companies and camps have been immortalized in this way. These roads ran through the properties of the giant industrial fiefdoms of Canadian Collieries (Dunsmuir) Ltd., Canadian Western Lumber Company Ltd., and many others.

An unknown island—a resource-based hinterland—lies west of the Island Highway. Generally, the Vancouver Island Mountains are inaccessible to the public, and much of the Island west of the highway is still controlled by logging companies. Their large private properties originated in the timber sales of coal magnate Robert Dunsmuir, who in 1884 received the massive Esquimalt & Nanaimo (E&N) Railway land grant of about 3,000 square miles (1,920,000 acres) from the federal government. Dunsmuir and his son James sold off their timberland to finance the Island's railway. With the notable exception of Strathcona Provincial Park, the mountainous interior of the Island is still either owned outright by the big logging companies or leased by them from the provincial government. Gates and restricted access roads still block the routes up to the Nanaimo Lakes, the Nitinat River valley, the entire Cruickshank River system, and most of the Vancouver Island Mountains. On maps these areas are referred to as "Private Managed Forest Lands."

THE EARLY 20TH CENTURY saw the accessible Douglas fir removed from lowland flats and river valleys between the Cowichan Valley and Campbell River. Many of these river valleys—including the Tsolum, Puntledge, and Oyster—extended 20 or 30 miles west and north into the Island's interior. Railway technology was central to this era, and powerful geared locomotives were put to work on the steeper grades everywhere, as historian Robert D. Turner showed in his book *Logging by Rail*. Most of the lowland timber in the Douglas fir and coastal western hemlock ecological zones was removed by logging railways between 1910 and 1950.

The basic pattern was clear to men who spent the early decades of the 20th century logging this level coastal plain. "This was the Garden of Eden for loggers," retired superintendent Bob (Robert) Filberg told journalist Pat Carney in 1967. "Almost solid fir, flat terrain, dense stands, five-foot fir on the stump." When he arrived in the Comox Valley in 1909, Filberg wrote in a 1968 article, "the terrain was mostly valleys and the stand of timber heavy. Less than 1 per cent of the vast stands of

WALTER GAGE PHOTO

In the early 20th century, sawmills of the Lower Mainland quickly converted the level forests of the east coast of Vancouver Island into marketable lumber. Shown here is a portion of the Island Highway between Courtenay and Campbell River, *c.* 1910, in the Oyster River area and within a stone's throw of the Strait of Georgia. About five miles of the highway went through Comox Logging's Block 29, and the timber on this part of the flats was all cut by about 1926.—MUSEUM AT CAMPBELL RIVER, 6950

Timber cruiser Eustace Smith (1876–1964), known to family and friends as Dan Smith, sits on a log with his notebook in hand, *c.* 1925. From his office in the Marine Building in Vancouver, Smith worked extensively for Comox Logging and other logging companies on the Coast and in the Interior. He "cruised" (surveyed) forest land for companies and calculated the species and quantity of timber for them. He was a member of a well-known Comox Valley family that included his brother Cecil "Cougar" Smith and sister Mary Pidcock.—Brian Ross & Diane Fleetham

WALTER GAGE PHOTO

Seeing the forest for the trees. A poster with this large Douglas fir's vital statistics serves as its death sentence, c. 1910. The sign reads, "Douglas Fir. Diameter 6 ft (Breast High). Circumference 18 feet. Total height 220. Contents 10,000 board feet." The Comox Valley presented Comox Logging with a Douglas fir bonanza; over 90 per cent of the original timber north of Courtenay was of that species. This photo was taken on the Campbell River Road around Oyster River, a photographer's favourite area on account of the accessible and continuous old forest. The origins of the sign are unknown.—COURTENAY & DISTRICT MUSEUM & ARCHIVES, D-926

Friends of the Holmes family somewhere in the Comox Valley, *c.* 1910 (photo from the Holmes family album). The fallers are standing on springboards—Douglas fir planks inserted into notches above the swollen and pitch-filled tree butts—and starting on the undercut. An advantage of logging in the Comox Valley was its proximity to established farms and towns, with women and families. Northern coastal camps were quite different in their resolute isolation and maleness. A peavey handle intrudes into the photo at bottom right.—MURIEL HOLMES

Pacific coast timber had been logged. It is small wonder everyone thought then that here, surely, was enough timber to last forever."

Formed in 1909, the Comox Logging & Railway Company ("Comox Logging") spent its initial energies cutting its accessible lowland holdings, which consisted, according to timber cruisers and surveyors, of over 90 per cent Douglas fir.[2] This property (Blocks 24–29) stretched for 30 miles north of Courtenay and included the lower stretches and tributaries of Browns River, Tsolum River, Black Creek, Oyster River, and Quinsam River. After 1910 Comox Logging operated as the logging subsidiary of the Canadian Western Lumber Company,[3] controlled by wealthy land promoters and railway magnates Alexander McRae, Sir William Mackenzie, and Sir Donald Mann. They owned some 60,000 acres of superb Douglas fir timber and possessed the capital required to build logging railways and buy locomotives, logging equipment, and large steam tugs. In 1911 Comox Logging became the first Canadian company to introduce highlead logging, then known as aerial logging, in the form of state-of-the-art, track-mounted Lidgerwood steam skidders. Steam tugs towed booms of logs to Fraser Mills (the name of a Canadian Western subsidiary as well as the associated company town, now part of Port Coquitlam). There it was sawn for export and for shipment by rail to the Prairies, where Canadian Western's lumberyards numbered around 200 in 1912. Fraser Mills had the enormous advantage of access to the Canadian Pacific Railway (CPR) main line, an essential national transportation corridor.

The forest industry, including its main components of logging, towing, and sawmilling, was the largest industry and employer on the British Columbia coast in the 20th century. It supported the whole coastal economy, directly employed tens of thousands of workers, and provided the structural timbers for those 1912 houses. Typically, the large companies formed distinct subsidiaries for all three components: Canadian Western, for example, operated the Comox Logging & Railway Company, the Canadian Tugboat Company, Fraser Mills Ltd., and the 200 lumberyards on the Prairies. Each subsidiary handled logs on a certain step of the journey: Comox Logging took the logs from stump to dump, or "hill to spill"; Canadian Tugboats took them from "spill to mill"; and Fraser Mills turned the logs into lumber, structural timber, plywood, doors, and sashes for Canadian and world markets. The term "vertical integration" is now used to describe companies, like Canadian Western, that not only own a resource but also control subsidiaries that transport, manufacture, and distribute the product.

> "Loading the Logs. Andersons Camp No. 1. B.C." was Walter Gage's caption for this *c.* 1910 photograph, taken at Union Bay. The third logging company in B.C. to have a railway show, P.B. Anderson logged timber owned by the Chemainus Company (Victoria Lumber & Manufacturing Company) at Union Bay. The railway flatcars are at left, and the maker's name is still fresh: "Designed & Built by Seattle Car Mfg Co," a reminder of the American origins of early railway and highlead equipment. A parbuckle system was in use, in which lines were stretched between the donkey engine and a block (pulley) on the gin pole at left. The logs were then dragged down the rollway and onto the flatcars. "They used a little short gin pole and rolled the logs up," recalled Joe Cliffe of Comox Logging in a 1980 interview. "They built all this expensive decking to the height of the car then more or less rolled the logs onto the car." At left are two Chinese workers, one with a bucking saw over his shoulder. Anderson's Camp was in level country about three miles into the forest straight up from the government wharf at Union Bay; it employed 50 or 60 men. Jake Bell (born 1895), who worked at the camp, recalled that Jeremiason was superintendent, Bill Kearns was timekeeper, and Dailey was bull bucker. Anderson's Camp was where 17-year-old Bob Filberg first worked when he came to B.C. in the spring of 1909.
—Babs Hind

The Edge of the World Economy

I have borrowed the evocative phrase "edge of the world economy" from historical geographer Cole Harris, whose ideas, in turn, owe much to the work of the influential Canadian scholar Harold Adams Innis (1894-1952).

Following Innis, Harris argued that, historically, the "staple trades" (or more accurately, resource industries that produced export staples) have acted as "the most vigorous motors of the Canadian economy." Harris wrote about the removal of silver and lead ore from the Selkirk Mountains above Slocan Lake in the West Kootenay, but extractive economies such as mining and logging had much in common. Staple trades were tied by capital, labour, and lines of transport—expensive railway lines—to distant markets.

The hardrock mining companies had to get their ore "from mine to mill to railway," Harris wrote, whereas coastal loggers had to get their logs from stump to dump to sawmill. An ore car at the end of an aerial tramway high in the Selkirk Mountains was not unlike a loaded flatcar at the end of a spur line in the Vancouver Island Mountains. Ultimately, these intricate and expensive local transport systems led from an isolated wilderness, and from transient work camps, to the markets of the world. "There was always the same essential challenge," Harris observed, "to connect isolated resources, tucked in corners of the province, to one of the corridors of modern transportation." *(See Harris, "Industry and the Good Life around Idaho Peak" and "The Struggle with Distance.")*

WALTER GAGE PHOTO

A very early skidder photo, *c.* 1911, printed as a postcard for the skidder-proud public in the Comox Valley. The maker's name, Lidgerwood, is still visible on the frame. "They got the skidders, six of them, which was the biggest accumulation of skidders on the West Coast," recalled high rigger Joe Cliffe in a 1980 interview. "I have been in all the camps in Washington and Oregon, and nobody had that many skidders. Quite a few camps had two or three, but Comox had six. They were twice as expensive as the donkey engines, more complicated, and an awful lot more rigging." Comox Logging would use these machines until 1946. Note that half the crew still have moustaches, a fashion that barely survived the First World War ("*No one* had moustaches on the skidder crews!" remarked Dick Bailey, who started in 1928).—Donna Edwards

An early two-drum donkey engine in the Comox Valley, *c.* 1910. The engine belonged to one of Comox Logging's predecessors. At left are two Chinese woodcutters, and seated at bottom left is George Cliffe (1892–1973), younger brother of Lu (Lucius) and Hughie Cliffe of Comox Logging. George Cliffe had started working as a skid greaser for Howard McFarlan on Denman Island in 1905–6, and in 1907 he joined Fraser River Sawmills (bought by Canadian Western in 1910). Cliffe then qualified as a donkey engineer and spent most of his career as a loading leverman for Comox Logging on Skidder 5 out of Camp 3. He retired in 1958 after a career that spanned the entire steam and railway logging era in the Comox Valley.
—Dorothy Tukham

Steam, steel, and grease: unidentified donkey engineers at Anderson's Camp, Union Bay, *c.* 1910. They are standing on a skid-mounted donkey engine with a vertical boiler, which stands at upper right. Donkey engines dominated Vancouver Island logging between their introduction in the 1890s and Comox Logging's purchase of Lidgerwood skidders in 1911. At right, a stack of wood has been split to feed the boiler.—Muriel Holmes

A Camp 1 skidder at work near Kitty Coleman Road about six miles north of Courtenay, 1914. The point of this photograph was to display the two functions of skidder logging: yarding and loading. At left, the skidder yards a log in by tongs while simultaneously, at right, the skidder loads a log with a single set of tongs. Later, chokers (short loops of line) replaced tongs for yarding, and after about 1925 loading was done with the help of a device called a hayrack boom. "The skidders were built with a fifty foot steel frame with two separate engines on them," said Joe Cliffe in a 1980 interview. "The front winch did the loading and until 1921 they used tongs for pulling the logs to the home tree." The foreman of Camp 1 at this time was Tommy Tame Sr. (1880–1947), an American from Chautauqua, Kansas, who had married Ruth Lever of Comox in 1912.—Babs Pelto

Canadian Western was only one of many big companies on the coast of British Columbia. Between 1900 and 1950, the east coast flats of Vancouver Island between the Cowichan Valley and Johnstone Strait provided the bulk of the timber for the large sawmills on the Lower Mainland. Hastings Mill (British Columbia Mills, Timber & Trading Company) on Burrard Inlet got its timber from Island regions served by its main camp at Rock Bay, on Johnstone Strait. Elk River Timber (known as International Timber Company until 1930), Campbell River Timber, and Bloedel, Stewart & Welch all logged inland from Campbell River, and all three companies dumped their logs at Menzies Bay and towed them to Lower Mainland mills. The Canadian Robert Dollar Company dumped its logs at Union Bay, and Victoria Lumber & Manufacturing Company got the timber for its Chemainus sawmill from the Fanny Bay, Copper Canyon, and Lake Cowichan areas. Industrial Timber Mills and the McDonald Murphy Logging Company were also active at Cowichan Lake. These and many smaller mills also bought timber, as necessary, from smaller producers on the open market.

"Vancouver Island logging drove the B.C. economy," retired forester Darrell McQuillan asserted.[4] "Farming and fishing were small fry in comparison." On the mainland, Vancouver's immediate timberlands were limited. The city was restricted by the rugged Coast Mountains to the north, the American border to the south, the Strait of Georgia to the west, and the heavily farmed Fraser Valley to the east—yet Vancouver Island was an easy tow across the strait.

The early-20th-century logging companies had much in common. Each controlled roughly 30,000 to 80,000 acres of timber, hired between 250 and 450 men, built camps identified by utilitarian numbers, and developed similar local patterns of economic activity based on similar combinations of resources and technology. They coveted the same lowland old-growth Douglas fir, cut their own portions of this level and accessible forest as quickly as possible, used similar railway and skidder technologies to obtain and transport the logs, and provided similar products to the same markets. In 1924 Comox Logging was what the *Comox Argus* called "the biggest logging concern in British Columbia," but this included Fraser Mills Ltd. and Canadian Western's other subsidiary, the Columbia River Lumber Company, in Golden, British Columbia. In 1926 Comox Logging's daily output was 500,000 board feet, and by the late 1940s it sent a million board feet a day to the log dump at Royston from Comox Lake and the forests north of Courtenay.

At its peak in the 1940s Comox Logging produced 200 million board feet a year, a figure which, over 50 years, would have amounted to 10 billion board feet. In fact, the total estimates approach this figure. In 1960 Filberg calculated that Comox Logging had produced 6½ to 7 billion board feet in the Comox Valley since 1909, and in 1985 retired surveyor Wallace Baikie estimated that Comox Logging and its successors had taken between 6 billion and 8 billion board feet from the Comox Valley. These huge figures resist comprehension. To help put them in perspective, a five-room bungalow requires between 5,000 and 6,000 board feet of lumber. So 6 billion feet of timber would yield enough lumber to build at least a million homes—enough to house everyone in British Columbia. And even when houses built in the early 20th century are demolished a century later, the timbers continue to benefit demolition firms, local builders, and the Duluth Timber Company.

MOUNTAIN TIMBER IS CONCERNED with Comox Logging's later and higher fortunes. As the company depleted its supply of coastal Douglas fir in the 1920s, it moved inland to log the Bevan sidehill, the shores of Comox Lake, and the upper tributaries of the Puntledge and Cruickshank rivers. Wherever it moved, the company had the same purpose: to find and cut

Life at the Edge

Coastal Douglas fir forests attracted capital and labour to the province, and logging activity left numerous transient settlements behind. Geographer Cole Harris used the term "work camp" to refer to these settlements which, in their basic form, are as old as Canada itself. Work camps appeared first in the cod fisheries and fur trades of the Atlantic coast and St. Lawrence River in the 16th century and later typified economies as diverse as the fur trade of Hudson Bay in the 18th century, the New Brunswick timber trade of the early 19th century, the Kootenay hardrock mining boom of the late 19th century, and Vancouver Island logging in the 20th century. "Underneath the detail, there are some consistent patterns in this country," asserted Harris. The work camp tradition continues today in the form of the company town or single-resource community.

Typically, resource companies built camps in isolated and unpopulated places and attracted men, and sometimes women, who worked seasonally. Harris noted, "All were connected to distant sources of capital, and by ties of family and culture to distant sources of labour. All were dominated by specialized economies and techniques, and by the transportation problem. All were vulnerable to international fluctuations in price and to local depletion of stocks. Often there was no alternative employment."

These communities tended to have a short shelf life. "Camps," Harris observed, "probably the most common form of settlement in the province, burst into existence, generated other settlements by their momentum, and then flickered and died, leaving a residue of people, settlements, and derelict landscapes."

Comox Logging's story, therefore, has more general connections to enduring traditions of Canadian work and scholarship. The connections between these older traditions and the story of Comox Logging are numerous, and this book will return to them from time to time. *(See Harris, "Industry and the Good Life around Idaho Peak" and "The Struggle with Distance.")*

Highlead logging dwarfed the little gin poles used in parbuckle loading. "It was all aerial logging: no fooling around dragging logs," said Joe Cliffe in a 1980 interview. On this postcard, high rigger Henry Nieminen sits about 140 feet up a head spar near Courtenay, *c.* 1911. The skidder sits on the forest floor at bottom right. Nieminen, a Finn who later changed his name to Henry Norman, had learned to rig on a ship coming from Finland to North America.
—Al Norman

mature Douglas fir forests. Whereas *Island Timber*, the first book in this series, was concerned with Comox Logging's slow but steady removal of timber on the level coastal plain north of Courtenay, *Mountain Timber* follows the company as it turned west, for survival, to untapped sources of wood in the mountains. Faced with the impending exhaustion of its original property, Comox Logging bought new blocks of high-elevation timberland in the 1920s and 1930s from the E&N Railway, the Rockefeller family, and smaller local timber owners. In 1927 the company introduced highlead skidder and railway logging to the mountains west of Courtenay.

Technological challenges accompanied the vertical shift. Geared locomotives ("locies") were required on the hillsides and in the mountain valleys because conventional rod engines were useless on the steep gradients. But even geared locies required expensive and impractical switchbacks on the steep terrain, and to avoid these measures Comox Logging kept its railways on the valley bottoms whenever possible and sent small mobile units known as cold deckers up the hillsides. In this way, the lower reaches of the mountain valleys were logged off, as Darrell McQuillan recalled: "The skidders were simply the spearhead of the company's logging. The best timber was in the valley bottoms, and rail logging limited the opportunities of reaching far up the hillsides. The high and steep timber could only be reached by truck logging, which came later." This book ends in 1946, with the abandonment of skidder and railway logging in the mountains west of Courtenay and the introduction of truck logging, which permitted access to even higher timber.

"STUMP TO DUMP" is an organizing principle that offers perhaps the most natural and convenient way to arrange the surviving photos, reminiscences, and basic contours of logging activity. It recognizes the natural pattern of log production and transport. I devised a "stump to dump" model for *Island Timber* and, finding it useful, continue to apply it here. Each logging chapter is a discrete geographical study of Comox Logging's progress through its forest properties.

In the vast space that was early British Columbia, capitalists faced common problems of distance, isolation, and transport. Much of the history of this province—from the fur trade to the fir trade—involves the acquisition of resources, followed by their sometimes ingenious, if shortsighted, exploitation and rapid exhaustion. And much of its history involves the physical act of transporting resources from trapline to ship's hold, mine shaft to smelter, hill to spill, gillnet to cannery, orchard to packing house, cattle range to stockyard, and then, for all of them, to market. Companies had a constant need to find fresh sources of wealth, to shift locations, to recreate themselves in new settings, to connect resources to markets, and to ensure adequate supplies of labour. Many communities in the province appeared, and disappeared, as a result of these processes.

Mountain Timber follows Comox Logging's flatcars as they make the transit from what geographer Cole Harris called the "edge of the world economy" to the log dump at Royston, a short tow away from Fraser Mills and the markets of the world. But it is Comox Logging's remote

> *Above*: Railways were the conveyor belts that moved the timber riches of the flats from stump to dump. Here, the One-Spot hauls a load of logs, 1912. The One-Spot was a Baldwin, built in Philadelphia in 1909 and for many years Comox Logging's main-line engine. In his book *Vancouver Island Railways*, historian Robert D. Turner called the One-Spot a "classic wood-burning 2-6-2."—VANCOUVER PUBLIC LIBRARY, 6046

> *Below*: Logging photographer Leonard Frank took this photo of a trainload of large old-growth logs being dumped from the trestle at Royston into Comox Harbour, 1912. The vertical boiler at left is the log unloader known as the "humdirgen," a donkey engine grafted onto the chassis of an old locie. The humdirgen has already emptied the two flatcars at left.—VANCOUVER PUBLIC LIBRARY, 6071

LEONARD FRANK PHOTO

LEONARD FRANK PHOTO

workplaces, its employees, and their work that are central to this book. Logging, the engine of the Vancouver Island economy, sustained logging camps as well as the surrounding farming and rural communities. This book focuses on the people who were drawn into the company's orbit. Four chapters (1, 5, 8, and 9), expanding on themes from *Island Timber*, present histories of logging camps and adjacent communities, both passing and permanent, created, sustained, and eventually abandoned by Comox Logging in the first half of the 20th century. The photographs and loggers' recollections in these chapters provide glimpses of the lives of people for whom central Vancouver Island was not an unknown hinterland, but a home. Two of the best-known contemporary Vancouver Islanders, Jack Hodgins (born 1937) and Pamela Anderson (born 1967), come from Comox Logging families, but tens of thousands of other people are also connected historically to various stages in this company's inroads into mountain timber. This is their story. ■

---

1 The name "Vancouver Island Mountains" was rescinded in 1948 and changed to "Vancouver Island Ranges." This book uses the former version to suit the time period covered.

2 These ventures are the subject of my book *Island Timber: A Social History of the Comox Logging Company, Vancouver Island* (Victoria, BC: Sono Nis Press, 2000).

3 Canadian Western retained control of Comox Logging until 1954, when Crown Zellerbach Canada Ltd. bought a controlling interest in Canadian Western. Crown Zellerbach sold its Canadian operations to Fletcher Challenge Ltd. of New Zealand in 1983, which in turn sold its coastal and forestry logging operations, including Comox Logging's original mountain holdings, to TimberWest Forest Ltd. in 1993.

4 Quotations attributed simply to the speaker, without additional details, are derived from interviews conducted by the author (*see bibliography*).

Three of the Hawkins girls of Comox at Elk Falls, *c.* 1928. Elk Falls was a popular place for car excursions and was the effective northern territorial limit of many Comox Valley families.—Myrtle Heron

A postcard dating from Comox Logging's first years of logging by rail, *c.* 1912, with the view through the uncut forest to a wooden railway bridge in the distance. The photo was taken on the main line a few miles north of Courtenay. Logging had not yet started, but this forest had already been visited by timber cruisers, surveyors, fallers, grade crew, steel gang, and bridge crew. The settings had been laid out and the spar trees selected.—Courtenay & District Museum & Archives, P200-1155

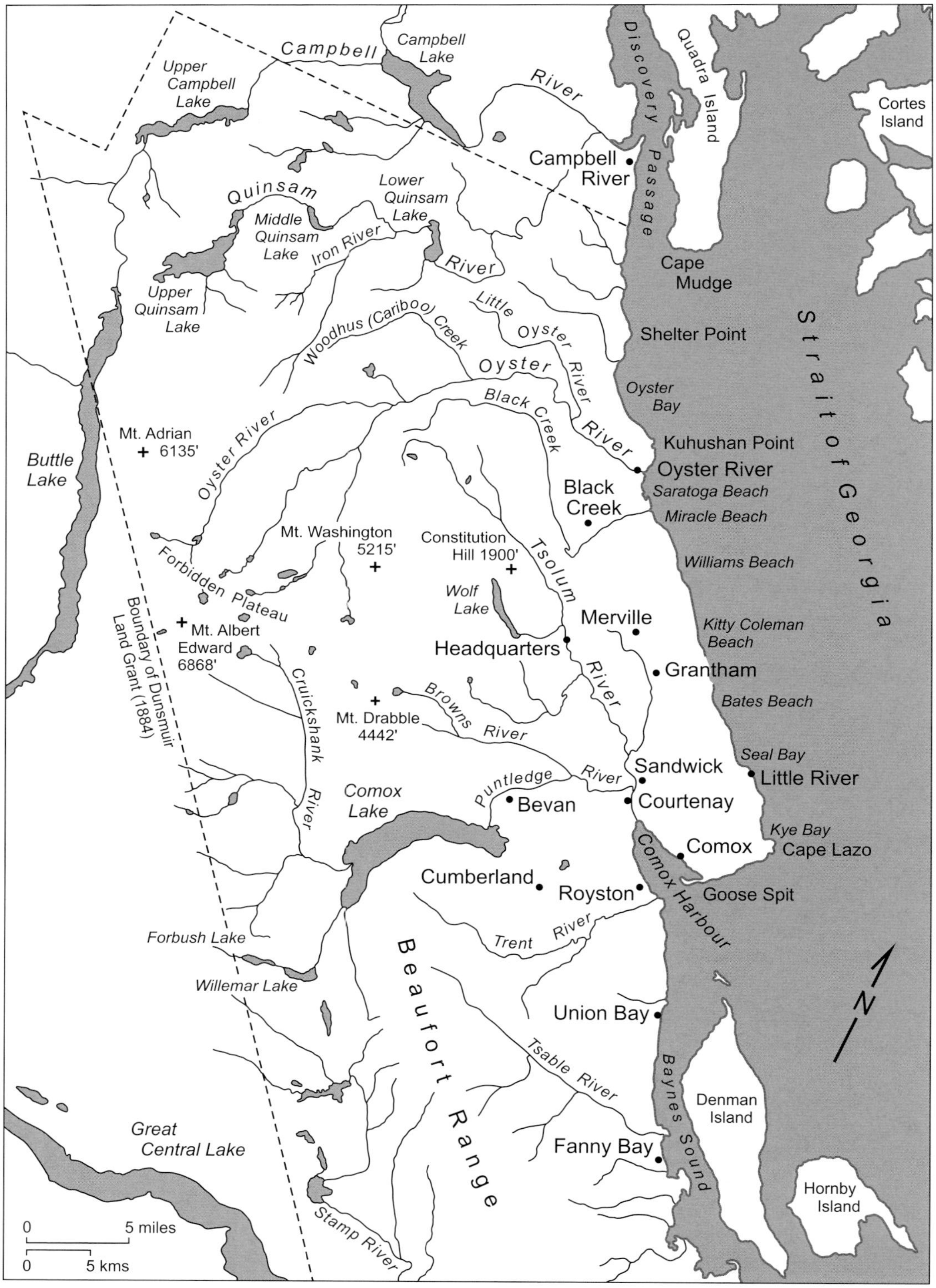

East-central Vancouver Island, 2009

British Columbia is the opposite of an advanced country & for it I am profoundly qualified.

—MALCOLM LOWRY, C. 1940

1

CAMP AND COMMUNITY

THE COMOX VALLEY, THE LARGEST FARMING DISTRICT ON THE coast of British Columbia north of the Fraser Valley, was opened to colonial settlement in 1862. Derived from a Kwakwala word meaning "plenty" or "abundance," the word *Comox* is popularly translated as "Land of Plenty," and early settlers welcomed the bounty of the soil, forest, and sea. The fertile alluvial soil of the lower Tsolum River valley extended for only six miles inland, and this narrow finger of good bottomland was quickly occupied by farming settlers from the British Isles, who within a few years produced oats, flour, potatoes, eggs, and butter for markets on the Strait of Georgia. Staple trades and urban development were not far behind. Three townsites were laid out in the next two decades: Comox, surveyed in 1875; Courtenay, surveyed in 1888; and the company town of Union, renamed Cumberland ten years later and surveyed for coal baron Robert Dunsmuir, also in 1888.

The next two major settlements in the valley were also company towns and major economic motors, and both dated to 1911: Canadian Collieries' town of Bevan, on the Puntledge River; and Canadian Western's town of Headquarters, some ten miles up the Tsolum River from Courtenay. All five communities, with their various service and transport industries, workers, and tradesmen, offered strong local markets for valley farmers. The last notable influx of settlers took place in 1919 at the Soldier Settlement at Merville, where 279 families arrived to farm 14,000 logged-off and agriculturally marginal acres of Comox Logging's Block 29.

The valley's population rose from 1,400 in 1893, to 4,900 in 1919, to about 6,000 in 1930, reflecting the growth of local logging and mining.

By the early 20th century, farming, mining, and logging coexisted, but coal mining, at first dominant, steadily dwindled until the Cumberland mines closed in the 1950s; as a result, logging increased in importance. Logging contributed a monthly payroll of $40,000 in 1913 and almost $100,000 in 1938. "In those days you worked for Comox Logging or you didn't work," recalled Bob Cliffe, though many loggers alternated seasonally between their farms and the forest. Back from the front lines of the skidder settings, Comox Logging's large operation stretched by main line, locomotive, and branch lines directly to the bunkhouses, cookhouses, and shacks of the logging camps and indirectly to the kitchens, living rooms, and bedrooms of hundreds of houses, farms, and stump ranches in the Comox Valley. In 1926 the company employed 400 men, about half of whom were married with families, and as a result at least 1,000 people depended directly on the monthly payroll. About half of the company's employees, and some 50 or 60 employees' families, lived in logging camps; the rest lived in the farms and small towns of the Comox Valley.

From a maximum of nine logging camps north of Courtenay during the First World War, Comox Logging operated only two camps, in addition to Headquarters, its industrial and transport hub and company town, between 1922 and 1926. Camp 1, in its final location behind Miracle Beach, closed for good in 1921. The two survivors were Camp 2, situated near the mouth of Oyster River, and Camp 3, a few miles away at Black Creek. Both camps operated within Comox Logging's main northern timber property, but Camp 2 closed in 1926, leaving Camp 3 to cut the remaining timber on the flats until its closure in 1942. Typically, logging camps remained open for only as long as timber remained available for exploitation. In this they resembled the hardrock mining towns of the Kootenays. "The closer the town to the resource base on which it depended, the more ephemeral it was likely to be," Cole Harris noted in "The Struggle with Distance."

WHEN IT BEGAN OPERATIONS in the Comox Valley, Comox Logging needed hundreds of men right away (600 men in 1911; 700 in 1913), and the problem was where to get them. The valley could not furnish a large and experienced workforce at such short notice. At first, the company did what all logging companies did: it engaged most of its workforce at Vancouver employment agencies known as "hiring halls." These halls, and the transient lives of men hired at them, feature in novels such as Martin Grainger's *Woodsmen*

Old Taffy and the Big Fir

Bob Harding told the story of a Welshman named Paddy or "Taffy" Wakeford, who settled at Little River:

Old Taffy, who was short and fat, came with his wife and a boy and girl. They got off the train in Courtenay and were met by an Empire Settlement fellow with a horse and wagon who took them out to their homestead at Little River. The fellow who picked them up said, "We've got to get your tools," so he picked up a seven-foot bucking saw, a couple of axes, shovels, rakes, and so on. That saw fascinated old Taffy because it was about twice as long as he was!

They had a shack there at Little River ready for them, built of shiplap that you could see out through the cracks of, and he dumped them off there. They had to clear two acres of land within a certain time, but it was covered with huge trees. Taffy said, "We never dreamed anything could be that big."

So with his boy, who was aged about 14 or 15, they sawed away at a tree. This tree was 4 feet at the butt and 150 feet high. Nothing happened, so they went around the tree and sawed at the other side, and then they sawed in a line right around the tree! Then it occurred to them that they didn't know which way it would fall. So they sawed and sawed away at it, but it still wouldn't fall. So they left it and went to bed. Well, that night the wind blew the tree down! Taffy said, "By the grace of God it didn't fall on our house."

They had no idea what they were doing. Old Taffy was absolutely the wrong person for that settlement scheme. But he could sure tell a tale, and if a quarter of it was true, it would be wonderful. His wife, though—she was a hard worker. ■

The Lure of Logging

From the industrial Ruhr Valley, Willie Adamschek and Fritz Passmann abandoned careers in coal mining in Germany and farming in Alberta to join Comox Logging's big skidder crews in the 1920s. Adamschek's niece Eleanor Keller (*née* Richter) described his decision to emigrate:

Uncle Willie Adamschek was the first of seven children in his family to come to Canada. The Adamscheks were coal miners in the Ruhr District. They were of Polish descent; their name had been spelled Adamczyk. The CPR sponsored the Adamscheks. The CPR man said, "You don't need money to go to Canada, but you'll have to work for a year on a farm." They departed from Hamburg, and Willie turned 21 on the Atlantic in July 1925.

They were on a farm at Hillcrest, Alberta, for less than a year; there was not much work in the coal mines of the Crowsnest Pass, so Willie went to check out Vancouver Island. He went to Headquarters and got work the next day at Camp 3. This was 1926, and he became a high rigger.

The Adamschek parents came in 1928. Eventually the whole family—parents and seven children—came from Germany. I have photos of the chickens that the boys bought for their parents when they came. Among the others were Rudy and Max Adamschek and Mrs. Passmann. The Adamschek family farm was on the corner of Howard and Merville roads: they had 60 acres with their first house.

Another Adamschek niece, Emmy Passmann (1929–2009), told a parallel account of her father, Fritz, who worked on the back rigging out of Camp 3:

My dad, Fritz Passmann, was a German from the Ruhr Valley. He was born in 1904. When he was about 14, in World War I, conditions were bad and he had no way to make a trade and there was no work, so he became a miner from the age of 14 or 15 until he was 21. My mum was from the same place and she knew the Passmanns from childhood. Mum's brother Willie Adamschek and my dad were good friends.

In Hamburg my mother's aunt saw a sign that said, "If you come to Canada as a farmer, a farmer will pay your fare." They had friends at Hillcrest, Alberta, so she went back and approached her dad. The result was that Uncle Willie and my dad decided to go to Canada. My dad was not yet 21. He worked for a year on a farm in Alberta, near Wetaskiwin, but the winter was bitterly cold and he was not used to it. They couldn't stand the climate! They were still dressed like the old country. The farmer in Alberta wanted to give him a piece of the property, but he declined because it was too cold. It turned out to be where they struck oil! Anyway, they paid the fare off and Willie came first to Comox, to the bunkhouse at Camp 3, and then Dad came and they met up here, and my mother came soon after. My parents got married in June 1927 in the Catholic Church in Comox. ■

A horse-drawn freight wagon at the top of Comox wharf, *c.* 1910. The driver is probably one of the Creech brothers—Bert, Dick, Harry, and Jim, sons of carpenter Samuel Creech, who came from England in the 1880s. Among the brand names on the wooden crates are Maple Leaf Flour of Winnipeg and Kellogg's Toasted Corn Flakes. The Lorne Hotel is at top right.
—Muriel Holmes

of the West (1908) and Roderick Haig-Brown's *Timber* (1942). Coastal logging literature is rich in images of colourful hiring-hall characters—itinerant men known as "bush apes," who moved between hiring halls, hotels, and bunkhouses. Writer Arthur Mayse, who worked in the camps in the 1930s, summarized the attributes of "your highlead logger" in a 1985 essay:

> He was a high-stepping, free-spending aristocrat among workers. He might, and did, endure with Spartan fortitude working conditions that made his job even more dangerous than its nature dictated. But he would not brook a personal slur or what he considered to be an unwarranted demand or criticism. His answer to such was to throw his gloves away and hit the grade—in other words, quit the job and camp forthwith. He would then enjoy a sojourn to town, unfailingly Vancouver, until he had blown his stake and was ready to sign on again. This he would do through Black's Loggers' Agency, which from its murky, smoky, and perpetually crowded Carrall St. base, dispatched loggers up and down the coast.

Transiency was central to the hiring-hall model. Some loggers spent so little time at any camp that they were known as "camp inspectors." Loggers with nicknames like Eight-Day Wilson, Boxcar Pete, Panicky Bell, Seattle Red, and One-eared Mike appear in the record. "The usual rule," the *Comox Argus* observed in November 1935, "is that the West Coast logger is a bird of passage." The nicknames suggest marginality and an absence of local roots, farms, women, schools, churches—these were yearned for, idealized, and sometimes attained. The camps were isolated, masculine, and temporary. Married quarters were not provided: married men left their wives at home. The coastal logging camp was a 20th-century version of an old Canadian tradition of remote and temporary work camps, a tradition that extended from 16th-century Basque whaling stations on the Labrador coast to late-19th-century Kootenay mines. For hiring-hall men, the past was usually a foreign country, the future was an expectation and an ideal, and the present was a chance to accumulate money on a coastal river valley or inlet.

In the first decade of its existence (1910–20), Comox Logging hired hundreds of men in Vancouver, chartered steamships, and sent them up to Comox. In doing so it followed the classic coastal employment approach. Stories from before about 1920 emphasize the transiency of the loggers, their bunkhouse culture, their dissipated lives, their "stakiness" and periodical binges, and the high likelihood of accident and death in the woods. Men needed little encouragement to quit or to be fired. "In those days," Comox Valley historian Ben Hughes wrote, "it was said that every camp

> Until the Esquimalt & Nanaimo Railway reached Courtenay in 1914, Comox was the main entrance to this largely farming district. The Comox wharf, built in 1874, was for many years the most picturesque and arresting feature of the rustic village. The Comox townsite had been laid out by surveyor George Drabble in 1875 and had gradually filled in since then. Here, an unidentified group waits as the steamer arrives, *c.* 1910. Boxes, barrels, lumber, and provisions await shipment, and a pile of hay for horses has been left on the wharf at right. In the background, at the top of the wharf, are the Elk and Port Augusta hotels; to the left of the wharf is the Robb farm, and in the distance, St. Peter's Anglican Church. This quiet 19th-century settlement would soon be overwhelmed by the capital and personnel of Comox Logging.
—Muriel Holmes

> The Riverside Hotel, the famous loggers' pub on Union Street (now Fifth Street), Courtenay, *c.* 1930. Built in 1889–90 and destroyed by fire in 1969, the Riverside stood in the centre of Courtenay, next to Comox Logging's main line. The car at right belonged to the proprietor of the Riverside, Richard Dixon, more often known as "Pa," "Dad," "Pop," or "Dick." Dixon, an American, had worked for Comox Logging as a donkey engineer before 1914 and later became engineer of the Three-Spot locie, but he left logging in about 1920 to buy the Riverside. As bartender he hired his old brakeman Buster Brown Sr. (J.N. Brown), who "claimed he got more in tips from the sawdust floor than in wages," according to Buster's son Bob Brown.—Harry Erickson

RIVERSIDE CAFE
HOTEL

had three crews, one coming, one going and one on the job. They packed their blankets and clothes ... on their backs."

The initial transiency of Comox Logging's workforce explains the anonymity of men in pre-1920 group photos at Comox Logging camps and skidders. Most of the men in these photos were not local men but birds of passage; they came from the hiring halls.

IN 1921, IN A SUDDEN REVERSAL of policy, Comox Logging started hiring local men. A surveyor's assistant named Wallace Baikie later wrote about this transition in *Rolling with the Times*:

> We were at Headquarters when the loggers came back to work that spring [1921]. They brought them up from Courtenay by train, fed them their lunch and took them on to camp afterwards. There must have been two or three hundred of them who had wintered in Vancouver and come up to Comox by special boat. I don't recall the Company moving the men in that manner after that time. Bob Filberg, about that time, realized (I heard him say it) that the company was going to be there for a long time and he was interested in establishing a permanent crew all down the line and not having to bring in so many of his crew from Vancouver. From that time on local boys had a preference. Bob was always encouraging and advising his men to get a hold of a piece of land and build a home, which a lot of them did. The Soldier Settlement at Merville turned out to be a good source of manpower; these fellows could work their land and also get a job with the Comox. This increased as the families started to grow up and look for work. From that time on the number of local men increased. When the company started logging on the Bevan sidehill and on the Comox Lake they took on quite a few Cumberland boys. Again when the Mennonite settlement came into the Black Creek area, another group of potential loggers became available, especially the younger generation. It was just a few years until all of the personnel of the Comox Logging were local residents.

Although Baikie attributed the local hiring policy to Bob Filberg, the real innovator might well have been Filberg's logging foreman, Jack McQuinn, whose son Howard stated that his father was the first to hire "a local crew of young fellows, but Filberg told him, 'For Christ's sake, leave them alone; they don't know anything!' Filberg preferred to get his help from Black's agency in Vancouver, but it wasn't ideal, so Dad got his own crew, all local people, and he trained them. And then when the camp closed down they just went home."

Jack McQuinn, who had come from a farm on the edge of the forest in the upper Willamette River valley of Oregon, simply legitimized an existing Comox Valley tradition, one that he understood

Travellers and patrons at the Lorne Hotel, Comox, *c.* 1910. Built in about 1880 and named after the Governor General, the Lorne has stood ever since as a hotel and pub. Two owners of the Lorne, the Cliffe and Fairbairn families, would provide sons to the ranks of Comox Logging, including the innovative logging foremen Hughie, Joe, and Harold Cliffe.—MURIEL HOLMES

Jack McKenzie Jr. (1895–1980) with a horse team at the top of Port Augusta Street, Comox, *c.* 1914, near his father's livery stable. To one side an early motor car kicks up a cloud of dust and augurs a new era. The son of Jack McKenzie Sr. and Maud Cliffe, "Big Jack" McKenzie would become a famous high rigger and later foreman with Comox Logging, and his brother, Bill McKenzie, would become a loading leverman on a skidder out of Camp 3. Like many of the Cliffe family, Jack was a capable athlete: in the 1925 loggers' sports at Courtenay, for example, no fewer than eight competitions were won by members of the extended Cliffe family: Edgar, Hughie, Joe, Percy, and Walter Cliffe; Bill and Jack McKenzie; and Andy Radford.
—Wilma McKenzie Collection, Comox Archives & Museum Society, A2000 19.49

Headquarters in the 1920s. Left to right: the Tsolum River, the hotel for single men, car shop, machine shop, roundhouse, and the row of married men's houses. Headquarters houses were painted white with green edging and windowsills, and many residents built extensions, sheds, or lean-tos behind their otherwise identical company houses. In the right distance is the Headquarters School. Note the vagrant cows grazing on the railway tracks in front of the roundhouse. Several Comox Logging employees, including Norman Battersby, Bob "Scratchy" Grant, and John Murphy, let their cows graze at Headquarters.—Frank Biss

Headquarters Townsite

An unknown writer penned the following recollections of Headquarters in about 1960:

In 1911, the Comox Logging & Railway Company planned Headquarters as a centre for their operations and in that year sent workers up and soon there was a street of 19 bungalows, an office, store, warehouse, machine shop and roundhouse for the logging locomotives, donkey engines and rolling stock. Soon families moved in from outlying camps and a fine schoolhouse was built. The Company provided houses for the married workers and their families. These were quite modern for 1911, had indoor plumbing, flush toilets, and hot and cold running water. To accommodate the teacher a tent was erected between No. 4 bungalow and the superintendent's house. Rent for the houses was $10 a month, $12.50 for the larger homes. The rents remained unchanged up to 1958.

The townsite was a thriving community of some 40 homes eventually, and later a hotel and dance-hall were built. Logs travelled by rail from here to the booming grounds near Royston, and then they were towed to sawmills in the Fraser River.

The Headquarters Hotel was built to house the single men. Room and board was charged at rates that varied with the current wages, eg. $1.35 in 1924 when wages were running around $2.00 a day. The hotel was often the scene of a poker game on the Saturday night after payday and continuing on until it was time to work on Monday morning. The hotel was closed down when the depression hit in 1929.

Those wishing to chance their money on a much smaller scale could stroll down to Nakamura's Store. For $.20 you could buy a go at the "punchboard," poke out one of the holes and a tiny piece of paper would tell you your luck—perhaps win a $1, $5 or $20 gold piece. There were lots of the gold pieces around in those days. Nakamura, the Japanese shopkeeper, also ran a taxi service. At first it was by horse-drawn carriage, but shortly after Mr. Filberg, the superintendent, bought his first motorcar, Nakamura had one too—a long touring model—to drive people around in.

In the winter the Tsolum River used to freeze solid and there was skating—into Courtenay or up towards Camp Three. At other times a group would borrow one of the pumpcars and ride the rails into Courtenay for a Saturday evening of rollerskating, saving enough energy, of course, to pump their way home under the stars. (Anon., "Headquarters, The Old Townsite") ■

The office staff at Headquarters, *c.* 1920. Back row, left to right: John McGarry, Bob (Robert) Filberg (logging superintendent), unknown (with dog), John Tukham (paymaster). Front row, left to right: unknown, unknown, Charlie McIver (accountant), Milton Campbell (timekeeper). These men kept the books, kept track of hours and finances, compiled statistics, and wrote and filed letters.—Dorothy Tukham

< Captioned "A Handful" by John or Signe Tukham, this *c.* 1924 photograph shows John Murphy, the postmaster at Headquarters, holding Anna Murtsell, who is holding Fred Tukham (born 1923). Murtsell had recently arrived from Sweden.—Dorothy Tukham

personally. He and his wife, Goldie, and their 11 children lived on a farm in Comox. "Jack knew about farming from his Willamette days," recalled his nephew Bob Hunt. "He had a farm in Comox, on Anderton Road, of 50 or 60 acres. He had a few cows and an orchard and a hell of a big garden." Bruce King remembered that McQuinn planned something similar when he was put in charge of Comox Logging's new show at Ladysmith in 1936:

> Comox Logging was a great old company! They called them the Comox Logging farmers, you know. All those guys in Courtenay owned a farm. When Jack McQuinn first came down to Ladysmith he bought a cow! Of course, Jack had a big family. In Courtenay they could pick up the whole crew locally. They didn't need bunkhouses because they all bought acreages; a lot of them had farms. In peak periods they might hire in Vancouver, but it was a Homeguard company. The advantage was they always had a crew. In bad times with layoffs they could always get them: there was always the basis of a crew there.

Filberg and McQuinn were not alone in their appreciation of the value of local farms and the farming ethos. Malcolm Morrison recalled that he went to ask logging foreman Bob "Scratchy" Grant for a job in the 1940s. Grant answered, "I like to hire kids from farms. They know how to use tools and they know how to think for themselves." Grant himself had a productive farm on the Dyke Road in Comox.

In fact, some local men had been hired by the company from its very beginning in 1910, and also by earlier logging concerns in the Comox Valley in the first decade of the 20th century—Victoria Lumber & Manufacturing Company, Fraser River Sawmills Ltd., and Fraser River Lumber Company. Between 1910 and the early 1920s almost every old farming family in the valley and on nearby islands contributed at least one member to the company's ranks. Among these families were Anderton, Baikie, Beech, Berkeley, Blackburn, Carthew, Carwithen, Cessford, Chalmers, Cliffe, Downey, Duncan, Grant (Scratchy Grant himself had started in 1911), Grieve, Harmston, Hawkins, Janes, Machin, McKenzie, McQuillan, Parkin, Piercy, Pritchard, and Radford. Most of these families had come to the valley in the 1860s or 1870s, and as the 20th century began, they had started to divide their farms among their sons. The company's huge payroll invited a measure of local involvement, and logging furnished plentiful and varied employment.

IN THE EARLY 20TH CENTURY Comox Logging attracted everyone into its sphere, including the grandchildren of labourers

Joe Cliffe (1897–1985) of Courtenay, *c.* 1918. Cliffe started as a faller in the Queen Charlotte Islands and, after army service in the First World War, joined Comox Logging in 1920 and eventually became logging manager of Crown Zellerbach's coastal operations. Cliffe had left school in Grade 3 "because he had to help out at home," recalled his daughter, Myrna McTaggart, "but he educated himself. He had a special interest in geology. He was completely self-taught, had a great influence on his brothers and helped send his younger brother Harold to university." He also taught his brother Roy Cliffe how to box. "Joe was a good fighter," recalled Bill Wilson. "I wouldn't put my dukes up to him." Joe worked as second hooker on a skidder for six months before going head hooking, and in 1921 started a 15-year career as a high rigger north of Courtenay and at Comox Lake before moving to Ladysmith as logging foreman in 1936. He once described himself as a "second-growth Englishman," a phrase that points as much to his English background as to his own adaptation to the forests of Vancouver Island. In his retirement he wrote invaluable logging memoirs (*see bibliography*).—MYRA BAIKIE & NANCY BROWN

Haying on the Robb farm west of the Comox wharf, *c.* 1910, showing the valley's characteristic mixture of farm and forest. Farms in the Comox Valley provided Comox Logging with its Homeguard employees in the first few decades of the 20th century. The Robbs leased this farm to Edmund and Nellie Davis, among whose children were two Comox Logging foremen: Clive "Trapper" and George "Pinky."—Muriel Holmes

Haying at Jack Hawkins's farm above Goose Spit, Comox, 1928. Hawkins (1863–1940), who arrived from Somerset, England, in the 1880s, married Isabelle Graham of Denman Island and had five daughters and two sons, including Jasper, who became a loader on a Comox Logging skidder crew at Camp 3. Three of his sisters married men who worked for the company: Amy married Wallace Piercy, Ella married Jack McKenzie Jr., and Frances married Clifford Janes.—Myrtle Heron

Ward Piercy with rifle, dog, and cougar shot on his farm north of Courtenay, *c.* 1925. Cougars were a major threat to livestock, and the government offered a substantial bounty for them. Son of New Brunswick immigrant and Comox Valley lumberman Harry Piercy, Ward (Alfred Warden) Piercy was born in 1896 and joined Comox Logging when he was 16. He worked as a farmer, blacksmith, and cougar hunter on his family farm, and as locie fireman, donkey engineer, and powderman for Comox Logging, mainly at Comox Lake. He married Nellie Booth from Lancashire, England, and died in 1980, aged 84.—Donna Edwards

A Comox school group, 1920s. The teacher is Rose Baikie, from Denman Island, who later married Bill McKenzie of Camp 3. The dark-haired boy in the middle of the photo is James Owen, son of Comox Logging faller Alexander Owen (Oñ), an Estonian. The girl at right is holding a lunch pail made from a Rogers' Golden Syrup tin.—Victor Owen

The school at Headquarters, early 1920s. The teacher here and at Tsolum for many years was Nellie (Helena) Cartwright, who had originally come to teach at Bowser. On the steamer from Vancouver she met another teacher, who was going to teach at Dashwood. They had never heard of Bowser or Dashwood (both north of Qualicum Beach) and didn't quite know where they were going. "Eventually, the two young women discovered they would be teaching in schools eleven miles apart," wrote Marguerite Holgate in an unpublished biography of Cartwright.—Dorothy Tukham

No frills. Florence McNeil and her children Sarah (later Verchère; 1906–2003) and Johnny (born 1907), at Camp 4, in what is now Merville, *c.* 1912. Comox Logging's transitory camps were captured in photographs and remain in memories. In *"For Our Children,"* Rose McKenzie (*née* Baikie) and Jean Feely (*née* Baikie) described "a typical logging camp bunkhouse" at Camp 3 as "long and narrow so it could be moved from place to place by train if the need existed. The outside was board and batten while the inside consisted of painted shiplap. The heater was located at one end of the building and three loosely fitted windows rattled whenever the wind blew." Married in Revelstoke in 1905, Florence MacDonald and her logger husband, Jack McNeil, were both Nova Scotians. When asked why so many left Nova Scotia, Sarah replied that "it was simply because things were so impoverished back there, particularly in the Antigonish area where Dad came from, that virtually all the young people, male and female, left to seek a better life elsewhere. Others had left ahead of them, and my father had friends and relatives in Vancouver. Besides, there were stories of riches to be found in the Klondike about the time he came west, and he made a disappointing trip to the Yukon before coming to Ladysmith." Two more McNeil children, Madeleine and William, would be born in the Comox Valley in 1913 and 1915.—Bill Verchère

brought to Vancouver Island on immigrant ships by the Hudson's Bay Company (HBC) in the 1850s. The island was colonized in 1849 by the HBC at the request of the British government, which insisted that the company adhere to the principles of "systematic colonization" that had been articulated by British colonial theorist Edward Gibbon Wakefield (1796–1862). Wakefield sought to reproduce in this colony the three main classes of British society: a landowning gentry, a shopkeeping middle class, and a working class of agricultural and industrial labourers. The HBC sent several hundred young labouring families on immigrant ships—such as the *Tory*, *Norman Morison*, and *Princess Royal*—around Cape Horn from Britain to Victoria and Nanaimo between 1849 and 1854.

Edna Marsh (left) and unknown girl, Cumberland, 1920s. Edna later married Hallie Dixon, a Comox Logging leverman, and lived at Comox Lake and in the Queen Charlotte Islands.—Fred Parkin

It is curious to find the influence of Wakefield at remote Comox Valley logging camps in the 1920s and 1930s, but it lingers in several of the valley's older families. Most of the HBC's Wakefieldian immigrants were employed on the farms around Victoria and in the coal mines of Nanaimo. These families are still on Vancouver Island, and descendants of at least four of them appear on the Comox Logging payrolls.

The Parkins of the Comox Valley are descended from Eliza Parkin (*née* Malpass) of Dove Creek, who had arrived in Nanaimo from England as a girl in 1854 on the *Princess Royal*. Her father, John, was a collier at Nanaimo. On her death in 1935, Eliza was survived by ten sons and two daughters. Among her logging grandsons were Alvin, Bill, and Norman Parkin, all of whom worked at Comox Lake in the 1930s. Another Comox Valley logger with HBC origins is George Yates Jr. Born in 1922, Yates is a great-grandson of blacksmith John Parker, from Kent, who arrived on the ship *Norman Morison* in 1853. Another is Neil Martin (1921–2008), whose ancestor Jonathan Martin also came on the *Norman Morison*. The large Stockand family of Cumberland are descended from William Stockand, another labourer who arrived on the *Norman Morison*. Ray and John Stockand both worked for Comox Logging. Other HBC families in the Comox Valley include descendants of Adam Grant Horne, who arrived as a labourer on the *Tory* in 1851, and his wife, Elizabeth Bate, sister of Mark Bate of Nanaimo, who arrived on the *Princess Royal* in 1854. Also in the valley and more or less connected to local logging are descendants of two Saanich pioneers, Henry Simpson (a baker with the HBC) and Duncan Lidgate (a carpenter with the HBC), both of whom arrived on the *Norman Morison* in 1853.

The migration of these families to the Comox Valley from Victoria and Nanaimo reveals a good deal about the underlying social and economic structures of Vancouver Island in the late 19th and early 20th centuries. The staple trades—mining and logging—

> Arrivals from the valley's late-Victorian farming era survived well into the logging era. Here, a group of Comox Valley oldtimers poses at Comox, *c.* 1935. Back row, left to right: Florence Sophia Radford (*née* Cliffe, 1876–1957), Joseph Burnard Holmes (1855–1938), Mary Anne Gage (1865–1956), Jack Hawkins (1863–1940). Front row, left to right: Mary Willemar ("the Archdeacon's wife"), Rev. Jules-Xavier Willemar (died 1949), Mr. Hartshorn, Ellen Holmes (1849–1942), Isabelle Hawkins (*née* Graham, 1871–1939). Hartshorn, whose wife, Maude, took the photo, was a British colonial official who had retired to Comox. Descendants of the Radford, Cliffe, Holmes, and Hawkins families worked for Comox Logging.—Muriel Holmes

The Bush Gentry

Immigrants from the British Isles made up local Anglo elites known elsewhere in British Columbia as the "bush gentry." Not as numerous as the bush gentry in the Cowichan or Okanagan valleys or on the southern Gulf Islands, the Comox Valley's version was a considerable local presence nonetheless. Aloof from local corporate structures, these settlers were quite outside the thread of capital that linked head office, Headquarters, and the camps, and they were unfamiliar with the stump-to-dump conveyer belt. Their connections were imperial, naval, metropolitan, and cosmopolitan. They brought with them what the CPR called "settlers' effects": household wares including furniture, silver, china, and family portraits.

The educated British fell into two broad categories: those with a private (external) income and those who worked locally for a living, often in the white-collar professions. Those with private or retirement incomes included Sidney "Dusty" d'Esterre, Arthur Dundas, Captain Patrick Guthrie, Sir Ernest Petter, John Petter, Commander Geoffrey Spicer-Simpson, and their wives. They tended to gravitate to houses on the Comox waterfront. In Cumberland, the bush gentry was represented by Lady Kathleen Villiers and her husband, Colonel C.W. Villiers, general manager of Canadian Collieries (Dunsmuir) Ltd.

Those from educated middle-class backgrounds who worked for a living included Vincent Bayly, Dr. George Beadnell, Maj. Allan Brooks, Geoffrey Capes, Reginald Carwithen, George Corfield, Winifred Crockett (*née* Gabriel), George Curtis, James Curtis, Rev. John Flinton, Chris Holmes, J.B. Holmes, Ben Hughes, Percy Osler, Theed Pearse, Reginald Pidcock, Ted Sackville, Cecil Smith, Eustace Smith, and Ronald Stewart. These families tended to intermarry. The second generation of these families generally went into the professions, typically teaching, and left the Comox Valley, but several families, notably Carwithen and Pidcock, joined the ranks of Comox Logging; some of their descendants are still in the coastal logging industry. ■

The English Immigrants

The completion of the CPR in 1886 made British Columbia much more accessible, and hundreds of thousands of British settlers poured in, often motivated by broad imperial sentiments or attracted specifically by work in the logging, mining, farming, and fishing industries in the booming young province. In the Edwardian years, B.C. was a favoured destination for prospective British emigrants, as the poet Rudyard Kipling noted in an essay entitled "Mountains and the Pacific": "Were I an intending immigrant I would risk a good deal of discomfort to get on to the land in British Columbia; and were I rich, with no attachments outside England, I would swiftly buy me a farm or a house in that country for the mere joy of it."

A good many British settlers came to the Comox Valley, both in the general course of immigration to Canada and specifically to take advantage of the Soldier Settlement at Merville and the Empire Settlement north of Courtenay. Many of these immigrants eventually joined the ranks of Comox Logging, the motor of the local economy. As well, local men returned from the Great War with British war brides.

By 1921, one of every three people in the province was British born, and most of these were English. English immigrants occupied every level in Comox Logging's structure, from superintendents (Arthur Hilton) and office managers (Chris Holmes), to accountants (Lewis Biss), timekeepers (Harry Steere-Clark), scalers (Alec Tilleard), locie engineers (Len Spencer), carpenters (Walter Rigler), flagmen (Jack Hawthorne), blacksmiths (Courtney Ingram), and fallers (Les Marshall). ■

Joseph Burnard Holmes (1855–1938) of Comox. Educated as a bookkeeper in Lincolnshire, Holmes came to Canada with the Hudson's Bay Company and worked at Kamloops before opening a store in Comox in 1884. Soon afterwards he returned to England and came back to Comox with fiancée Ellen Bubb, a nurse from London. Ellen was five years older than he. In February 1886 the men of Comox staged a famous drunken charivari—a noisy mock serenade—for the couple, probably because of their age disparity. Despite this, Ellen had three children and lived to the age of 94. Known to everyone in Comox as "J.B.," Holmes owned the Port Augusta Hotel and was active in the Anglican Church and Masonic Lodge. His daughter, Nellie, married the Comox game warden and ornithologist Ronald Stewart, and his nephew Tom Holmes worked for Comox Logging in the office at Headquarters.
—Muriel Holmes

The Comox home of merchant J.B. Holmes and his wife, Ellen, *c.* 1940. The house was built by their son Willie (Bill) Holmes, a carpenter and building contractor who was killed in action on the Western Front in April 1917.—Muriel Holmes

Our Kind of People

While most English immigrants readily embraced the novelty of life in the Comox Valley, some lamented the loss of familiar class structures. Pat Williams (*née* Holmes), recalled her mother's stubborn refusal to adapt to the new society. Mary Holmes was a war bride who came to the Merville settlement in 1919, to a homestead that was "an awful mess with stumps." Mary Holmes was close to other English settlers, such as the Sackville and Capes families, but she refused to embrace the community at large, and her children had a much wider circle of friends, as Pat (born 1917) recalled:

My mother tried to bring me up as a pre-First World War English child. When I was two, she said that I had perfect table manners! She associated only with a certain group of people in the Comox Valley. Mother had a very, very limited circle of friends. She would say about certain people or families, "They're not our kind of people." And I would say, "Why not?" Or mother would say, "Oh, that isn't done!" And I would say, "Yes, it is!" or "Everyone else is doing it!" Mother was not in touch with the New World. And she never took out her Canadian papers, so when her British passport ran out we could never take her on any trips anywhere. ■

Rural Transfers

The English novelist Thomas Hardy might have appreciated the story of Jack Ware (1875–1961), as recalled by his son Jack Ware Jr. (1918–2003):

Jack Ware was my dad's name too. A Ware is someone who looks after fish in a stream—it is the same name as Weir. My dad was born in August 1875 at Bidbeare, Winkleigh, near Torrington, Devon. My grandfather, another John Ware, was a farm labourer, and my grandmother, Emily Ellis, was a servant in Winkleigh. They were married in 1873. Emily could sign her name, but John could not.

Dad grew up on a farm. In those days they cut hay with a scythe, but when the first mowing machine arrived, the workers went out and smashed it! Dad used to tell the story of them smashing that first mowing machine because it was going to put them out of work. But he left the farm and joined the army because you got paid in the army. He had to lie about his age. I'm not sure of his army career, but he said once he was lucky not to be shipped to Khartoum, in Egypt, where everyone was massacred under General Gordon. Later he was sent to the Boer War in South Africa. He and his partner were shot off their horses and were rescued by a Dutch boy. This was in 1900 or 1901.

He left the army in about 1902, trained on streetcars, and became a streetcar conductor in Bradford, Yorkshire. Then in 1910 he immigrated to the United States and got a job right away in a stove factory in Kalamazoo, Michigan, and later he ran the interurban streetcar between Detroit and Port Huron, along Lake St. Clair.

American neutrality bothered him so he crossed the border into Ontario in 1916 to join the Canadian Army. He was attached to the 72nd Highlanders. By this time he was over 40 years old, and they wouldn't send him to France because he had been injured in South Africa. He met his wife in England, my mother, Lily Mantle, and she was from Buckinghamshire, and I was born at Littlehampton, Sussex, in 1918. When I was a newborn they were dropping bombs on Devon, and my mother hid me in a hedge. I was torpedoed in 1945, so I can safely say I have been in two world wars!

I came to Canada when I was 18 months old and I've never been back to Littlehampton. We came to Victoria first and then, because Dad had been in the Canadian Army, we moved up to the Soldier Settlement in Merville. My mother was not very pleased when she first saw Merville! But she was sure glad her kids grew up there. Dad ran the farm at Merville, and in the 1920s he also delivered mail between Sandwick and the Headquarters area. He never starved because he'd been raised on a farm. In later years he worked for Comox Logging as sparkcatcher and night watchman, and he died in 1961. My brother and my nephew still have the Soldier Settlement land at Merville. ■

Merville Memories

In 1918 the Land Settlement Board bought thousands of acres of logged-off lands in Block 29 north of Courtenay from Comox Logging's parent company, Canadian Western Lumber. Most of the land was covered with huge stumps and was unsuited to farming, but the Land Settlement Board divided it into 50- or 60-acre farm lots and awarded them to returned veterans, many of whom later quit their marginal stump ranches and went to work for Comox Logging. One of them was Chris (Cedric Chrishop) Holmes, an Englishman who had come to Vancouver Island before the war, and who returned in the summer of 1919 with his war bride, Mary (*née* Osborne), and daughter, Patricia, who was born in England in 1917. Patricia (later Williams) described her arrival in Merville:

My only memory of arriving in Merville is of lots of white canvas and a woman and two girls just older than me waiting outside one tent. They were Helen Capes and her two daughters, and they were welcoming us to the tent which would be our home for the rest of the summer. I knew the Capes girls quite well. That was one of my earliest memories—I was just two at the time so I only have a few rather disjointed ones—and we had moved to the farm on the Headquarters Road before I was four. One of my most vivid is of sharing a blackberry bush with a black bear!

When we were in our mid-teens the Capes girls and I used to go camping up the Forbidden Plateau when very few people did. It was a full day's hike to reach the best camping sites, and the gear was much heavier in those days, but we loved it. Later, when I taught school at Comox Lake, in the summer a few of us would go fishing or swimming and in fine weather one could always go for a walk along the train tracks, but no one had time or energy to climb mountains. All you would have seen would have been more mountains! ■

< The Prince of Wales on his visit to the Merville Soldier Settlement, September 1919. This is the future Edward VIII, the king who would abdicate in 1936. The prince also visited Merville School, St. Joseph's Hospital in Comox, and Lewis Park in Courtenay, where he was greeted by Courtenay's mayor, William Duncan. "Everybody got dressed up for that day," recalled Duncan's daughter Barbara Marriott. "My father agonized over what to wear. Finally he wore a proper suit with tails, but when the prince arrived he was wearing a comfortable tweed suit, and Father felt overdressed. Father had been telling the prince about the salmon in the river, so the prince had him stop the car on the way through the Duncan farm in front of the barns. They got out of the car and stood on the bank of the river and watched the salmon spawning in the river below. The river was just full of fish. The prince picked up a pebble and threw it in the river and the fish scattered."
—Dorothy Tukham

An unidentified girl on the small lawn in front of the Riverside Hotel, Courtenay, 1920s. In the background are the Fifth Street bridge across the Courtenay River and the "B & K" feed store. Victoria-based Brackman & Kerr had feed stores in farming districts across southern B.C.
—Fred Parkin

Coming to Canada

Comox Logging employees came from many backgrounds and brought a number of skills with them. Former Bevan faller Bill (Bror Eugen) Lyttback (1908–2003) recalled his reasons for coming to Canada:

I was born on February 4, 1908, in Kronoby, Finland, and came here in 1926, when I was 18. Kronoby is a farming community in the far north of Finland. It was a Swedish settlement with Swedish schools. In those days we didn't have to learn Finnish. I only have a few words of Finnish. I was raised on a little farm. There were 12 children, and when I was ten my mother died of cancer.

The land was poor. The farms were small, though there were a few big ones. We had horses and cows and sheep and pigs. We raised them and sold them. We grew everything; grew what you had to eat. We had one sawmill in the whole community. We did our own logging and sold our logs to the mill—whatever was not needed—but it was not enough to employ a lot of men. Yes, I did fell trees in Finland. I worked in the woods but it was nothing compared to what's here. I pulled the Swedish fiddle for 35 years, and now I've got long arms!

I decided to leave and I got my passport in three weeks. If I stayed I had to go into the army for 12 or 15 months. There were five of us young fellows from the same place. We took the train to Helsinki and then a boat to Stockholm, and a train from Stockholm to Göteborg, and a boat from Göteborg to Halifax. The ship took seven days—and I had three meals in seven days! I was so seasick. I swore I'd never go back on a boat, and I never have, only on the airplane four times. Then it took another seven or eight days to get to Vancouver on the immigrant train. Quite a bunch of Finns stayed around Port Arthur in Ontario, but some moved here later. Port Arthur was too much like the old country, too cold!

All five of us were from Kronoby and we all came to Vancouver. They were Viking Osterblad, who changed his name to Johnson here; Irska Laderstrom; and the two Sundquist boys. Viking came from the same place in Finland and we worked together here too. His father was a shoemaker in the old town. The Sundquists settled around Vancouver, and Irska Laderstrom went to the States and married a Swedish woman who had a brother there.

All my brothers and sisters are dead now. They sold the farm when the old man died. There was not much to do on the farm. I'm 94 years old now! When I'm 100 I will walk to the Saltair Pub. ■

Bob Larson with his English setters, c. 1933; one dog was named Queenie. Larson was a sawfiler at Camp 2, Quadra Island, Bevan, and Comox Lake.
—Bob Jennings

The Wandering Sawfiler

Born in 1885 to Swedish parents in rural Minnesota, Bob Larson worked as a boy on the family farm and in the woods. As a teenager he left Minnesota in disgrace after siring an illegitimate daughter, whom he never met.

Larson became a typical wandering logger of coastal lore—one for whom Comox Logging would provide a refuge from his past and a welcome escape from the transiency of the hiring halls. He arrived in British Columbia in 1910 but worked on both sides of the border for the first decade. In 1918 he was working for Comox Logging, but he soon left for the Capilano Timber Company and the Mainland Cedar Company of Jervis Inlet. He returned to Comox Logging in 1923 but during the shutdowns stayed in a hotel in Vancouver near the hiring halls. A year later he was back at Camp 2 as a sawfiler. In February 1928, from Camp 2 on Quadra Island, he wrote to his mother and sister in Minnesota:

It is now three weeks since I came back to camp was in Vancouver one month had a hell of time spent 500 boxs and bouth two English pup dogs thies I paid $100 for but they are fine dogs ... I am sure taking great care of my self and I like this country so I think I will stay with it.

Bob. Heriot Bay BC Comox Log.

PS this is all I have time for I cam here on the boat coming and we onlly have one mail day a week.

Larson subsequently worked at Headquarters and Comox Lake as a sawfiler. In the 1920s he was active in loggers' sports and invented a saw guard and a log undercutter. He committed suicide in February 1935 at his home three miles south of Campbell River. "He was comfortably off and there is no reason known why he should have been led to take his life," commented the *Comox Argus.* In 1997 his grandson Bob Jennings of Switz City, Indiana, placed a stone on Larson's unmarked grave at Sandwick Cemetery, Courtenay. ■

Nestor Hagg

(Karl) Nestor Hagg (pronounced "Haig") was a faller and champion bucker for Comox Logging. A Swede-Finn, Hagg (1901–84) was also known as sawfiler for Esther Forsman, ladies' world bucking champion, as his son Roy recalled:

Dad came from Finland in 1923, which was when the big migration was. His father had gone to the U.S. quite a few years before, but he lost him: he was killed and robbed in New York when he was on his way back to Finland! He was going home with the money he had made. He was killed and robbed. So Dad had to look after the place in Finland for a while, but they were looking for people here. At first Dad worked in a steel mill in Hamilton, Ontario; and then he worked his way across the Prairies threshing hay; and then he moved into the Interior of B.C. making railway ties for the CPR; and then he came to the Coast, where he fell trees on Burnaby Mountain. He always said you could see his work from downtown Vancouver!

At Comox, Dad did most of the sawfiling for Esther Forsman, and also filed the competition saws for Allan Woodrow and others. He was sawfiler at Headquarters with Jabe Day, and then he went falling, but he fell at Camp 3 before he became a sawfiler. He had been at ERT [Elk River Timber] *too. He'd worked all around. When Ladysmith opened up, quite a few guys went down there and he went there for a while, but he came back.*

Dad bought a small farm, like most of the Finns. Filberg promoted this to keep the crew happy. Comox Logging always had a stable crew, even in the 1950s when I started. People stayed for years!

Bob Muir (born 1931) remembered one of Nestor Hagg's more spectacular performances:

Nestor Hagg was a pretty tough guy! He was Roy and Runnar's father. I remember once at the Courtenay Hotel, and Ivan Randall, who was Roy Randall's brother, had a bad toothache. Nestor said to him, "Show me which tooth," and Ivan opened his mouth and pointed to it. Nestor was hiding a pair of pliers, and he reached in suddenly and gave the tooth a twist with the pliers, and out she came! It bled a little, but with enough beer he didn't feel it. Nestor was a faller for Comox at this time. ■

HARRY STEERE-CLARK PHOTO

Timekeeper Harry Steere-Clark's caption for this photo was "Happy? You bet." It shows three of Steere-Clark's colleagues at Camp 3 in 1921. Left to right: Ray Dawson, Eugene Croteau, and Reid Good. Dawson and Good were skidder levermen (engineers), and Croteau was then a log scaler. "Ray Dawson was one of the old-time skidder fellows, by golly!" said Davey Janes. "I always thought that Croteau was fired because he made beer on the side," recalled Alan Stewart of Comox, "but this is only hearsay." In 1930 Croteau would build a camp on Forbidden Plateau at what became Croteau Lake. He closed the camp 12 years later because, the *Comox Argus* noted, "he is over 80 and can't do the packing any more."
—Noëlle Clark & Roger Emery

were labour intensive, and the old HBC-era families gradually worked their way northward with the Island's resource frontier to take advantage of work opportunities. Persistence, mobility, and adaptation characterized the second and third generations of Island colonists.

FAMILY CONNECTIONS IN THE COMOX VALLEY were deep and sometimes complex, as these HBC family stories suggest. Already by 1910, two or three generations of mixing had occurred. The valley contained old farming and mining communities with many family interconnections, as Allan Pritchard (born 1928) wrote in a letter in 2000: "When I was growing up in the Comox District one usually became aware when connections existed between families but often did not understand exactly what the connection was. Trying to figure out clans like the Cliffes and the Grants was too much even for natives and the complete despair of newcomers."

Many farming families of the late 19th century adopted and embraced woods work. They had always bridged the farms and the forests. Farming required versatility and an active engagement with the forest, portions of which could easily take up half of the most productive Comox Valley farms. Large agricultural properties, which originated in the late 19th century as 160-acre preemptions, often included considerable stretches of mature forest that, before 1920, many farmers sold to Comox Logging. In turn, the company spent much of its first decade extending spur lines into farms and logging the "farm timber" of the lower Tsolum River valley. Since many farmers were already part-time loggers, working seasonally for the logging company was a continuation of old habits and an opportunity to work locally with family and friends. And abundant local employment meant that local men married at a younger age and stayed around home, which helps explain the remarkable persistence of 19th-century Comox Valley farming names to the present day.

The Comox Valley was not unique in Canada in offering settlers a seasonal blend of farming and logging. Historical geographer Larry McCann identified a similar "alternation of farm and forest" in the Maritime provinces in the early and mid 19th century. Logging offered cash, whereas farming offered subsistence, stability, and domesticity. In other words, woods work on both coasts of Canada was grafted onto an existing agricultural economy without ever quite supplanting it. It may be more than a coincidence that many Comox Valley settlers in the 1870s came from the Maritimes, having logged there too. The Berkeley, Carthew, Grieve, and Piercy families had come from Britain to Nova Scotia or New Brunswick before moving to Vancouver Island.

Oswald William Watson Harmston, *c.* 1910. The grandson of early settlers in the Comox Valley, Harmston (1901–75) started with Comox Logging in about 1920 and worked in the following decade on the flats at Black Creek, Oyster River, and Dove Creek. He became a skidder engineer at Camp 2 in the late 1920s, Cat driver at Comox Lake in the 1930s, foreman at Cruickshanks Camp in the 1940s, and superintendent of Crown Zellerbach's Courtenay operations in the 1950s. Raised on the family farm at Mission Hill, Harmston was known for his athleticism and prodigious strength. This "burly logger," noted the *Crown Zellerbach News* on his retirement in 1960, "could bend a six-inch spike into a horseshoe shape with two hands, then bend the two ends together with one hand." He was an outstanding foot racer at the company's loggers' sports and once climbed a 90-foot rope to ring a cowbell at the top in 33 seconds. He married Alice Phye (1900–79).—GLORIA TWAMLEY

The legendary Jimmy McIvor of Oyster River with an unknown visitor, *c.* 1930. McIvor was a Cape Breton Scot who farmed at the mouth of Oyster River from the 1880s to the 1940s. He figures in the memoirs of many Comox Valley and Campbell River settlers and visiting fishermen and hunters. "His belt was three quarters of an inch thick," recalled Fred Parkin. "It was made of a horse harness. Jimmy McIvor was my grandfather's uncle on the Marsh side. He said he had been shipwrecked at Oyster River. He bought his land for $1 an acre. We used to go up and see him. All he ate was fried flour, salt, and water, except when my grandparents visited. We would take him a sack of flour, a bag of salt, and a needle. He had beautiful animals, and he would sleep with them—he'd sleep with the horses. We went out there when he died. His cabin was a mess and we had a bonfire. He had a box of blasting caps and a pile of mattresses in the corner full of shotgun shells. We auctioned it all off: horses, rifles, furniture. It'd all be he rlooms now. He had dogs and cats. He was a tough old bugger! Old McAlpine, the Campbell River cop, found him, and then his relations came out of the woodwork because they thought he'd buried money. When they were through it looked like they'd been digging fence posts! Maybe old McAlpine got the money. He left no will, so the government took the property, and then Filberg bought it because he thought the airport would go in there, but it didn't. It was a lovely piece of property. We used to go there to pick mushrooms."—Fred Parkin

Drew Berkeley with a bear cub at Camp 3, June 23, 1917. Camp 3 was then situated in its original location west of Dove Creek Road. Bear cubs were popular mascots of Canadian troops in the First World War, and this cub was adopted by the 102nd Battalion, which trained at Comox. The most famous wartime bear mascot was that of the Fort Garry Horse Regiment, "Winnipeg Bear," who was later donated to the London Zoo and who became the inspiration for A.A. Milne's Winnie-the-Pooh stories of the 1920s. Drew Berkeley, born in 1896, was the son of logger John Berkeley, who came from New Brunswick to the Comox Valley in the 1870s. Drew was high rigger on a skidder and also worked on the rigging-up goat. He died of a heart attack in the woods outside Headquarters in June 1942, aged 46, and was survived by his wife, Maude; his twin daughters, Bernice and Beryl; and his son, Bob. Drew's brother Lance was a blacksmith for Comox Logging, and his nephews Jack and Eddy also worked for the company.—Bob Berkeley

An example of an early local hiring is Clifford Janes, who started work in 1917. Comox Logging was then active at Grantham, near Merville, where Janes's father, Fred, had a farm. "I started logging practically on my back doorstep," the *Crown Zellerbach News* quoted Clifford in October 1965. "I was 15 years old at the time and the company was logging on the property that adjoined my father's. I went to see the foreman and was hired as a signalman, and stayed with the company until the time of my retirement." Clifford's three brothers—Howard, Norman, and Davey—also joined the company. These men tended to have logging jobs; that is, they worked on the cold deckers and skidders rather than on the falling crews, grade crews, steel gangs, or boom crews, all of which typically hired men from northern and central Europe. Similarly, Mack (Albert McKay) Grant (1902–2003) started working for Comox Logging in 1927, and his brothers Harold, Arthur "Butch," and Bob "Scratchy" also worked for the company. When asked whether the company hired men from Vancouver, Mack replied, "Not very much. They got most of the men locally. The skidding machines were made up of pretty near all local help." Naturally, men employed by Comox Logging would try to find work for their brothers, brothers-in-law, and cousins, of whom local men had a readier supply than newcomers.

Wallace Baikie's observations about hiring local men apply mainly to the large yarding and loading divisions of the company; hiring-hall men were still obtained after 1921 for several occupations. In June 1969, in the *Crown Zellerbach News*, Harold Cliffe described the changes in logging since he started as a boom man at Royston in 1927:

> When I started out, we used to move logs from the woods to water by rail. It meant lifting and laying about 50 miles of steel a year. The gandy dancers (rail crews) were usually from mid-European countries. The fallers and buckers were Scandinavians; the riggers were Finnish; the cookhouse crews were Chinese; and the yarding men were native-born. Because of language difficulties, the groups didn't mix much. But there was one big thing that brought them all together in the bunkhouses—poker. Cards gave them an international language.

Cliffe's recollection that the yarding crews were "native-born" is supported by the evidence. Skidder and cold-decker photographs show the same local faces year after year. In the 1920s and 1930s it was virtually impossible for outsiders to get work yarding or loading on the cold-decker and skidder crews, which were the bailiwick of men with strong family ties in the Comox Valley. "The best-paying jobs were reserved for the Anglo-Saxons," Don McIver observed.

Horse logging continued for several decades after Comox Logging introduced modern methods and equipment. In 1923 horse logger Clement Burchell logged "Fairholm," Chris Holmes's farm on the Headquarters Road near Courtenay. This photo was taken between Fairholm and Portuguese Creek. Burchell and his team of ten are yarding logs down a skid road to a landing, where they will be loaded onto trucks. Burchell (1885–1956) emigrated from England before the First World War.—John Holmes

Hiring-Hall Hands

In 2007, Erik Piikkila described by e-mail his grandfather Toivo Koski's emigration from Finland and his early work on the British Columbia coast:

In the post First World War era of the Russian Revolution and Finnish Independence, two young men from west central Finland left forests of Scots Pine and Norway Spruce and flat land for the steep hillsides and unbelievable timber of coastal British Columbia.

My grandfather, Toivo Hanhikoski (shortened to Koski in Canada), fresh from serving two years in the Finnish Army, returned home to be told by his father that as the third son and youngest of five children, there was nothing for him at the family farm. This was in addition to his mother's death when he was eight years old and the death of his two older sisters from tuberculosis a few years later.

His distant cousin and childhood friend, Waino Jylha, left due to similar bleak circumstances. Waino was the eldest surviving child of 16 children, where the roost was ruled by an evil step-mother who saved the most and best food for the second brood and gave the dregs to children of the first brood.

The boys left Finland on November 10, 1923 bound for Vancouver, British Columbia. The trip took them about a month, with stops in Germany, Denmark, the Kiel Canal (which separates Germany and Denmark), and Hull and Liverpool, England. The North Sea was quite rough, but the Atlantic was flat calm.

They arrived in Vancouver by train on December 1st, 1923 and started their first logging jobs at Youbou on Cowichan Lake on January 1st, 1924. They had taken the boat from Vancouver to Victoria and a train to Cowichan Lake. They arrived unannounced and without a reference from either of the placement agencies in Vancouver. They had heard from other Finns in Vancouver that the Youbou

The Eight-Spot ploughing snow at Headquarters, *c.* 1935. Built in 1911, this powerful Heisler engine worked at Golden, B.C., until 1929 with Comox Logging's sister company, Columbia River Lumber. —Elwood White

The Two-Spot on a trestle near Headquarters, *c.* 1918. " The men in the Headquarters shop were kept busy keeping the steam locomotives in good repair," an anonymous writer noted in about 1960. "In 1926 there were seven locomotives ... When one of the locies was brought up for a major overhaul, it was up to the slimmer, more agile crew member to squeeze himself through the opening of the dome and down into the boiler itself to back-up the riveter. The job often took two or three hours and left the man's ears ringing."—Elwood White

Comox Logging as a faller in 1933 at Bevan, recalled, "When I started, half the outfit—maybe more—was from Courtenay." The company, according to Buck Pedersen, was known in the 1940s as "a school for the young and a home for the old." Stories abound of men who worked until their mid-70s or later as whistlepunks, bunkmakers, sparkcatchers, watchmen, and hostlers. Bob Harding summarized some of these themes:

> They used to call Comox Logging the "Old Man's Home" because they'd keep them on so long. There was no Canada Pension then, no UIC [unemployment insurance]. It was "We, us, and the company." They were *all* part of the show, even if it's only the guy who greases the skids. They'd ask, "Who do you work for?" and you'd answer, "The Comox Logging," and that answered most of your questions! And because you worked for them it meant you had money; this was the hungry '30s. It got to the point where the grocery store gave Comox Logging employees 10 per cent off. Comox Logging was the only place in the district with anything like a payroll. Cumberland? You might as well be at Cranbrook as Cumberland! It was isolated, and they were miners: a different race.

> The Seven-Spot moving Camp 3 bunkhouses from Black Creek (Schulz's store) to the new location at Northy Lake in 1922. "Besides hauling logs," wrote railway historians Cummings, Oliver, and Anderson, "the railway often hauled construction materials and sometimes a whole camp, cabins and bunkhouses being moved wholesale." These same bunkhouses would be destroyed in the August 1925 Camp 3 fire. The notion of community formation is challenged by the spectacle of logging camps on wheels, but it is worth noting that before vanishing, the camps sent logging money to homes in Comox and Courtenay, and to many farms in the district. Therefore, in a logging context, a more pertinent term than "community formation" might be "community maintenance."—Noëlle Clark & Roger Emery

Men like Clifford Janes and Mack Grant were known as Homeguards. The standard meaning of this term comes from logging poet and engineer Bob Swanson, who in *Bunkhouse Ballads* defined a Homeguard as "a man who works for the same company and never leaves his job." This definition certainly applies. "Fellows spent practically their whole life working for Comox Logging. They treated fellows real good," recalled Bus Griffiths (1913–2006). When asked if he knew the term "Homeguard," Eddy Berkeley (1920–98) answered, "That's what I was! When you went to work for Comox Logging you stayed there. You spent your whole life there." Eva Bailey (*née* Manson) had similar recollections. "They called them Homeguard when they stayed in one place all their life. It wasn't meant as an insult. They knew their cheques were coming in every month, which was a good thing."

> A Homeguard platform. Riggers Drew Berkeley (left) and Norman Janes pose in front of Skidder 5, north of Courtenay, *c.* 1925. To the left is a row of empty flatcars. "Depending on the grade, about seven flatcars were brought into a setting at a time," recalled Walt Edwards. Janes started working for Comox Logging in 1916. "Norman was 16 when he started," said his brother Davey. "He started working right at Tsolum School and he could run home for an apple at lunch. He spiked his age by a couple of years. He was giving hand signals with flags as a whistlepunk." He then quit the logging company to serve in the First World War. In the early 1920s Berkeley and Janes were members of Comox Logging's rigging-up crew, or "flying corps," a mobile crew that went from camp to camp rigging spar trees ahead of the skidder crews, and by the 1930s both Janes and Berkeley were head riggers attached to Camp 3.
—Bob Berkeley

In Comox Logging usage, "Homeguard" took on an additional meaning: it referred to employees with viable family farms or marginal stump ranches, and strong family ties in the Comox Valley. In the 1920s Homeguards were primarily of British descent because at that time almost all the settlers in the Comox Valley were of British ancestry. Longevity in the Comox Valley was a main criterion of Homeguard status. To Homeguards, the land was a retreat, a place of safety from the seasonal rhythms of logging and the absence of welfare programs. Men like head rigger Norman Janes woke up early to feed and milk their cattle and do the farm

HARRY STEERE-CLARK PHOTO

chores before going to work. "He had to get up in the morning at 5 a.m. to milk the cows, and then he'd go logging," recalled Janes's daughter Phyllis Hatfield. Homeguard employees were a natural feature of a company whose territory overlapped with one of the Island's older, stable, rural communities.

By the 1930s the Comox Valley was known for its resident population of farmer-loggers who provided Comox Logging with a stable local workforce. In the late 1940s the company placed an ad in the *Argus* with the following caption:

> One pioneered with the crosscut saw and the axe and the other with the plough and harrow. Gradually as the years went on they merged and became the breed that now live in the valley, the man who works in the woods with the power saw but comes back to his home in Courtenay, and all over the district, or at night to the farm. When snow in the winter closes down the camps and heat in the summer, they go back to the farm.

Stump ranches were vital to this rural economy. Newcomers to the district and younger Homeguard sons with no claim to the family farm bought property of marginal quality on logged-over land for the seasonal shutdowns. The stumps themselves could be removed with blasting powder obtained through a government program or procured illicitly from Comox Logging's grade crew. Hay was grown as feed for livestock, and most stump ranches featured a garden, orchard, and poultry. Stump ranches provided a way of avoiding what John Holmes called the "endless cycle" of feast and famine: "You'd save your money and then spend it when the work ended. You saved a bit, and then it was fire season or snow. If the fire season didn't get you, the snow did." Similarly, Harry Erickson, whose Finnish parents farmed at Merville, observed, "Dad was on a stump ranch. The thing was, there was no UIC, and they said that an extreme winter, a cold winter, would be followed by a dry summer, so you needed to be able to support yourself." Some of the soldier settlers of 1919, unable to make a living on their stump ranches, gravitated into Comox Logging's orbit, while others left the district altogether and sold their tarpaper shacks on the burned-over logging slash of Merville's glacial moraines.

Many people credit Bob Filberg with encouraging employees to buy small farms and stump ranches. "Most of them bought small farms. Filberg promoted this to keep the crew happy," said Roy Hagg. *The Timberman* noted in September 1935, "For those who want their own acres to cultivate, he [Filberg] makes it remarkably easy for them to acquire them," and many have echoed this connection. Historian Ben Hughes wrote that Filberg "encouraged men to take up farms in the Comox Valley on which they could

> Comox Logging trains near Black Creek, c. 1920. At right the Four-Spot waits at a siding with flatcars of logs. At left, the Five-Spot passes with at least 20 men on flatcars. This is the lunch train, as Wallace Baikie wrote in his 1985 memoir: "In those days [1921, at Camp 1] nobody packed a lunch. When noon time came the locomotive with a couple of flat cars would be waiting, pick up the different groups as it came along and wheel us into camp. The dinner bell would be ringing and we would charge in, eat our meal in 15 minutes and the locomotive would be whistling to take us back out by one o'clock. They did that for years." Davey Janes recalled, "At Comox Lake, we had to sit on a flatcar to go to work. There was not even a crummy! In a howling southeaster and in any weather." Janes started working for Comox Logging when he was a boy in 1927, but by the 1930s men were usually transported in converted boxcars known as crummies.
—Donna Edwards

CL & R Co
4

work when they were laid off by a bad market for logs or for fire hazard." Indeed, in March 1943 Filberg himself told *West Coast Lumberman* that "men who work in the woods nowadays have their homes within easy reach of the logging operation. Many of them have families and have a permanent stake in the country where they work, owning their own homes and in some cases small farms."

Proof of the success of the stump ranch model was its appeal to the Swedes, Finns, and other relative newcomers, many of whom had come from countries where farming and logging similarly overlapped or competed seasonally. Land, usually of marginal quality, requiring solid labour and best suited to grazing, could be had for cheap in the Comox Valley. "Here, most loggers had farms or little ranches," recalled Stan Hodgins. "Quite a few Finns and Swedes had farms: two or three of them would buy a farm together, split up the land between them, build a house on each property, get a couple of cows between them, and turn a few acres of land into a garden." Barbara Marriott (*née* Duncan) had similar memories: "It was sometimes referred to as a family company because they all knew each other and most lived locally, and the Swedes and Finns followed suit with the others. The fact that they had land helped stability. Most of the loggers didn't live at Headquarters but on their little farm places, their acreages." Once they had settled down and even married locally, newcomers were sometimes called, paradoxically, "Swedish Homeguards."

The benefits of Homeguarding also accrued to Comox Logging. Hiring locally was cheaper than finding crews in Vancouver—workers who might be green, itinerant, unreliable, or radicalized—and paying agency fees and passage to the Island. Local hiring meant both security and cost savings. "They didn't have to run camps as much," said John Holmes. In this highly seasonal industry, local employees were simply contacted at their farms and stump ranches at the end of the winter and fire seasons. News of the woods' opening was passed quickly by telephone and word of mouth—what Neil Martin called the "Salal Telegraph"—through rural districts. Year after year, Comox Logging got familiar, trained, and reliable crews. "Comox Logging always had a stable crew," said Roy Hagg. Local men also provided a resourceful and versatile workforce. The well-known inventive streak of a family like the Cliffes, for example, was turned to the company's advantage.

These employees, whether they originated locally or came from the hiring halls of Vancouver, whether Homeguards or recent immigrants from Sweden, Finland, or Britain, manned Comox Logging's assault on the timber of the flats north of Courtenay in the 1920s. ■

Bill Lyttback and the Man Catchers

Bill Lyttback arrived in Vancouver in 1926, aged 18, from a farm in Finland. He recalled his first jobs in the woods of coastal British Columbia:

I went to Black's agency on Carrall Street in Vancouver. My first camp was Billy Hanson's camp, which was way up the coast. I took the Union steamboat. I was fireman for three months, then second boom man. I was 16 months at Hanson's; I was 20 years old when I left there and went down to Vancouver. In 1928 it was easy to get work. They even had man catchers on the street trying to get you to work. They'd ask you right on the street.

Then I went to work for Bloedel, Stewart & Welch around Campbell River. They were a big outfit with seven or eight locies at Camps 3 and 4. I worked three or four years at Bloedel. Every July and every Christmas, Bloedel sent a Union boat up for the holidays and brought the whole crew down to Vancouver. I worked for Bloedel till I started to work for Comox in 1933.

[Did Bloedel, Stewart & Welch have a good safety record?] *Jesus Christ, no! One fellow who was second loading got killed in the morning, but they had no time to take him down till the evening, and they just laid his body down. One year just about every day someone was hurt or killed.* ■

Home Truths

Doris Walker (*née* Marvin):
Filberg did something for logging: he instilled in his workers how nice it would be to have a little ranch or chicken farm, because they were shut down in winter for the snow, in summer for fires, and sometimes in between for strikes. No money was coming in at these times. So they listened to Filberg and bought property and did what he suggested. They had chickens and cows, so when the place shut down they had food anyway. So when the logging shut down, everything didn't go to pieces; they had their farm to run and a place for their kids to work. Why they didn't do that years ago I don't know! They started small farms, instead of going out and boozing it all, which loggers did at one time.

Bill Woods:
Comox had pretty good timber, but they shut down for fire season and in winter. You'd just get all your bills paid by the end of June and then it's fire season! You'd just start to get out of the hole when the summer layoff came, and the same when the winter layoff came. Filberg would say to the loggers, "If we can sell our logs, you have a job; if not, no job. If we can't sell our logs you're out of work." Filberg's philosophy was you should have a two- or three-acre plot of land, and some guys had four or five acres. He encouraged the loggers to buy small farms so you'd have some chickens, grow your own veggies, grow apples and potatoes, have a milking cow and a beef cow, get your own meat, so that you can live if there's no work because you never knew what the market would be. ■

Cows at Camp 3, 1930s; the view is west towards Mt. Washington. Unwilling to leave their rural habits behind them and mindful of Depression austerity, at least three Comox Homeguards—Arthur "Butch" Grant, Norman Janes, and Drew Berkeley—kept cows at the camp. Rural expressions filled the conversation of many logging families. A favourite greeting of cold-deck engineer Dave Aitken, according to his nephew Ken Aitken, was "Who left the barn door open?!"—George Yates Jr.

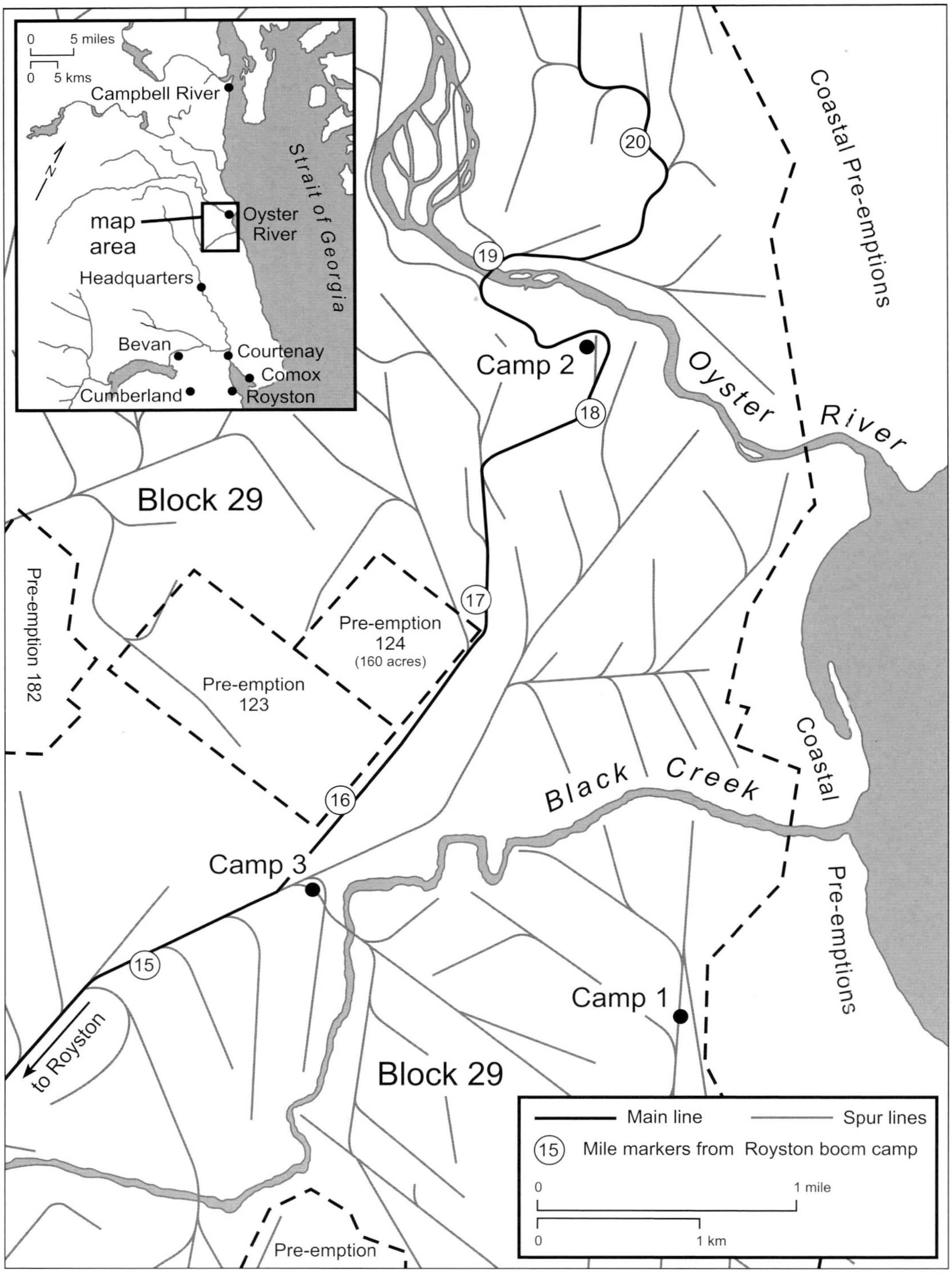

Comox Logging's Camps 1, 2, and 3, *c.* 1921, near the mouths of Black Creek and Oyster River. Major Harry Steere-Clark, timekeeper at Camps 1 and 3 between 1920 and 1925, compiled an impressive photographic record of this forested coastal region. Spur lines protruded everywhere into the uncut forest property of Block 29, and skidder setings were situated at the end of the railway spur lines.

—Redrawn from a map at the TimberWest office, Courtenay

It will not be many years now before the easy logging on the flat land near the coast will be a thing of the past, as it already is in the northern Pacific states across the line.

—*COMOX ARGUS*, OCTOBER 11, 1928

2

LEAVING THE FLATS

THE FINEST COMOX LOGGING IMAGES FROM THE 1920S ARE THE recently uncovered, heretofore unpublished photographs of Major Harry (Henry) Steere-Clark, an English-born veteran of the First World War and a timekeeper at Camps 1 and 3. Steere-Clark's photos show huge Douglas fir trees being felled, yarded, loaded, and hauled on the level flats near the mouths of Oyster River and Black Creek. They document the dramatic removal of the original lowland forest between Courtenay and Campbell River with crosscut saws, double-bitted axes, and steam- and railway-driven technology. This coastal lowland was ideally suited for skidder and railway logging; indeed, Block 29, the main property, "was as flat as this table," according to surveyor Tony Turner. The vast northern property of Block 29 contained 31,360 acres—49 square miles.

Steere-Clark's photos, and those of other early photographers, disclose the extent of that long-gone coastal forest and perfectly illustrate the main aspects of early logging on the flats north of Courtenay. They open a window onto this logging heyday: men, methods, machines, and the whole 1920s log production and transport system from stump to dump. *Island Timber* presents details of this system; the photos here summarize it and set the stage for the rest of *Mountain Timber.*

This chapter is, in a sense, an obituary for the fine old forest cut in the 1920s at the mouths of Oyster River and Black Creek and captured by Steere-Clark in its dying days. It is also a chronicle of a local environmental disaster. Fortunately, a small fragment of this forest remains at Miracle Beach Provincial Park, the remarkable legacy of two lawyers: Theed Pearse (1871–1971), Courtenay mayor, scholar, and ornithologist, who preserved the land in the 1930s and 1940s; and Frank Pottage (1885–1959), who donated the land as a park in 1950.

Steere-Clark's photographs are astonishing for the size of the trees. Vancouver Island's Cathedral Grove and Vancouver's Stanley Park give some idea of the scale of the coastal forest he depicted. The photos present a stark contrast to Comox Logging's later shows at higher elevations, where the trees were smaller, the ground generally steeper, and the momentum westerly instead of northerly. In the October 1978 *Crown Zellerbach News*, Mauno Pelto, who started logging at Lake Cowichan in 1933, estimated that the Island's "bottom lands" contained from 50 to 100 per cent more timber per acre than land at higher elevations:

> The big trees grew in the main drainage areas along the coast, and these were the most accessible areas, the ones first logged off. The early loggers began with the big fir and spruce, and worked inland and upward to higher levels over the years. At the higher levels, the runoff of moisture is faster and the soil is shallower compared to the lower drainage areas. There is less for trees to hang on to at higher mountain levels, making for generally smaller trees.

The riches of the coastal flats would never be repeated. As the company went higher, the trees—and profits per acre—got smaller. Steere-Clark captured Comox Logging's last years in the coastal "loggers' Eden" before the company left for higher and more challenging mountain timber.

IN 1926 *THE TIMBERMAN* PRESENTED the following bare facts about the Comox Logging & Railway Company. The company required 400 men to move its products from stump to dump. It operated four sides (logging operations), producing a total of 500,000 board feet a day. Equipment included 6 skidders, 6 donkeys, 4 rod locies, 3 geared locies, 20 tank cars, 8 speeders, 200 flatcars, and 60 miles of track.

These are the dry facts, the basic inventory which, for clarity, can be meaningfully rearranged into a "stump to dump" scenario and recreated visually with Steere-Clark's images. Coastal logging companies followed established and predictable steps. Comox Logging's 400 men were divided into large and small crews, each with a distinct and specialized expertise. These crews moved constantly from one 50-acre to 80-acre setting to another, and one side to another, as the forest was converted to a waste of logging slash. Some of the men on this log production line never touched a

A Lucrative Occupation

Future Canadian diplomat Hugh Keenleyside (1898–1992) spent the summer of 1916 as a timber cruiser on the British Columbia coast, as he later described in *Hammer the Golden Day:*

This occupation was a highly developed and lucrative one in British Columbia. The best of those employed in it became amazingly skilled at estimating the quantity and quality of lumber that could be taken off a given forest area. In more or less open country, and when exact figures were not required, the timber cruiser could do this by a quick optical survey. But when he worked in the dense rain forests of the coast and when large sums of money were at stake, greater precision became essential. Then the area to be cruised would be divided into blocks of, say, a square mile each. The cruiser would fight his way across the block, moving in straight lines, identifying and measuring every tree directly on his route. He would make five or six traverses from north to south and a similar number from east to west; he could then put together a remarkably accurate estimate of the number and size of each species of tree, and on this his employers could confidently base their plans of action.

Timber cruising was a hard physical operation because a cruiser's reputation depended not only upon the accuracy of his estimate but also upon the time involved in collecting data and preparing his report. We often started at five or six o'clock in the morning and seldom finished before the summer dusk was making work difficult. ■

The Flying Major

Harry (Henry) Steere-Clark came from a family of craftsmen that for several generations had designed and manufactured lace at Nottingham, England. His father, Albert, invented the world's first machine for producing coloured lace, and in the 1870s he went to Austria-Hungary and Germany to promote this machine. Albert settled down in Wiener Neustadt, Austria, but his wife returned to Nottingham each time she was pregnant so that her children would not be claimed as citizens of the Austro-Hungarian Empire. Harry, their oldest child, was born in Nottingham in 1880 and was educated in Wiener Neustadt, Prague, and Vienna. The family returned to Nottingham in the 1890s and Harry completed his education there.

In 1899, when he was 18, Harry Steere-Clark joined the Sixth Inniskilling Dragoons of the British army in Dublin, Ireland. "Dad didn't want to go into the family business, so he went into the army," recalled his daughter, Madge Herbert. He served through the Boer War and later in Egypt and India. In 1912 he left the army and came to British Columbia, where his sister-in-law and her family had settled and where Steere-Clark worked as an accountant. Meanwhile he had married May Thompson, and their three children were born in Egypt, India, and Vancouver.

In 1914 Steere-Clark went overseas with the Canadian Expeditionary Force and served in France and England during the entire war. Fluent in German, Arabic, and Hindi, he also spoke passable French, skills that suited him for the positions of liaison officer, general staff officer, and brigade major. He was mentioned in dispatches and ended the war in hospital after being gassed on the Western Front. While recovering, Steere-Clark patented a design for pushing railcars along railway tracks using an airplane engine and propeller, a device that he hoped would prove useful in the colonies.

In 1919 he returned to B.C. and found a restorative outdoor job as scaler at Comox Logging's outer camps. He was then timekeeper at Camp 1 near Miracle Beach from 1920 to 1921, scaler at Headquarters in 1921, and timekeeper at Camp 3 from 1922 to 1925. He was known as the "flying major" because he was always going up to Campbell River to fly-fish in his old Ford, and he was known for his potent homemade "near beer." He also developed strong interests in photography and woods safety, and he wrote frequently for *B.C. Lumberman.* In 1925 he was appointed safety director of the British Columbia Loggers' Association, with responsibility for the association's Safety First program, and lectured widely on woods safety with lantern slides and moving pictures. "Dad's nickname was 'Old Man Safety,'" recalled his daughter. In 1931, when the province took over the Safety First program, Steere-Clark was transferred to the Workmen's Compensation Board. He died suddenly in September 1932, aged 52, of a heart attack, just as he was due to become head of the Workmen's Compensation Board.

His photographs, taken with a craftsman's precision, have been preserved by his family and form the basis of this chapter. ■

ARCHIBALD BRIDGMAN PHOTO

Major Harry (Henry) Steere-Clark, *c.* 1932. His daughter, Madge Herbert, described this as "the last photograph taken of him before his sudden death. It was taken by the top photographer in Vancouver, Archie Bridgman, who was a close friend of Dad's."—Madge Herbert

log: those who prepared the ground and built railways, bridges, and trestles; maintained the logging infrastructure and machinery; or performed clerical, managerial, or menial roles in the company's offices and camps. This was a logging company, but it can accurately be called a railway company, a construction company, a bridge-building company, and a log-booming company.

However, about 340 of the company's men were somehow involved in producing and transporting logs: that is, in the specific assignments of felling, bucking, scaling, choking, yarding, loading, hauling, dumping, sorting, booming, and towing. Their job was to perform, repeatedly, a single step in transforming a mature forest into uniform 40-foot "merchantable" logs suitable for the sawmill at New Westminster.

First came the timber cruisers, usually literate men from reputable firms in Vancouver. The cruisers assessed the quality and quantity of timber in a given area. They produced maps and cruise reports of their findings. They spent their entire careers working on remote timber holdings for timber speculators and logging companies. They worked well beyond the industrial "edge of the world economy." They visited and reported on remote properties that might be logged in five or ten years, depending on the rate of cut and the larger market demand for lumber.

Comox Logging hired Eustace Smith for much of its cruise work between the wars. Born in England and raised at Black Creek, north of Courtenay, Smith ranged far and wide through the uncut forests of British Columbia from his office in the Marine Building in Vancouver. His brother was the legendary cougar hunter Cecil "Cougar" Smith, who made a living hunting cougars in the abundant second-growth clearings left behind by Comox Logging's skidder shows.

The timber cruisers were followed by the company's engineers, also loosely known as surveyors, who worked in three- to five-man crews consisting of an experienced engineer and several apprentices or assistants who acted as chainmen, axemen, and draftsmen. Comox Logging's chief engineer in the interwar years was Hughie Stevens. The surveyors merged the cruise maps with topographical maps to produce their own detailed blueprints showing how properties would be entered and logged. Stevens, who like Smith was born in England and raised in Canada, decided on the route of the logging railway's main line and laid out dozens of branch lines. He numbered the skidder settings in the order they would be logged and then passed the maps on to the grade foreman and the logging foreman. Among the many assistant surveyors were Bob Filberg from Seattle, whose career advanced spectacularly after his 1916 marriage to Florence McCormack, daughter of the

HARRY STEERE-CLARK PHOTO

The grade crew at work, Camp 3 area, 1920s. In *Rolling with the Times*, Wallace Baikie described the grade crew on the Oyster River over the winter of 1920/21: "John Nylander was the grading foreman and he and his eleven man crew were on contract, all hand grading, and the contract price was thirty-five dollars per one hundred feet. For that price they would have to fall and buck, blow the stumps and grade three hundred feet a day, to earn eight dollars per day per man, less the cost of the powder. They were all Swedes and hard workers, snoose chewers and big eaters."—Noëlle Clark & Roger Emery

HARRY STEERE-CLARK PHOTO

Making the grade. The grade crew with shovels and mattocks clearing the right-of-way and building a platform of cedar logs as a culvert, Camp 3 area, 1920s. The object was to prepare a level platform suitable for laying ties and steel. Stan Hanham recalled the grade crew at Camp 1 in about 1918: "They were mostly northern Europeans and Italians. They had horses and two-handled scoops. They used powder to blow stumps for the right-of-way, and then it was just hard work." —Noëlle Clark & Roger Emery

HARRY STEERE-CLARK PHOTO

Grade crew clearing a right-of-way, Camp 3 area, 1920s. Swedish faller Sven Ell, who helped cut rights-of-way, recalled, "They blazed the right-of-way in 100-foot sections. The bull bucker, the fallers' boss-foreman, would get orders to open up a certain right-of-way, and then he'd send you out. They sent in the grading crew after the fallers, and then the steel gang came after that, and they laid the track. After the fallers, the powdermen came in and blew stumps. Then the grading crew came in and removed the logs. They'd just push them aside, and the cherry picker came in and cleaned up as far as they could reach." Powdermen like Bill Nairn did essential work for the grade crew. "Bill was a powderman," said Davey Janes, "blowing stumps on the grade and in a blowdown."—Noëlle Clark & Roger Emery

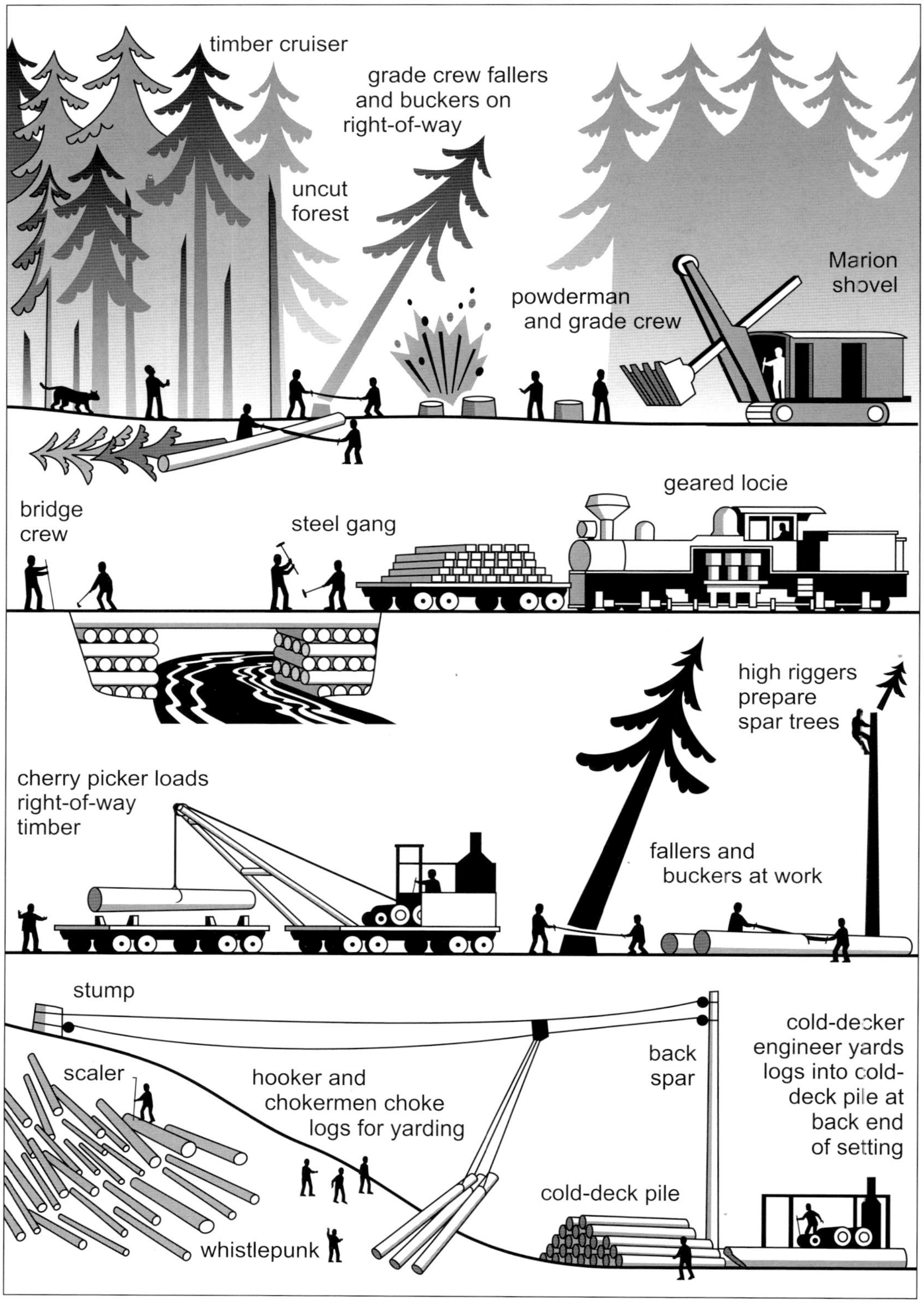

Logging the flats: a schematic portrayal of the stump-to-dump production and transit of logs by Comox Logging, *c.* 1925, a layout that required the specific and repetitive expertise of 400 men.

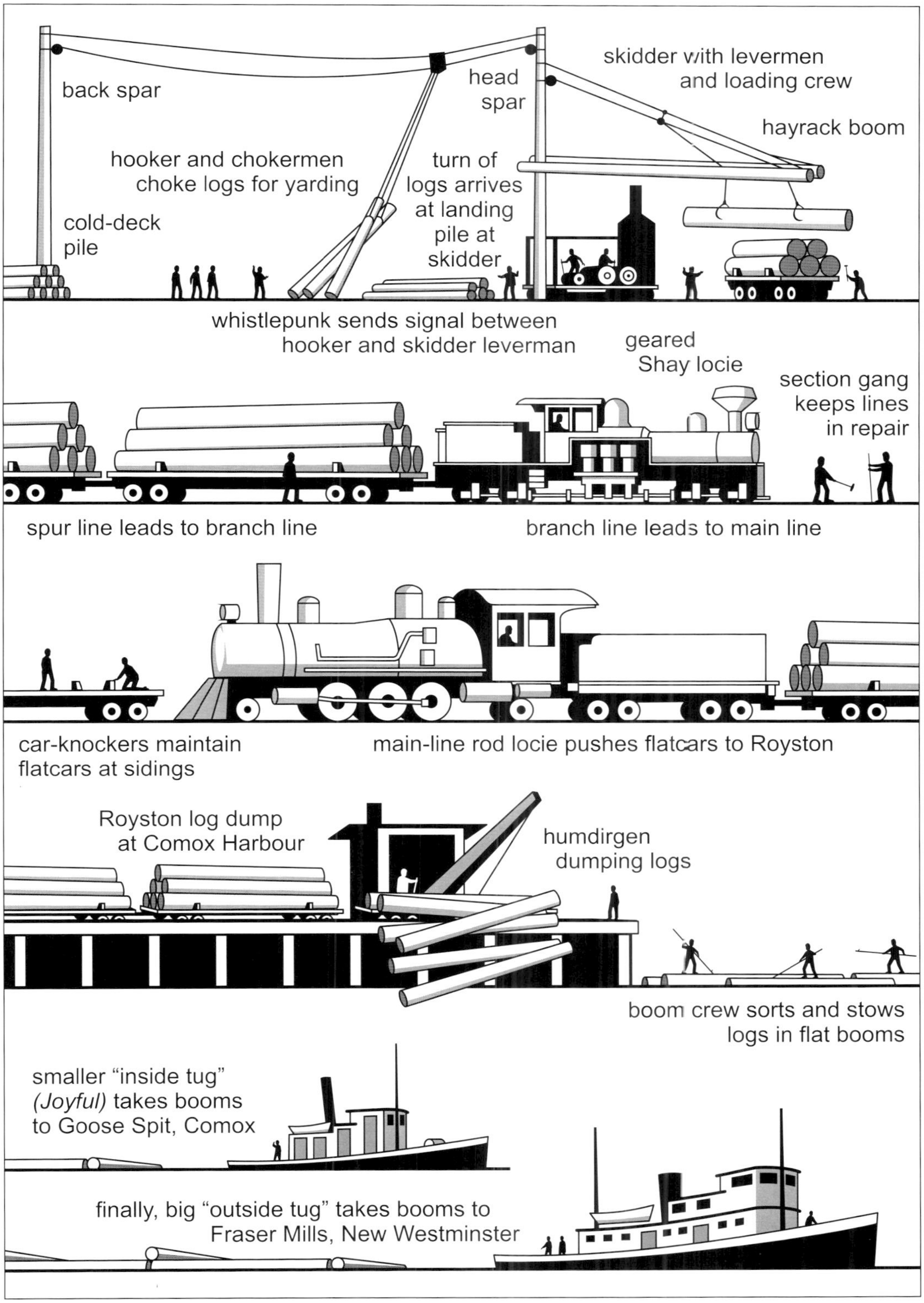
back spar
head spar
skidder with levermen and loading crew
hayrack boom
hooker and chokermen choke logs for yarding
turn of logs arrives at landing pile at skidder
cold-deck pile
whistlepunk sends signal between hooker and skidder leverman
geared Shay locie
section gang keeps lines in repair
spur line leads to branch line
branch line leads to main line
car-knockers maintain flatcars at sidings
main-line rod locie pushes flatcars to Royston
Royston log dump at Comox Harbour
humdirgen dumping logs
boom crew sorts and stows logs in flat booms
smaller "inside tug" *(Joyful)* takes booms to Goose Spit, Comox
finally, big "outside tug" takes booms to Fraser Mills, New Westminster

vice-president of the Canadian Western Lumber Company; and Wallace Baikie of Denman Island, who married Myra Cliffe and later published a memoir, *Rolling with the Times*, valuable for its chronicle of Comox Logging's shows and personalities of the 1920s.

Next came the grade crew, a multi-purpose gang responsible for falling and bucking the trees on the right-of-way, blowing stumps, and grading. They cleared an 80-foot right-of-way for the railway main line which was, essentially, a tunnel into the forest. The crew consisted of between 10 and 30 men, usually recent immigrants from northern and eastern Europe but also Irish and Italian immigrants, men usually hired originally in the hiring halls of Vancouver. Indeed, the crew was jocularly known as the "League of Nations" in the 1930s. The foreman was then an Italian named George Padula. The grade crew often stayed in its own remote grade camp consisting of mobile bunkhouses and tents. Originally they used picks, shovels, and wheelbarrows, but in the mid-1920s the company bought two electric-powered Marion shovels and entrusted them to Homeguard Tom McQuillan. Shovel operators and their assistants (the mat-hookers or swampers) also usually came from Courtenay or from farms in the Comox Valley.

Now came the men of the steel gang, under French-Canadian foreman Joe Asselin (1887–1944). They "pulled steel" rails and lifted ties from logged-off settings and laid them in new ones. Known as gandy dancers or tie tampers, the 20 or 30 men on the steel gang tended to be recent immigrants from central and eastern Europe, such as Croatians and Russian Mennonites, but they also included (in 1934) a Finnish bootlegger, a German estate manager, a shoemaker from Ontario, and a one-eyed Englishman. Like the grade crews, they were a transient force obtained from the hiring halls. Theirs was brutally hard work, as Mennonite Henry Schultz recalled: "I had often sweated while working, but never as hard as I sweated when I worked on the steel gang." Joe Asselin, known by a nickname that closely resembled his surname, could be hectoring and tyrannical. "I can hear him hollering!" said a man who worked within earshot in the 1930s. "You could hear it everywhere! There were lots of Europeans, bohunks we called them—not a good name. He was shouting orders to his men."

When a bridge was required, the steel gang laid the tracks right up to the edge of the river or ravine in advance of Bill Eastman's bridge crew. Eastman's equipment included a donkey engine mounted on a flatcar for lifting hewn timbers into place. The bridge crew consisted of five or seven highly trained and well-paid men, all Homeguards, who ran the donkey engine, climbed spar trees as needed, and hewed bridge timbers with broad axes and adzes. For a large

HARRY STEERE-CLARK PHOTO

Unnamed steel gang foreman at end of steel, Camp 3 area, 1920s, with steel rails and tie timbers. These ties were not hand hewn but sawn at a local mill. The aim of logging by rail was to construct a horizontal platform over the swamps, creeks, rivers, and hills of the flats. Geographer Cole Harris has noted the central importance of railways in the staple trades and transport systems of B.C. in the late 19th and early 20th centuries. Railways provided artificial corridors that replaced the vagaries of the surface of the earth with lines of machined steel, and the irregularities of topography with engineered grades. Their artificiality gave them their power, allowing steam engines to haul by day or night, summer or winter, in greatly varied terrain.

—Noëlle Clark & Roger Emery

Poling Stations

In 1929 the San Francisco-based K & K Pole Company opened an office in Courtenay. The company removed cedar poles from Comox Logging's properties and sold them as poles to carry telephone cables and electric lines during the rapid expansion of electricity in the United States. K & K paid a stumpage fee to Comox Logging—so much a linear foot. If K & K produced, in a year, 10,000 poles, each one 60 feet in length, stumpage payments were calculated on a total of 600,000 feet of cedar poles.

Buck Pedersen (1921–2007), son of K & K's Danish-born foreman Al (Axel) Pedersen, recalled the K & K's operations:

We moved to Courtenay in 1929, when I was seven. My dad moved there as hooktender for K & K Poles. Before that he'd been a K & K teamster at Camp 4, McNabb Creek, on the side of Howe Sound, up from Sechelt, where they horse yarded cedar poles and sent them down a flume to the beach.

The K & K had poling camps at Tsolum and Camp 3, another at the top end of Comox Lake, and one at Bevan hill almost where the lodge is. All this was between 1929 and about 1932, when K & K went broke. At that time K & K used horses for poling. The teamsters and horses brought the poles to the trackside and decked them—you'd clear a landing and put skids down for yarding. For lift, a line went up to a block on a gin pole, and you'd spot a flatcar in front of the pile.

They liked Percherons—not Clydesdales, because their feet were big and clumsy. Clydesdales were good for pulling heavy wagons, but Percherons were lighter on their feet. They had to step over windfalls and so on. I'd go out and feed them when I was a boy. They went into the bunkhouse every night, and they worked every day except Sunday, when they were shoed and groomed. Those horses hardly ever saw a field! If it was a long winter the teamster would put them in a field. I remember Alec Rennie, an old-timer with horses, an old Irishman.

The K & K used portable camps and bunkhouses in those days. You could get 12 men in a bunkhouse, and you could load the bunkhouses on flatcars. There were two important things about camp life: first, a teamster who abused horses went down the road in ten minutes. Second, a cook who abused food went down the road in ten minutes. Down the road he went! The best food in the world was in a logging camp. It was plain but it was cooked just the way you wanted it. I always preferred camp food. I went to the best restaurants in the world when I visited head office in San Francisco, but camp food was always better. ■

HARRY STEERE-CLARK PHOTO

Harry Steere-Clark's caption for this photo was "Cedar poles Camp 3." These poles were cut in the 1920s by the K & K Pole Company, which sent crews of fallers, buckers, peelers, and horses into the woods as soon as Comox Logging's grade crews were through. At this stage of logging only K & K's poles have been removed from the standing timber. Steel has not yet been laid. The pole company did not require an active railway initially: it used horse teams and gin poles to pile its small cedar poles beside the grade. "They used horses and cold decked [the poles] into great big piles next to the railroad tracks; then they'd spot a flatcar next to them," recalled Davey Janes.—Noëlle Clark & Roger Emery

HARRY STEERE-CLARK PHOTO

HARRY STEERE-CLARK PHOTO

Steel and timber in an uncut setting, Camp 3 area, 1920s. This man, possibly the bull bucker, appears to be carrying bottles and cans of coal oil (kerosene) out to the falling gangs for use as saw lubricant.—Noëlle Clark & Roger Emery

< Manpower: the steel gang at work, Camp 3 area, 1920s. The tie tampers, also called gandy dancers, often came from continental Europe. The steel gang included Croatians, Russian Mennonites, and Italians. "You hardly ever saw Swedes on the steel gang—they were mainly Italians, Europeans," recalled Swedish faller Sven Ell. "The odd Croatian was a faller, but there were few of them, and they preferred the section, grade, and steel gangs." Between 1910 and 1950, wrote railway historians Cummings, Oliver, and Anderson, "the Company constructed approximately 27 miles of railroad a year, main line and branches." Rails were taken up and reused constantly, and ties were reused as long as the timber was still sound.—Noëlle Clark & Roger Emery

> A trestle on Comox Logging's main line near Camp 3, Northy Lake, 1920s. "When a bridge was to be built, the steel gang had already laid the rails right to the bridge," recalled Hi Churchill. "Most of the bridges consisted of big long fir logs. It's hard to recognize locations of bridge photos because when bridges were put in, the timber was still standing around them, but later the timber was all gone." —Bob Berkeley

The Four-Spot Heisler moving flatcars loaded with gravel, *c.* 1920. The grade crew used gravel for railway beds, and the steel and section gangs dumped lesser amounts of it between the wooden ties of the completed railway. Heisler locies did the many everyday tasks of railway logging: moving gravel cars, delivering empty flatcars to skidder settings, pushing around cherry pickers, moving loads of ties and steel, picking up loaded flatcars, and hauling crummies.—Donna Edwards

Fireman on a Cherry Picker

Son of Bob and Mildred (*née* Pritchard), Neil Martin (1921–2008) attended Headquarters and Tsolum schools and then got a job on the Camp 3 cherry picker:

When I was 17 or 18 I worked on a cherry picker. Joe Ducca ran it out of Camp 3 and I fired for him. There was a second cherry picker at Comox Lake and we competed! Cherry pickers were wood-burning donkeys. They were actually on skids on the flatcar; they sat on the car—they worked right from the flatcar even though they were on skids. They didn't move through the bush like yarders. I fired the donkey and I also pulled line. The donkey was fired by bark mostly; there was no time to cut wood, and there was lots of fir bark around.

There was no swing boom on the cherry picker so we used tongs to bring the logs right into the middle of the track for loading. Sometimes we put lines out, and when we did we had to pack tongs out with us. We cherry picked wherever good logs were close to the railroad. It was a matter of economy: these were cheap logs, and it was cheap to log this way, to get logs close to the railroad. We also logged in nuisance areas, for example between the track and a lake, where it was not handy to set up a skidder. We also cherry picked spar trees from logged-off settings: by that time the spar trees had been dropped and they sent me down ahead to buck up the spar trees into lengths. They kept a crosscut saw on the cherry picker for this purpose. Luckily I had done a lot of bucking on the farm! When our flatcars were loaded, the spotting engine—the Two-Spot—came and pulled them out. John McLoughlin was running the Two-Spot, Tom McKay was brakeman, and Bill Defty was fireman.

Martin later became a scaler, safety coordinator, and bull bucker for Comox Logging and its successors. ■

Once the grade was in and the steel laid, the cherry picker came along to collect the right-of-way timber left by the grade crew's fallers. The grade crew had simply pushed the track timber aside when making the railway. Here, a loader pauses, c. 1930, with the cherry picker behind. The cherry picker leverman (far left) is probably Jim Mathers Jr. Cherry pickers were donkey engines secured to flatcars and fitted with heel booms for loading. "A heel boom comes from the foot," said Bruce King; by contrast, a hayrack boom was attached to the spar tree (some 30 feet up) and suspended horizontally above the flatcars. Cherry pickers were usually the first and last loading machines to visit settings. "They generally go in before a setting is logged," explained Sam Telosky. "The cherry picker goes along and gets all the ones it can reach. It tidies up. We used to say they were cheap logs because you only needed 3 or 4 men on a cherry picker, instead of 21 on a skidder." The cherry picker would return again when logging was over to retrieve logs from nuisance areas and to pick up logs that had fallen from flatcars.—Laurie Mathers

A Swede for the Section Gang

Walter Anderson described his first six months of work for Comox Logging:

I arrived from Sweden at Headquarters on July 22, 1924, and my sister Signe Tukham met me. I had just turned 17. My father [Fred]*, an electric welder, worked overtime till 9 o'clock, so my sister took me right to the hotel and introduced me to the bullcook, known as "Hardtack," who showed me to a room in the hotel. My sister told me I had to get up at 6 o'clock in the morning and that breakfast would be at 6:30. "You've got to start work tomorrow morning—you have to work on the section gang," my sister said. She had to burn all my clothes because the train had been full of lice!*

The section gang boss was an Irishman named Fred Griffith, a very nice guy. The section gang repaired the old railways: they changed ties and maintained the railway. There were four men on the gang, two and two: two carried ties and two tamped them. There were four workers

and a boss. Griffith walked around checking ties. Comox Logging had another two section gangs, each with its own boss. They each had their own certain length of railway to look after. The territory of the Headquarters section gang ranged from Headquarters to Dove Creek and then halfway up to Camp 3. That was our territory. I stayed on the section gang for three weeks.

In September 1924, I was moved to the steel gang. Joe Asselin was the boss. They were building a new railway to the north and west of Camp 3, and Camp 2 was still open at that time at Oyster River. Men came and went on the steel gang. They'd say, "There was one crew coming, one crew going, and one crew working." Joe Asselin would pick up small rocks and throw them at the guys and say, "Come on, come on!" But he'd never hurt them. The steel gang provided recruits for the skidder crews. You'd see a guy in the bush a lot of times: he was the woods foreman out in the bush watching the crews to see who was working hard, and he'd take them to the skidder crew. I saw the bush moving and didn't pay attention to it.

Then, after only three weeks, I went setting chokers on a skidder crew out of Camp 3. This lasted from September until December 1924. I got the job when the hooker got sick; the second hooker got his job, and there was a vacancy at the bottom. This often happened. They were still firing the skidders with wood at this time. They hadn't yet switched to oil. I remember they just left any log in the bush that was two feet or less at the butt. They never even cut them down! They left the snags and the small trees standing.

There were 300 to 400 men working for Comox Logging at that time at all the camps. There were a lot of fallers, who worked in four-man gangs: two fallers and two buckers. In the whole bush at this time there were mostly Swedes, Finns, Norwegians, Russians, and Germans. ■

HARRY STEERE-CLARK PHOTO

Lifting and laying steel near Miracle Beach, *c.* 1921. Harry Steere-Clark captioned this "Track-laying machine at Camp 1." This crew is lifting steel from a setting that has already been logged. Frank Davison recalled that Pete McLoughlin ran this "steel-laying machine" on the steel gang. Soon after, Davison said, they got rid of the steel-laying machine because it was faster for men to push the steel rails off a flatcar and onto the grade than it was to lift and place the rails with this contraption: "It was twice as fast!" The locie in the background is a Heisler, the Four- or Five-Spot, with a crummy behind it.—Noëlle Clark & Roger Emery

bridge, such as the Tsolum bridge built in 1921–22, up to 25 men were required.

Meanwhile the engineers had been busy marking the skidder settings in the forests beyond the bridges. They blazed large X marks on head spars and back spars—that is, on large, strong Douglas firs left standing and used for what was called highlead, skidder, or aerial logging. Highlead logging was introduced to Canada by Comox Logging in 1911 and was used continuously by them until 1946. In it, a block, line, and carriage system working on the same principle as a clothesline was stretched between a head spar and back spar, which were about 1,100 feet apart. At every setting, powerful skidder engines pulled logs from back spar to head spar and loaded them onto flatcars.

But first came the falling gangs, consisting of fallers and buckers under the management of a foreman known as the bull bucker. Fallers were paid by how much they cut, not by the hour, which meant they took only the marketable timber and left behind a characteristic landscape of worthless old trees, snags, and saplings. Buckers then cut the logs into 40-foot sections, and scalers measured the bucked logs. Between 1920 and 1945 two of Comox Logging's long-serving bull buckers were Eric Flinton and Clarence McCormack. Flinton (1899–1978), who started his Comox Logging career in 1919 as a timekeeper, was son of the Anglican priest in the Comox Valley, Rev. J.W. Flinton—an indication of the ability of Comox Logging to attract everyone into its engine of employment.

Each falling gang consisted of two fallers and two buckers until the Great Depression, when one bucker was let go to form three-man falling gangs. Most fallers and buckers were Swedes, Finns, and Swede-Finns (Finns of Swedish ancestry) engaged originally at the Vancouver hiring halls. Comox Logging employed about 30 three-man gangs. The Swedes and Finns were part of a substantial 1920s migration to coastal British Columbia. Some of them brought their wives with them or met Scandinavian women locally or in Vancouver, and both Merville and Bevan had considerable Finnish communities, complete with saunas and subsistence stump farms for the off-season. A few also met and married local women when they moved to the Comox Valley. The Finnish influence was considerable. "I could almost talk Finn at one time! All the guys in the boxcar talked Finn and I picked up the odd word," recalled Art Beech, a local man who started working at Headquarters in 1936.

The Finns were also a considerable presence on the rigging-up crew, or "flying corps," a crew that went from setting to setting preparing the head spars for highlead logging. This crew, numbering five or six men, consisted of a high rigger assisted by a couple of second riggers and by the engineer and crew of a small donkey

HARRY STEERE-CLARK PHOTO

In about 1925 Comox Logging bought some Marion shovels. They were used for making railway grades and for loading gravel cars. This 1929 photo shows (left to right) Jimmy Sheasgreen, grade foreman based at Camp 3; an unidentified operator; and Tom McQuillan, shovel operator.—Noëlle Clark & Roger Emery

> A four-man falling and bucking crew, Comox Logging, 1920s. "The fallers were off by themselves and I didn't see much of them," recalled Hi Churchill, who worked as a chokerman, loader, and brakeman between the wars. Skidder leverman Mack Grant said, "We rode to work on the same car as the fallers. You could hear them working: they worked pretty close around. They cut the trees always out ahead, but the fallers and the skidder crews more or less worked together. You wouldn't leave the logs in the bush for too long."—Noëlle Clark & Roger Emery

Playing Second Fiddle

The great majority of Comox Logging's fallers and buckers were Swedes and Finns. The handful who were of other backgrounds included Les Marshall, an Englishman whose sister (Alice) had married Lu (Lucius) Cliffe; Austin Blackburn, who grew up on a farm near Merville; and Lin (Lilburn) Hunt, an American who had married Birdie McQuinn, sister of Camp 3 boss Jack McQuinn.

Generally, these men prided themselves on being able to "keep up" with the Swedish and Finnish fallers, but the Scandinavians sometimes resented the favouritism shown to the few Anglo-Saxon fallers. "Hunt's gang got the best quarters because he was McQuinn's relation. He always got the best chunk of timber," recalled Gunnar Jonsson (1909–2006), a Swede who worked as a faller out of Camp 3 from 1937 to 1944. Finnish fallers Toivo Koski and Waino Jylha had similar memories, as Koski's grandson Erik Piikkila wrote by e-mail in 2007:

After several years of working at Camp 3 [in the 1930s], *my grandfather, Toivo Koski, noticed that the good falling areas were always allocated to the same select few. My grandfather often found himself with the nasty cold deck yarder settings at the fringes or limits of the main railroad setting, in the very back, in a long corner, behind a hill, or at the bottom of a slope. These were often covered with scattered old growth, unmerchantable species, dense immature stems, poor quality, dead, blown-down or fire-killed timber. On top of all the stand difficulties, the groundcover might be very thick and very tall which would help slow down production as well. All of these factors made it very difficult for a falling and bucking crew to achieve their daily production rate or scale requirement of 50,000 board feet of felled and bucked timber.*

Having seen way too many poor settings and falling quarters, my grandfather spoke to Bob Filberg about this issue, and thereafter the good wood and crap wood were shared and dispersed among all the falling and bucking crews. ■

HARRY STEERE-CLARK PHOTO

engine known as a rigging-up goat. Originally Comox Logging's high riggers were Finns like Herman Anderson, Emil "Sailor" Lehtonen, and Henry Norman, but by the mid-1920s they were more likely to be Homeguards like Joe Cliffe, Jack Downey, Bob "Scratchy" Grant, Norman Janes, or Jack McKenzie Jr. Accustomed to working on the front lines of the whole operation, the high riggers were often selected for promotion, first as head riggers on skidders, and later to managerial positions.

Next came the logging crews, who transferred the felled and bucked timber onto flatcars. "Logging" referred to two central tasks: yarding and loading logs. As Roderick Haig-Brown wrote in 1942, "Loggers [are those] who work around the machines, distinguishing them from trainmen, fallers and buckers, graders and trackmen ... he [a rigging man] is quite likely to insist that only men of his particular calling are really entitled to the title 'loggers'—and that the others, including even the fallers and buckers, are not loggers at all." Yarding meant pulling logs from where they were cut to where they were loaded; loading meant lifting logs and placing them on flatcars for transport to the boom camp. These were vital and highly visible steps in the process, but all the steps were vital, even if photographers rarely bothered with them. Yarding and loading were performed by crews, usually Homeguards, attached to two machines: cold deckers and skidders. Cold deckers, also called yarders, were small and mobile donkey engines that yarded logs from the bush and piled them up in temporary "cold-deck piles" to await further yarding or loading. The large trackside machines, known as skidders, were glorified donkey engines that yarded logs to them and then loaded the logs onto flatcars. Yarding crews comprised riggers, hookers (hooktenders), rigging slingers (on Clyde yarders only), chokermen, and whistlepunks. Landing (loading) crews comprised yarding and loading levermen, firemen, loaders, chasers, and bunkmakers (*see glossary*).

Cold deckers and skidders were the engines of highlead logging. First, a cold decker yarded and piled distant logs into an accessible cold-deck pile beneath the back spar. Then the yarding leverman on the skidder yarded the logs 1,100 feet into a landing pile at the head spar, and the loading leverman loaded them onto waiting flatcars. Until 1921 Comox Logging had three camps going simultaneously, with two skidders at each camp.

Skidder crews numbered about 21 men each, and cold deckers had crews of 7. Each setting, therefore, required about 30 men. In the 1920s and 1930s Comox Logging always had four skidders and four yarders going simultaneously, which meant that logging crews accounted for about a quarter of the 400 men employed by the company.

> ***Top*: Camp 3 fallers on springboards work on the undercut of a tree at the back end of a setting, 1920s. The grooves and scoring on the near side of the tree were caused by friction from a cold decker's receding line.**—Noëlle Clark & Roger Emery

> ***Bottom, left*: Camp 3 fallers, 1920s, standing on springboards and cutting the undercut of the same tree with double-bitted axes. Springboards could be turned in almost any direction, recalled Neil Martin, and fallers would adroitly kick them around beneath them to find the best angle.**—Noëlle Clark & Roger Emery

> ***Bottom, right*: Sighting a tree. A faller at Camp 3, 1920s, uses his double-bitted axe to determine where he will fall the same tree. "The undercut has to be dead level," explained retired faller Keith Cardriver (*né* Pirie), "and the back of the cut gives you a basic sighting plane." In 1922 Comox naturalist Mack Laing recorded how his neighbour Norman Pritchard taught him to fall and buck trees: "I learned how to notch-in to mount the foot-boards on which we stood, to put in the under-cut, to sight along the straight haft of the axe, to drag the narrow-bladed falling saw, to use the wedge here or there to correct alignment of the falling giant. It is a game for the exercise of skill. Norman had worked in the logging camps where, if you stay on the pay roll, you must drop your logs without much breakage. And there are fallers who will gamble with you to break your watch for you if you place it on the ground."**—Noëlle Clark & Roger Emery

HARRY STEERE-CLARK PHOTO

HARRY STEERE-CLARK PHOTO

HARRY STEERE-CLARK PHOTO

HARRY STEERE-CLARK PHOTO

HARRY STEERE-CLARK PHOTO

Above: The thin end of the wedge. Camp 3 fallers demonstrate the use of the falling saw and the steel wedge, 1920s. "There is not yet an undercut on the other side of the tree," said Keith Cardriver. "The wedge is to hold the cut open so the saw doesn't bind in the cut. The wedge is to loosen the cut." Note that the falling saw is thin and flexible, whereas a bucking saw is thick and rigid.—Noëlle Clark & Roger Emery

< Harry Steere-Clark's caption for this photo was "Cedar 10' dia on stump. Two more scarfs [wedge-shaped gouges removed from the undercut] had to be chopped after this snap was taken. 1922." This photo was taken in the Camp 3 area. "A wedge-shaped piece is cut from one side of the trunk about waist high or lower," wrote John Gough in *The Story of British Columbia.* "Properly done, this will cause the tree to drop where the loggers wish." Finnish faller Bill Lyttback, who started working on the coast in 1926, recalled that "they called the falling saw the Swedish fiddle. They had big seven-foot saws. The falling saws were nine feet originally, then seven."—Noëlle Clark & Roger Emery

Top: Fallers from Camp 1 chopping the undercut on a large Douglas fir near Miracle Beach, 1921. This tree was 500 to 600 years old. Note the pitch oozing into the undercut from above and the fire scarring on the bark, which in Douglas fir could be as thick as ten inches. The fallers have removed the bark at the sides of the tree to expose the cutting surface. Two experienced fallers might take two and a half hours chopping out the undercut on a tree of this size. —Noëlle Clark & Roger Emery

Centre: Camp 1 fallers cutting down the same Douglas fir near Miracle Beach, 1921. About this photo, retired faller Keith Cardriver said, "This is a perfect picture of a side-notch. You need a side-notch to allow for the length of the saw and the size of the tree. This is probably a nine-foot saw and an eight- or nine-foot tree. They've removed the bark from around the tree to prevent the saw handles from hanging up. The undercut is on the far side of the tree." It could take two experienced fallers 90 minutes to saw through to the undercut. —Noëlle Clark & Roger Emery

Bottom: Harry Steere-Clark's caption for this photo was "A big Douglas fir (side-notched) going over Camp 1 1921." This is the same Douglas fir pictured in the previous two photos. "As the saw approaches the undercut, from the other side, the monarch of the forest begins to tremble," wrote John Gough in *The Story of British Columbia*. "Then it crackles at the butt, leans toward the earth and plunges groundward." The standing trees in the background, considered unmerchantable, are about to be smashed. Wallace Baikie was scaling at nearby Camp 2 when these photos were taken. In his 1985 memoir, Baikie wrote, "It was beautiful fir timber, straight and on level ground. I scaled one tree that the bucker cut five forty-foot logs from. That was the only time in my lifetime that I have seen that happen."—Noëlle Clark & Roger Emery

HARRY STEERE-CLARK PHOTO

HARRY STEERE-CLARK PHOTO

HARRY STEERE-CLARK PHOTO

The Finnish High Riggers

Before about 1930 Comox Logging was known for its Finnish high riggers. The three best known were Henry Norman (Nieminen), Herman Anderson (Hyytiäinen), and Emil "Sailor" Lehtonen. Other Finnish riggers were Vic Ahoe and Mauno Pelto.

Various theories have been advanced for the prevalence of Finnish high riggers with Comox Logging. "The Finns made good high riggers because they'd been on sailing ships," asserted Walter Anderson, who started working for Comox Logging in 1924. Norman and Lehtonen had, indeed, been crew members on sailing ships. Finns were also famous for their physique and work ethic. "On the average, the Finnish loggers were big, strong men who worked hard and complained little," recalled Sailor Lehtonen's son, Larry. "Consequently, they soon earned the respect of the local loggers. Dad made around $7 a day in the 1920s, and that was a very good wage for those times."

Emil Lehtonen, born in Finland in 1893, started working for Comox Logging in about 1919. "Sailor Lehtonen was fast; fast up the tree," recalled Walter Anderson. He was known for standing on his head on top of the spar tree. Certainly his sense of balance was superb. Art Beech remembered that Sailor Lehtonen "used to put a full glass of beer on his head in the beer parlour and he'd walk around with the glass on his head! I've seen him do that many times."

Henry Norman, born in 1885, joined the crew of a windjammer when he was 15 and arrived in New Orleans in 1903, having learned to rig on the ship coming to America. "Then he left his ship to go rigging on the skidders in the great alligator-infested swamps of Louisiana," wrote logging historian Bob Swanson in *Whistle Punks and Widow-Makers*. "The system had just been developed and riggers were as scarce as hen's teeth." Henry Norman told his son, Al, that he'd worked with skidders in Washington State before coming to Vancouver Island to work for Comox Logging in 1911. "I brought the skidders into Canada," he told him. Soon afterwards he was lured away from Comox to Chemainus to rig for the Victoria Lumber & Manufacturing Company, but he spent most of his career as a rigger for ITM (Industrial Timber Mills) at Lake Cowichan, where he trained another Finn, Mauno Pelto, as a high rigger. Norman married the daughter of a Finnish miner at Ladysmith, Ester Solmie, whose sister Rachel married Herman Anderson. In 2007 Henry's grandson Rob Norman donated his grandfather's belt, spurs, and topping axe to the B.C. Forest Discovery Centre in Duncan. ■

HARRY STEERE-CLARK PHOTO

Barking up the right tree. A high rigger, probably Emil "Sailor" Lehtonen, starts up a head spar, Comox Logging, 1920s. "It was the rigger who climbed with belt and spurs to cut the top off a carefully selected standing tree which served as the spar tree," high rigger Mauno Pelto told the *Crown Zellerbach News* in October 1978. The ideal spar tree was about four feet across at the base. Lehtonen will climb the tree, limbing it as he goes, before topping it at about 120 feet. Dangling from his belt is a specially made short-handled, double-bitted axe known as a "topping axe."
—Noëlle Clark & Roger Emery

HARRY STEERE-CLARK PHOTO

A felled setting, 1920s. The Douglas fir trees have been felled side by side to prevent breakage, and in the direction of the skidder to speed up yarding. These logs will soon be bucked into 40-foot lengths; 40 feet was the length of a railway flatcar. In the background are the young and dead trees—saplings and snags—left standing. "Everything in the background was unmerchantable," pointed out Keith Cardriver. "It will all be mown flat by the big trees!"—Noëlle Clark & Roger Emery

HARRY STEERE-CLARK PHOTO

Fallers at Camp 3, *c.* 1922. "A splendid Douglas fir, Camp 3," wrote Harry Steere-Clark in his caption. "Four 40-foot logs were cut from it." When shown this picture, Keith Cardriver pointed to the "perfect bucking cut" at right. "When you're finished the cut, the cut should separate cleanly. Usually they flop down inert on the forest floor, but they can do something unexpected, say if a stump 40 feet away puts torque on the tree." —Noëlle Clark & Roger Emery

Finnish faller Oskar Aaltonen (1887–1968) rests on a huge Douglas fir somewhere north of Courtenay in the 1920s. He is sitting on the side-notch. The rounded axe marks of the undercut are visible at right, while the smooth surface left by the crosscut saw is at left. The jagged wood to the right of Aaltonen is what retired faller Keith Cardriver called the holding wood, "which acts just like a brass hinge when the tree falls." Forester Bill Grange, who surveyed in the Oyster River area in the 1950s, remembered in a 2005 correspondence "the large Douglas fir stumps, which in some cases took three strides to get across." Aaltonen's daughter (Bertha) married into the Mathers family, and his descendants are still in the Comox Valley.—Laurie Mathers

Riggers at a head spar, *c.* 1930. Left to right: Drew Berkeley, unknown, Carl Sundin. "The high rigger was a special breed of woodsman. They were paid top dollar because they had to deal with high winds, dense forests, and constant danger," wrote journalist Don Lyons in 1985. The swelled butt indicates that the head spar is a standing tree (one used where it stood rather than being brought in and raised). The men are standing on a shovel mat. The bucked log in the background suggests that these riggers were preparing the setting for logging. They have already attached the hayrack boom (top left). "Drew Berkeley was head rigger: he rigged all the spar trees on the rigging-up goat," said Hi Churchill. "The goat was never attached to one skidder. If No. 5 was finished a setting first, they would move to the next tree, and the rigging-up crew would move on and prepare a tree for Skidder 6." Carl Sundin was a Swede who also worked as a faller, scaler, and boom man. "He once told me that if he had to do it again he would have stayed with the rigging," recalled Neil Martin.—Bob Berkeley

HARRY STEERE-CLARK PHOTO

Topping the head spar, north of Courtenay, 1920s. The top 60 feet of the tree crashes to the forest floor. "The job did have its hazards, but as long as you kept your mind on the job you were all right," high rigger Mauno Pelto told Don Lyons of the *Nanaimo Times* in 1985. Once the tree had stopped swaying, and if a photographer was present, the rigger might stand on top of the spar tree. Otherwise he might just sit on top and enjoy a smoke. Note that this setting has already been laid out: the grade and steel gangs have completed the logging railway, and the right-of-way timber has been bucked into 40-foot logs and pushed aside to await the cherry picker. A handcar has been pulled off the track below.
—Noëlle Clark & Roger Emery

A gas-powered cold decker at a landing north of Courtenay, *c.* 1920. This unit is probably on its way into the setting. At left, loaded flatcars await the journey to Royston. Cold deckers were overhauled periodically at Headquarters. "The skids don't have much taper to them; they had the choice of the best timber," observed Davey Janes. Douglas fir was the best material. Note the inlaid bolts that held the two skids of the sled together. These sleds were so sturdy that when worn out, they were taken to Royston and used in the log boom breakwater. This photo belonged to foreman Jack McQuinn.—Dick McQuinn

Raising a tree around Camp 3 or Bevan, *c.* 1927. If a suitable "standing tree" was not available in the right spot, or if it turned out to be punky or unsuitable, a tree had to be brought from elsewhere and raised as a head spar. This was known as a "raised tree." This tree is between 120 and 140 feet tall. A rigging-up donkey, out of sight to the right, is pulling the tree upright. According to Neil Martin, spar trees had to be at least 130 feet high and 19 to 20 inches across at the top. "Raising and rigging a wooden spar tree 150 feet in height, with a double set of guy lines, main line, haul back, blocks and pulleys, is probably a lost art," wrote logging photographer Wilmer Gold in 1979. In a 1980 interview, former high rigger Joe Cliffe described the rationale behind raised trees: "Later on, when the big trees got scarcer we used to haul out trees. We'd get them where we could, good big trees, then we'd haul them to the place where we wanted to put them up. Then we could put the logging jewellery on the tree down on the ground before we raised it. Then we raised it up—all it meant was a little adjusting here and there, tighten the guy lines and all that. You never knew it was raised of course, unless you looked. That way you got away from the swelled butts and you got them exactly where you wanted them close to the track. It helped the loading."—Babs Pelto

Donkey crew out of Camp 3, *c.* 1935. The emulsion is peeling from this tiny snapshot, a reminder of the fragility of the photographic record. Left to right: Stan Hodgins, Bert Woodruff, Jack Marsden. Woodruff and Marsden were brothers-in-law. Woodruff's caulk boots are visible. "Worn-down caulks mean a sprained ankle or an injured leg," advised "Old Man Safety" in the *Comox Argus* in May 1928. "Look after yer caulks, fellers—the woods ain't no ballroom!"—Tom Marsden

HARRY STEERE-CLARK PHOTO

HARRY STEERE-CLARK PHOTO

< Skidder 6 at Camp 3, *c.* 1922. The engineers, wearing overalls, are most likely Reid Good and Clifford Janes. Skidders 5 and 6 were attached to Camp 3. "Of course, when they had the three camps there were two skidders in each camp," explained Joe Cliffe in a 1980 interview. The closure of Camp 1 in 1921 meant that two skidders were moved to Camp 3, as Howard McQuinn recalled: "At Camp 3 there were four skidders at one time. The other camp was Camp 2, around Oyster River."—Noëlle Clark & Roger Emery

< Harry Steere-Clark's caption for this photo was "Skidder 5 Camp 3." The photo shows loading in about 1922. At far left is loading engineer George Cliffe, who from 1907 until 1958 worked for Comox Logging or companies connected to it. ("All engineers in those days wore those bib overalls," said Davey Janes.) Powerful logging machines, skidders were hard on their wooden frames, and Comox Logging experimented with different materials in frame construction. The first frames of 1911 were made of square timber, but the early 1920s saw the introduction of simple rounded log frames, shown here. These were apparently too heavy or unstable because the company returned to square timber before switching finally to steel frames in the mid and late 1920s. As a result of this experimentation, no two skidder photos were quite the same: timbers soon went missing from a new frame and were replaced with unmatching timbers or were bandaged together with planks. The same skidder, therefore, could appear in four or five different sets of clothes between 1911 and 1946 while remaining identifiable by the crew members. George Cliffe, for example, was always with Skidder 5.—Noëlle Clark & Roger Emery

Lord of the woods in a straw hat. Hard hats did not come into use until the 1950s. This is an unidentified crew member of Skidder 4, probably the head loader, 1920s. "You couldn't put a greenhorn on a skidder; they were highball!" said Sam Telosky. Notice the gleaming drums and lines on the skidder behind and the gouges in the cork-like bark of the Douglas fir spar tree. The gouges were caused by friction from moving and tightening the lines attaching the skidder to the spar tree.—Donna Edwards

HARRY STEERE-CLARK PHOTO

< The back rigging crew fastening a guyline from a spar tree to a stump, 1920s. "It could take two weeks to rig a tree for a skidder, not including the back spars, which came later," recalled Hi Churchill. "Usually they had to bring a tree in and stand it up. And they had to notch all those big stumps! Everyone was handy with an axe. They had to make a notch [all the way around] the stumps that was six to eight inches wide so they could take two wraps with the guyline, and then they would spike the lines to the stumps with railway spikes ... It was a *lot* of work." Keith Cardriver recalled that lines were still spiked in the 1960s. "You'd run the line around the stump and then the donkey would tighten it. You clinched the lines with the top of the spikes: you never bent them." The men are wearing gloves to protect their hands from jaggers (broken strands on worn line). "Never take hold of wire rope with your bare hands," advised "Old Man Safety" in the *Comox Argus* in May 1928. "It'll tear them to ribbons some day and you'll be lucky if you don't get blood-poisoning. Always wear your gloves, and good ones at that, and save your hands. And when you do get a jagger in your palm, give it first aid just as quickly as ever you can."—Noëlle Clark & Roger Emery

Between the Lines

Line was essential to highlead logging, and manufacturers in Vancouver turned out thousands of miles a year of different sizes. It is no accident that line splicing was the main event at the annual loggers' sports. Line tied the entire logging show together: cold deckers, skidders, and spar trees yarded, loaded, and supported logs by means of it (only the two-inch-thick skyline was called a cable; the smaller sizes were called line).

In the 1940s Angus McMonnies worked for Vancouver Barge Transportation, which brought Comox Logging's heavy equipment up the Courtenay River on barges. He recalled the huge rolls of skyline cable required:

They brought in big rolls of skyline containing the big two-inch skyline cable. These were huge things wrapped around great big wooden drums which were eight feet tall! I remember that "British Wire Rope" was written on most of the drums. It was quite a job getting them onto the truck. Edgar Cliffe had a hand winch, and he'd go from the front to the back of the truck turning the winch a click at a time—click-click-click—parbuckling it slowly onto the truck.

Leverman Mack Grant explained how he kept a careful eye on the skidder drums:

Most problems on a skidder involved lines breaking. Lines would weaken by use. You worked in close proximity to them at any time, and the lines came right in the drums and you could touch them. The lines usually showed signs of breaking—they wouldn't just snap—there'd be fraying, so you'd see it. The lines that pulled the logs were an inch and an eighth thick. They were made of woven cable. When the lines started to fray, you'd have to replace them—that was quite a common occurrence. They gave out quite often and they'd have to be spliced. ■

HARRY STEERE-CLARK PHOTO

Hooker and chokermen pause as they choke two logs for yarding, Camp 3, 1920s. Yarding logs required the attention of these workers as well as the yarding leverman and whistlepunk (signalman). The whistlepunk relayed messages from hooker to leverman by means of a whistle wire strung between saplings.—Noëlle Clark & Roger Emery

Death at Camp 2

Born a farmer's son in the small town of Saarijärvi, central Finland, Juho Hyytiäinen arrived in Canada in 1909, aged 19. He married Rachel Solmie from Ladysmith's Finnish mining community in 1916 and later joined Comox Logging as a high rigger. Like many Finns, he changed his name to something more digestible to the locals, and Juho Hyytiäinen became Herman Anderson.

On March 5, 1925, the *Comox Argus* reported on the accident that fatally injured Anderson:

Crushed in the writhing folds of a guy rope jarred loose from its moorings, Herman Anderson of Camp 2 of the Comox Logging Company, was so badly hurt yesterday afternoon that he died in the hospital at Comox this morning. His wife, who lives with his eight-year-old son at Ladysmith, sped up the island by car to see the last of her husband but arrived at 2:30 on Wednesday morning, an hour and a half after the head rigger had died.

Anderson was the acting head rigger at Camp 2 and was one of the best men at his job in British Columbia. He and his crew were tying a guy line on a new road and were fastening the inch and a quarter line around a stump when the back spar tree came down and crashed on the guy line, jerking it off the stump. The line came back at a terrible speed and with terrific force and writhed itself around Anderson, who was astride it at the time. Like Laocoon, the line wound around his head and legs, crushing him fatally. Anderson's body acted as a buffer and only one of his crew, a man named Erkalla, was hurt and he only slightly. Anderson was brought down on the train to Courtenay, where the ambulance was awaiting him and he was driven with all care to the hospital but he never recovered consciousness and he died at 2:30 [1:00] *yesterday morning. An inquest was held last night at Courtenay by Dr. Butters and a verdict of accidental death was returned.*

Anderson's great-granddaughter is the actress and animal rights activist Pamela Anderson, born in Ladysmith in 1967. ■

VINCENT RUSSELL PHOTO

Tight lines. Finnish rigger Herman Anderson stands on a fully rigged and operational head spar above a Lidgerwood skidder at Camp 2, Oyster River, *c.* 1920. Anderson, great-grandfather of actress Pamela Anderson, was killed a few years later. This photo illustrates the extensive preparations for highlead logging, which required, Eustace Smith recalled, "a spar tree rigged up with heavy yarding blocks placed at a high level, guyed for the heavy strain" (quoted by Wellwood, "The Smith Saga"). Mauno Pelto described the preferred dimensions: "The rigger would top the tree at about 160 feet—where the diameter would be 20 to 28 inches—and then the tree was rigged for logging. Rigging of the tree involved installation of blocks and guylines to hold it firmly in place while it served as the anchor point for the yarding of logs" (*Crown Zellerbach News*, October 1978). At the base of the tree is the wood-fired skidder with a sparkcatcher over its chimney. "Skidders used to spew out sparks like you wouldn't believe. Any time you fired wood in there, sparks were going out," recalled Bill Wilson. —Al Norman

The logging foreman was Jack McQuinn, who had been raised on a tributary of Oregon's Willamette River. Respected by his crews, the hard-nosed McQuinn had gone to work in the woods as a whistlepunk when he was about 12, came to Comox as a rigger with the first skidders in 1911, and was headhunted in 1919 by Bob Filberg, with whom he had worked as a boy in Washington State. McQuinn and Filberg were two of only a handful of American-born loggers to remain in the Comox Valley. While most members of McQuinn's crews were Homeguards, his crews typically included a few newcomers—usually Swedes, Finns, Germans, or Croatians, capable men who had come to the notice of the grade crew or steel gang foremen or who had married into local families.

In charge of cold-decker and skidder crews were the high or head riggers. These top men of logging crews climbed, topped, and rigged the big fir trees used in both yarding and loading. Their spectacular daily work figures prominently in logging photography. They have, literally, figuratively, and justifiably, a high profile in the logging literature. Their work is glorified in the paintings and comics of Bus Griffiths and in much popular logging history.

Once the logs had been loaded onto flatcars, they were the responsibility of the train crews. Every locie (that is, locomotive or engine) was manned by an engineer, a fireman, a conductor, and one or two brakies (brakemen), usually local men. The rail division included eight to ten car-knockers, who repaired and maintained (in the car shop at Headquarters) the fleet of 200 much-battered flatcars; the sparkcatchers, who patrolled the rail lines in hot weather to extinguish fires caused by sparks from the locies; and the flagman in downtown Courtenay, who held up the road traffic several times a day as logging and crew trains made their way to Royston and back. The entire rail network was maintained from every camp by a section gang, a six- to ten-man crew under the general supervision of a genial Irishman named Fred Griffith. The section gangs repaired the railways after floods, accidents, or other disrepair. About 45 men made up Comox Logging's rail division, which was managed by a New Brunswicker named Clay (Clarence) Walsh, known as the "train boss" or "locie foreman." Walsh had married Edith Cliffe in 1915; local respectability and influence were side benefits of marrying into the Cliffe family. Walsh was also, technically, assistant superintendent to Bob Filberg—a measure of the importance of the railway to the logging company.

The train crews took the logs on the final stage of the transit from stump to dump. They hauled them to the Royston boom camp, a few miles south of Courtenay, and dumped them into Comox Harbour. Here, the boom crew took over. The logs were sorted by species, shoved into rafting pockets, and boomed and stored in

HARRY STEERE-CLARK PHOTO

Logging out of Camp 3, 1920s. Suspended from the carriage (above), three choker lines make their way from the skidder and head spar, at left, to pick up a turn of logs somewhere in the setting. Joe Cliffe called choker lines "slings." At 21 feet, they were long enough to encircle a log six feet in diameter and keep it in a tight grip ("choke"). "The principle of hauling logs by sky-line is somewhat similar to bringing in washing on a line which runs out to a tree in the garden and back to the window," wrote journalist Renny Englebert in 1947. "In this case it is obviously far more complex because the logs have to be picked up from the ground. To do this, a carriage or trolley runs along the main skidding line down to the ground equipped with hooks or a noose of steel cable called a choker." The carriage could reach speeds of 60 miles per hour as it flew between the skidder and the back spar. Davey Janes, who worked at Comox Lake in the 1930s, recalled that some guys would ride the choker lines from the back spar to the head spar.—Noëlle Clark & Roger Emery

Greasy McQuinn

From a farm in Oregon, Jack (John Grover) "Greasy" McQuinn (1892–1963) was first hired by Comox Logging as a rigger on the skidders in 1911. He took charge of Camp 3 in about 1920. McQuinn was tough, blunt, efficient, resourceful, practical, abstemious, and loyal—qualities that endeared him to the Comox Logging Homeguard. "Over a long period of years, no one ever associated with the Company has worked harder or more successfully in the Company's interests than you have done," superintendent Bob Filberg wrote to him in 1933. "McQuinn taught by example," recalled Frank Biss. "If there was a problem, he wouldn't *explain* how to do it. He'd get in there and *do* it." He got his nickname from his mechanical streak. "That's how old McQuinn got his name: if they were setting up a skidder and there was a pile of blocks, he'd be under there getting greasy," said Art Beech.

He had sole control over woods hiring and firing at Camp 3 in the 1920s and 1930s. "Jack McQuinn, not Filberg, ran the day-to-day operations of the company," recalled Melda Buchanan of Comox. "You couldn't argue with or oppose McQuinn. He was a very powerful figure." Bob Muir related a story that emphasized McQuinn's ability to size up new employees:

My uncle Dick Bailey was a real admirer of Jack McQuinn. He said that when guys came in, McQuinn would look them over and if they had small feet he'd figure they were catty and quick, and he'd say, "You go second loading!" and if they had big feet, he'd figure they were slow, and he'd say, "You go on the chokers!" It was an old saying in the woods: the smaller guys are faster.

Others have less charitable memories. Men who worked on the falling crews and grade and steel gangs, immigrants from northern and eastern Europe, remember him as a callous, arbitrary, and unfeeling boss, and no friend of the unions that tried to get a foothold in Comox Logging in the 1930s. According to Eddy Murtsell,

McQuinn was an awful guy—he didn't care for anybody! Once, someone got killed at a skidder, and they brought in a stretcher. They asked McQuinn where to put him. McQuinn said, "Just put him down by the spar tree there—we might have two or three more by the end of the day."

By the late 1920s, however, McQuinn was actively involved in the company's safety committee, and he was transferred temporarily to Ladysmith as foreman in 1936. He and his wife, Goldie (*née* Fredenburg), brought up their 11 children on a farm on Anderton Road, Comox. ■

HARRY STEERE-CLARK PHOTO

When the steel had been laid, the logging foreman came in to assess the new settings. Harry Steere-Clark's caption was "Jack sizing up a new setting. Camp 3." Camp 3 boss Jack McQuinn is shown here on a speeder, 1920s. The Model 551 Casey Jones speeder, built by the North Western Motor Company in Eau Claire, Wisconsin, was powered by a Ford motor and claimed a maximum speed of 40 miles per hour.—Noëlle Clark & Roger Emery

the shelter of Goose Spit to await the final journey to Fraser Mills. The boom crew, numbering 25 or 30 men, and the boom camp were managed until 1927 by American Pete Donnelly and then by oversized Homeguard Hughie (Samuel Hughes) Cliffe, known as "High Pockets" or "Big Hughie."

These crews totalled about 345 men. They were distributed among Headquarters, the boom camp, and the three outer camps, the last of which each housed a range of stump-to-dump experts in falling, bucking, scaling, choking, yarding, loading, and hauling logs. The remaining approximately 50 employees were mostly based at Headquarters. At the shop they included mechanics, welders, machinists, lathe men, boilermakers, blacksmiths, sawfilers, carpenters, night watchmen, hostlers, sleighmakers, timbermen, handymen, and locie oilers. The white-collar staff included superintendents, engineers, scalers, timekeepers, paymasters, storekeepers, accountants, and postmasters. The Headquarters Hotel required cooks, flunkies, and bed makers, all of whom were Chinese.

The woods crews did their jobs all too well, and by 1926 the accessible coastal timber was exhausted. Logged-over land was useless for continuous profit-making; replanting was not started until 1943. Comox Logging needed new timber properties to survive, so Camp 2 was picked up and moved inland by rail—lock, stock, and barrel—to the former coal mining town of Bevan and new reserves of mountain timber. Until the abandonment of skidder logging in 1946, Camp 2 went through later and higher incarnations at Quadra Island, the shores of Comox Lake, and Cruickshanks Camp. But the men of Camp 2 would always look back nostalgically to their years among the big trees on the coastal flats. ■

WALTER MONTGOMERY PHOTO

The "edge of the world economy": the railway has been extended right up to the uncut forest in the distance. "Comox Logging's railway was really important," recalled yarding leverman Mack Grant, "maybe more important than anything else because it took the logs from way back halfway to Campbell River and brought them down here to Royston." At left, the engineers of Skidder 6 yard and load logs out of Camp 3, *c.* 1930. On the skidder the yarding leverman brings in a "turn" of three choked logs, while the loading leverman swings, by tongs and hayrack boom, a log beneath the hayrack boom onto a flatcar. "Man's inventiveness has made the skidder the king of the woods," wrote journalist Renny Englebert in 1947. "It does three operations at the same time. One drum brings the logs to the landing, another runs the miles of cable and lines, whilst the third loads on to rail cars."—LAURIE MATHERS

> *Detail of photo above*: A view down the empty flatcars towards the loaded flatcars. Two loaders pause from their work while the photo is taken. "North of the old camp at Headquarters on comparatively level ground, logging is being done with tree-rigged Lidgerwood skidders," commented *The Timberman* in September 1935. Seven or eight men were stationed at what was called the "front end" of a skidder setting. They were known as the landing (or loading) crew. Note the bark and other debris on the flatcars and the cheese blocks on each side to support the log loads.—LAURIE MATHERS

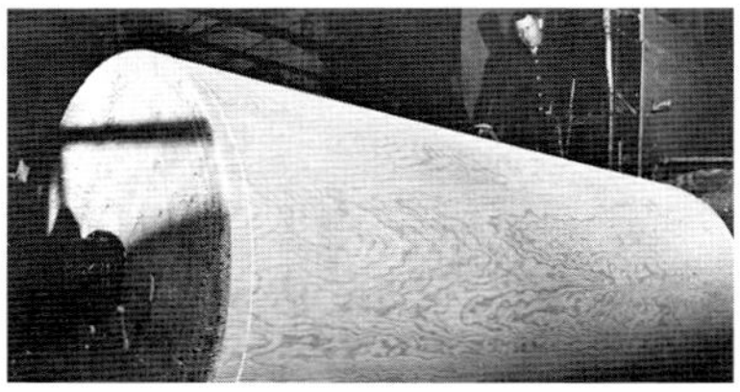

Peeling thin sheets of veneer off a Douglas fir log at the plywood plant at Fraser Mills, New Westminster, *c.* 1940. "A peeler was a veneer log," recalled Al Knappett. "Plywood in those days was very good stuff!" In 1913 Fraser Mills became the first Douglas fir plywood manufacturer in Canada. "Sheets of veneer are glued together, grain in each layer at right angles to adjacent sheets to give added strength," explained a Crown Zellerbach publication in the 1960s. "As many as seven layers, or plies, are pressed together under heat." According to Ted Nikiforuk (*Coquitlam 100 Years*), the glue used in this process was made from the blood of slaughtered cattle from Swift's Canadian Company's slaughterhouse in New Westminster.—RAY KNIGHT

WALTER MONTGOMERY PHOTO

The Skidder Skyline System

In "Early Methods of Logging," Joe Cliffe described the basic principle of highlead skidder logging, a system introduced to Canada by Comox Logging in 1911:

Many logging companies in Washington, Oregon, and British Columbia used the skidder skyline system of railroad logging. Simply explained it was a large cable hung on a spar tree at the track and also hung in a tree at the back of the felled timber. These back spars were spaced approximately 200 feet apart all around the setting. The skyline was threaded through a carriage. When the skyline was tightened the carriage travelled back and forth. Logs were lifted up clear of all stumps and landed at the track without any hang-up trouble. This system was especially suited to rough ground with ravines and side hills. These machines were more expensive than the highlead yarder and required an immense amount of steel cable as well as a well-trained crew of about 20 men. This system was favoured by government foresters as it did less damage to the young forest. ■

WALTER MONTGOMERY PHOTO

Yarding logs across the Oyster River, *c.* 1927. A major river was a minor obstacle in highlead logging. Burned in the 1922 fire, the lower Oyster River region still contained much fire-damaged but salvageable timber. The chokermen, barely visible among the slash at right, have just sent three large logs across the river to the skidder landing (out of view to the left). Near the top of the photo are the skyline and carriage, from which the choker lines descend. Snags are still standing in the burned-over landscape. At first, Comox Logging did not clearcut its forests but rather left standing the saplings and snags, which constituted a hazard to the men working among them. As Sven Ell recalled, "In the old days they left the snags standing, and the accidents were pretty awful because they would pull those turns among the standing snags. We were looking at an old photo of a loading crew, and right around the yarding machine were snags. Filberg said, 'When I look at that picture I get cold shivers!'"

—Eddy Berkeley

HARRY STEERE-CLARK PHOTO

***Above*: The magic of highlead: suspended from the carriage (bicycle), two 40-foot logs trundle back to the spar tree in the Camp 3 area, 1920s. "Lidgerwoods were all skyline," recalled Sam Telosky; "the logs hardly touched the ground." According to leverman Mack Grant, "They'd pull the logs up to the skyline. They'd pull them up on an angle to the bicycle, which ran on a cable; it would pull back and forth on a cable, and then they'd pick one end up and drag it along." The genius of highlead yarding, as Eustace Smith put it, was that "the nose of the log was lifted clear and would ride over all obstructions without being held up en route, which so often happened with ground yarding through logged-over areas" (quoted by Wellwood, "The Smith Saga"). These logs were then dumped at the landing pile next to the skidder before being loaded onto flatcars.**

—Noëlle Clark & Roger Emery

Hughie (Samuel Hughes) Cliffe, shown *c.* 1925, became foreman of Comox Logging's boom camp at Royston when Pete Donnelly retired in 1927. A towering six foot six, Cliffe was known as "High Pockets" or simply "Big Hughie." Born in Comox in 1889, he was a son of Comox pioneers Sam and Florence Cliffe, brother of Lu (Lucius) and George Cliffe, and uncle of Comox Logging foremen Joe and Harold. "As a youngster," his daughter, Lorna, related, "Dad was farmed out because there were too many kids in the Cliffe family. He lived and boarded with a blacksmith in North Vancouver and he was blacksmith's assistant." Later he worked as a logger, hunter, and fisherman, and during the First World War on the wooden shipbuilding program in B.C. Boom man Donnie Haas, who worked under Cliffe at Royston in the late 1920s, recalled that Cliffe was "a good boss, good to the crew. In the summer, when the tide went out, we only worked four or five hours, but we were paid for eight hours. Hughie would say, 'You've worked hard,' and he'd let us go home, and we got paid just the same wage." In the 1950s Cliffe ran Comox Logging's sister company, the Canadian Tugboat Company, and he is best known for his invention of the boom or "dozer" boat. He died in 1964.—WILMA MCKENZIE COLLECTION, COMOX ARCHIVES & MUSEUM SOCIETY, A2000 19.46

HARRY STEERE-CLARK PHOTO

Supplying a skidder setting, Camp 3, *c.* 1925. The Six-Spot, a Shay engine, prepares to unload flatcars containing tanks of oil and water and to deliver, behind the spar tree, a cold decker on a flatcar. Skidder 6 is parked on a tiny spur beneath the head spar. Tanks of oil and water appear in most skidder photos. In "Early Methods of Logging," Joe Cliffe explained the need for water and oil: "These log yarders and skidders when working used a tremendous amount of water which sometimes had to be piped for miles and quite often in summer months had to be hauled in steel tanks by a locomotive ... Around 1920 all railway logging companies started converting their skidders and yarders to burning bunker fuel oil. This gave a constant hot, even fire eliminating 90 per cent of fire hazard from red hot exhaust cinders and eliminating the old steam saw and wood splitter." —NOËLLE CLARK & ROGER EMERY

The Six-Spot, a coal-fired geared Shay locie, pushing its load through a Comox Logging camp, *c.* 1917. In the distance is a water tower with spout. Shays were ideal for hauling on steep grades. Built in 1913, the Six-Spot was later sold to Pacific Coast Terminals at New Westminster. It was scrapped in about 1939.—WHITFORD & CRAIG, *FORESTS OF BRITISH COLUMBIA* (1918), PLATE XVII

HARRY STEERE-CLARK PHOTO

A locie hauls three loaded flatcars through Camp 2 at Oyster River, *c.* 1923, down the main-line conveyor belt that gradually emptied the riches of the flats. Second-growth forest is arising spontaneously on each side of the main line, but fire-blackened snags still dominate the horizon. Natural regeneration was the norm until the first replanting in 1943. At right is one of the characteristic portable bunkhouses of Camp 2, with tarpaper roof and skids beneath. "A railway line wound inland to one or more woods camps, haphazard collections of white buildings flung down in a desolation of burned-over logging slash," wrote ex-logger Arthur Mayse in 1985 of woods work in the 1930s. At bottom right is a "Wee McGregor," a mechanical bucker used to saw logs into firewood.—NOËLLE CLARK & ROGER EMERY

Fireman Jack Armstrong with the Three-Spot main-line locie, *c.* 1930. Jack (John Bell) Armstrong fought at Ypres with the Canadian Artillery in the First World War. He married a war bride, a Londoner named Doris Grey, and took up land at the Soldier Settlement at Merville, where he ran cattle. Like many Merville settlers he found secure work at Headquarters. "He was a big man, holy cow! He had to duck going under the door," recalled Eddy Murtsell. In the 1930s he moved with Comox Logging to Ladysmith.—DOROTHY TUKHAM

The Three-Spot preparing to pull a load from the bush to Royston, *c.* 1935. Left to right: unknown, Jack Armstrong (fireman), Bob Martin (conductor). In cab: unknown. "When the Three-Spot arrived [in 1911]," wrote Comox Logging railway historians Cummings, Oliver, and Anderson, "she was the largest engine on Vancouver Island, and remained so for several years. She was considered the elite of logging locies." In the early 1920s the Three-Spot engineer was Len (L.F.F.) Spencer, who had worked previously with the Canadian Pacific Railway. Originally from England, Spencer had worked for several other logging companies on Vancouver Island, including Merrill & Ring and Elk River Timber. He married Margaret, one of the Williams girls of Williams Beach, and lived at Camps 1, 2, and 3 in the 1920s.—TRUDY MOONEY

BOB FILBERG PHOTO

Comox Logging's Seven-Spot backing at least 20 flatcars full of logs onto the trestle at Royston, *c.* 1938 (view is to the east). "This is the original trestle and log dump," recalled Al Knappett of Royston. "The piledriver is at right. That's probably bridge timber piled in front: they were always repairing that trestle. Some timbers were creosoted, but most were green and teredos would eat them out in no time."—LORNA CLIFFE COLLECTION, COURTENAY & DISTRICT MUSEUM & ARCHIVES, 2000.158.54

The Three-Spot at Courtenay Crossing pulling two loads of cedar poles in front of a regular log load, *c.* 1940. "They always backed up to Royston with the locie *backwards* and with the tender leading the way," explained George Yates Jr. "I never asked an engineer why they did this. I always meant to! I think it had to do with weight. The tender was lighter than the locie itself." The Riverside Hotel is at left and the Native Sons' Hall at far right. The locie engineer is Jack Carthew and the flagman is Dick Delin, formerly brakeman at Camp 3 and Comox Lake. Yates recalled that the Three-Spot would take 60 flatcars at a time to Royston on the main line, and the watchman stopped the traffic when the train reached the Riverside Hotel. Buster Brown Sr. (J.N. Brown), a brakeman on the main-line locie, used to tell his son Bob, "We'd jump off the train at Courtenay and run up to the Riverside for a beer—it was only 20 steps to the Riverside. We'd go in for a beer and come out in time to get on the last empty flatcar!"—ELWOOD WHITE

CHARLES SILLENCE PHOTO

A "perfect load" of six logs at Royston, *c.* 1925. "In the early days a giant tree made up a car load, but towards the end of rail operation, as the timber had to come from higher levels, it took more and more logs to fill a car," noted historians Cummings, Oliver, and Anderson. The man is water scaler and boom camp foreman Pete Donnelly. He is holding an L-stick—a device used to calculate the board measure of timber. Donnelly retired in 1927 at the age of 70 and was succeeded as foreman by Hughie Cliffe.
—BOB BERKELEY

A load of the K & K Pole Company's cedar poles at the Royston boom camp, *c.* 1930. Just visible between the cars are K & K's Danish-born foreman, Al (Axel) Pedersen and his daughter, June. These poles stretched over two 40-foot flatcars. The lengths are written on two of the poles: 95 feet and 75 feet. "Comox Logging took the cars on the main line to Royston and left the logs on a spur there and the CPR picked them up," recalled Al's son, Buck Pedersen.
—BUCK PEDERSEN

Forest meets field as the Three-Spot pulls a load of flatcars to the Royston log dump, *c.* 1923. "The No. 3 was the main-line locie. They'd line up log cars with small locies and when they had enough they'd take them down to Royston," recalled Ralph Dalby. According to Ralph Harmston, "Everybody would watch the train go through Courtenay. Fifty to sixty loads of logs! Twice a day they went through." Many people on the Comox Logging main line relied on the empty returning flatcars for a supply of long-burning Douglas fir bark.—Leona Shackleton

Empty flatcars being pushed up the main line between Courtenay and Headquarters, *c.* 1930. The toiling locie is just visible in the cloud of steam in the distance. An old Swede named John Anderson used to tell his nephew Eddy Murtsell about his days as fireman on the One-Spot. "Uncle John said those wooden flatcars were heavy as hell! They were firing with coal, and he had to shovel coal all the way from Royston to Headquarters. It was uphill and there were a lot of cars." Note the adjustable cheese blocks, which held logs on the flatcars. Comox Logging, wrote railway historians Cummings, Oliver, and Anderson, "is a distinguished hauler among logging railways for they do not use the customary skeleton type of log car, but instead a full deck flat car. This has been standard for many years."—Donna Edwards

Booming and Farming

In 1995 Don Berkeley tape-recorded his father, Jack (1913–96):

I was born at Comox hospital [St. Joseph's Hospital] *on September 29, 1913. At that time we lived up the Lake Trail Road and Dad had a farm up there, and I lived there till I was six years old. This was roughly 1920. And then they bought a farm down at the end of 29th Street. At that time it was called Embleton Road, and we had a 50-acre farm in there ... But my dad had to work up at Quadra Island—the camp moved up there, so in those days you had to follow the work, so he had to go up there. I had to stay home and look after the farm, put the crop in, look after the cattle, and so on, so that terminated my school days. This is still 1929. I worked as a whistlepunk and in 1931 I got a job down at the boom camp, down at Royston. Hughie Cliffe was the boom foreman.*

[Don: He was a relation of yours?] *Oh God, yeah, I'm sorry to—oh, strike that!*

[Don: Was he behind you getting a job there as well?] *Not really, no. He was instrumental in why I got laid off! No; through the old man they were distant cousins, I guess. The old 42nd cousin deal, you know. This guy, he was about six foot four and he weighed around 220, not an ounce of fat on him, and he was the only man I ever saw who couldn't get his hand down an auger hole. Couldn't put his hand down the* [four-inch] *auger hole! Hughie was a real bull-of-the-woods type like they had in the early days, you know. God!*

[Don: No favours going your way because of it?] *No, not a bit! I can tell you what he said a few times, but it might not be appropriate, you know!*

[Don: And you lived at Royston?] *Single men had to stay there. That was one provision—to hold your job you had to stay in camp, even though I paid board at home. The first year I worked there* [1931], *it was maybe six to seven months, I just handed the cheque over to Mum, and she gave me spending money, and that was it; I also had to pay board at camp. And even part of the next year I did that too. That seemed to be the standard thing in those days.* ■

BOB FILBERG PHOTO

A view to the south along the trestle at Royston, *c.* 1940. The humdirgen rides the rails, and beside it a tug is moored to the trestle. At left is a completed eight-section boom surrounded by a line of boomsticks with swifter logs pulled over the top. Boomsticks were the logs that surrounded a boom, while swifters held the boom down in rough seas. Swifter logs were attached to the boomsticks with a boom chain at each end. Ray Stockand recalled that flat booms were about 30 logs across. In the distance is Canadian Collieries' wharf at Royston.—Lorna Cliffe Collection, Courtenay & District Museum & Archives, 2000.158.34

Captioned "Comox from air—August 1934. H.M.S. Norfolk." The photo shows, from top to bottom, the village of Comox, the wharf, the *Norfolk* riding at anchor, Comox Logging's flat booms tied up in the shelter of Goose Spit, and the tents attached to the navy's firing station and rifle range. Log booms were stored here until required at Fraser Mills. —D'Esterre Collection, Courtenay & District Museum & Archives, 998.219.102

An "outside tug," probably the 85-foot, coal-fired *Gleeful*, tied up inside Goose Spit, Comox, *c.* 1930. This vessel was added to Canadian Western's fleet in 1914. "Prior to the arrival of the outside tugs," wrote Charlie Nordin Jr. in a 1999 letter, "Captain C.E. Nordin received orders from Fraser Mills by his home telephone on how many sections of Douglas fir, hemlock, cedar, peelers, etc, would be allotted to that particular tug. When an outside tug arrived, the extensive yarding by that tug, and the SS *Joyful*, began which would take several hours or more."—Dorothy Tukham

The sturdy tug *Joyful* leaving the Comox wharf, *c.* 1940, with Comox and the Carthew farm behind. Between the vessel's arrival in 1912 and its retirement in 1948, the *Joyful* pushed and towed over three billion board feet of logs between Royston and the boom tie-ups at Goose Spit. For most of these years her crew consisted of only two men: skipper Charlie Nordin Sr. and engineer Jock (John) Smith.—Charlie Nordin Jr.

Booming and Dancing

Bill Walker, Comox Logging boom man, was born in Scotland in 1906. His father, a machinist in Glasgow, was a "dedicated Communist," according to Bill's wife, Doris. "[My father-in-law] used to say, rolling his Rs in the Scottish way, 'We should be living in R-r-r-russia! R-r-r-russia is the only place to live!' " He lost his job in Scotland for organizing machinists' unions.

The Walkers immigrated to Canada when Bill was a boy, and he was raised at Union Bay, where his father was a boilermaker for Canadian Collieries—and a union organizer. Bill worked first for the Esquimalt & Nanaimo Railway when he was very young, and then he went to work for Comox Logging at Royston. He did a number of jobs, but mainly he was a boom man. "He was a good boom man; he was in demand all over the Island," recalled Doris. He was at Royston in the mid-1920s, Quadra Island in 1928, and Deep Bay in 1930 with the Canadian Robert Dollar Company. In 1933 he was back with Comox Logging at Comox Lake, and in 1934 he was at the Royston boom again; in that year he married Doris Marvin. Doris recalled:

The Depression had been hard on our family. We had moved from Saskatchewan to Royston in 1927. My parents owned the Royston Beach Auto Court on the Island Highway. I had trained as a nurse but I was working there when I met Bill. I chopped kindling, carried water, changed beds—there was lots to do. I joined clubs in Courtenay and also taught first aid. In the Depression we lived off the beach: we ate clams and oysters, and shot deer in the backyard. Bill and I used to dance at Royston Pavilion, which had a lovely spring dance floor. Loggers came from as far as Camp 3 for dances there. There was lots of dancing when the loggers were not up in the woods. The guys were in the logging camps, the girls were free, and there were dances to go to, so what do you expect? Later, when we were married, Bill put up the logs that held the sign over the entrance, and I painted a big sign reading, "Royston Beach Auto Court."

In 1937 Bill and Doris moved from Qualicum to Ladysmith, where Bill also worked on the boom. A year or two later he retired from Comox Logging and went into the furniture business in Duncan. "He'd had enough of booming work," Doris said. "It was hard on the legs and the back." In the war he joined the army and left Doris behind as matron of Shawnigan Lake Boys' School. Bill Walker died in 1984; Doris Walker died in 2000. ■

Bob Filberg photo

The double-ended bunkhouse at Royston, *c.* 1940. "It was pretty up to date," recalled Ron Bowen, who stayed there in the 1920s. "They used both ends of it. In the middle was a big hot-air furnace. A bell rang in the morning and we went over to the cookhouse on a wooden sidewalk." The bunkhouse provided accommodation for the boom crew and also for woods crews after the closure of the Headquarters Hotel in 1929. A train left Royston early every morning for Headquarters and took men from Royston to work in the woods. One of them was Mauno Pelto, then a cold-deck engineer, who arrived in 1938. "Uno Saari and his wife ran the cookhouse," Pelto said. "There were just four of us staying at the bunkhouse, but the boom crew was 35 or 40 men. It was all pike poles then, no dozer boats. Men were making up booms there; it was a *very* efficient operation." Another was faller Sven Ell, who was at Royston in 1941. "I worked for a while in the Courtenay area but I only did one quarter before going to Ladysmith," he recalled. "I stayed at Royston; I boarded in the bunkhouse and took the train to the woods. It would go right through town."—Lorna Cliffe Collection, Courtenay & District Museum & Archives, 2000.158.33

A Douglas fir boom at Royston, *c.* 1940. The boom men have finished securing 72-foot swifter logs over top of these flat booms. The company tried to get its booms to the fresh water of the Fraser River as soon as possible to avoid the pestilent teredo worm, a wood-boring marine mollusc. "We also blew the teredos from standing booms," recalled Sam Telosky, who worked at Royston in 1935-36. "If a boom was in the water for more than two or three months, the teredos got in, and the boomsticks would sink, so you'd have to blow the teredos on the boomsticks. You used half a stick of powder with a short fuse. It was the shock that killed the teredos—it just killed the teredos and didn't damage the boom. When I was at Royston, we'd always get a salmon when we blew the teredos! The salmon would be lost and running around under the boom when we blew it."—Lorna Cliffe Collection, Courtenay & District Museum & Archives, 2000.158.35

When push comes to shove. The humdirgen log unloader dumps a flatcar full of logs into Comox Harbour, *c.* 1930 (view is north towards Courtenay). In 1935 the Royston booming grounds could process 10 million board feet of timber a month. "The tracks on the Royston wharf were always canted a little bit to allow the logs to fall," recalled Bill Wilson (1912-2007). Down the track from the humdirgen are loaded flatcars and, in the distance, a locie, while below them, in the "bull pen," a boom man walks on the Douglas fir log at far right. This photo was taken before the log dump and booming grounds were rebuilt and reconfigured in the late 1930s. At that time the trestle was extended and logs were dumped on the inside (west) of the trestle, rather than on the exposed outside (east).—Donna Edwards

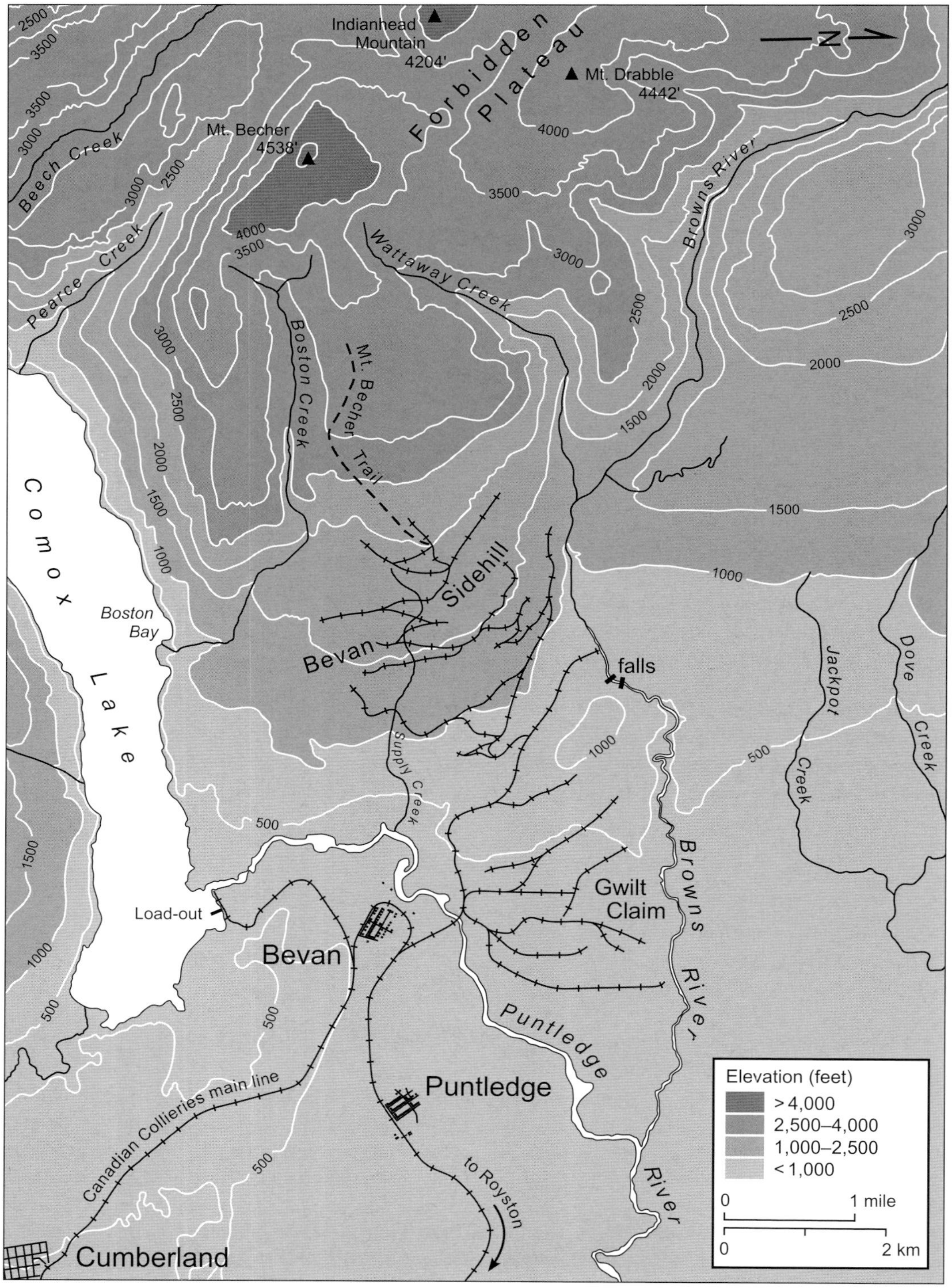

Comox Logging's railway show at Bevan, situated on the mountain slopes beneath Forbidden Plateau and bounded by the Puntledge River and Browns River. Camp 2 was kept busy at the Bevan sidehill between 1927 and 1933.

It is only within the last twenty years that this upland country has been known, the early Comox settlers being too busy making a living from the soil to do much exploration. Though they always lifted up their eyes to the hills, their way there was barred by the dense and heavy undergrowth in the huge forests surrounding the valley.

—ERIC DUNCAN, *FROM SHETLAND TO VANCOUVER ISLAND*, 1934

3

EYES TO THE HILLS

AFTER IT HAD EXHAUSTED THE TIMBER IN THE LEVEL COASTAL areas, Comox Logging turned to the Vancouver Island Mountains. Camp 2 on Oyster River was closed in May or June 1926 and the site was abandoned. Everything was moved onto flatcars and taken out; even the big cookhouse was moved. "When they took Camp 2 out," recalled Herbert Pidcock, whose family farmed nearby, "they took the old bunkhouses out, and we ran cattle there in the 1930s. There were no buildings left there then." The personnel and equipment of Camp 2 were transferred to Bevan, three miles west of Courtenay, where the company planned to log in a mountainous region more challenging than the flat "loggers' Eden" of Block 29 north of Courtenay. On February 4, 1926, the *Comox Argus* chronicled the project. "Arrangements have been completed whereby Camp 2 of the Comox Logging Company at Oyster River will be moved to Bevan at the latter part of May or the beginning of June. This will mean the transference of about a hundred men from fourteen or fifteen miles north of Courtenay to three miles west. Work in the woods at Bevan will commence this month in preparation for logging operations."

In the mid-1920s Bevan was a nearly defunct mining company town. "Bevan had a good mine known as the No. 7," recalled Stan Hanham, who lived there in 1929, "or so they thought, but it didn't pan out as well." Constructed in 1911 by Canadian Collieries, Bevan was designed as a workers' town for the No. 7 mine near the Puntledge River. Canadian Collieries built 150 houses in 1911–12 as well as a hotel and a store. Houses at this company town were rented to mining families, most of whom came, at first, from Scotland.

The unexpected demise of the mine, according to Dick Isenor's *Land of Plenty*, arose "from an excess of sulphur in the coal and a deterioration of the coal seam." In 1918, 100 Bevan houses were moved to Cumberland, where the coal deposits were more secure. The mine closed in 1922, after producing 1.15 million tons of high-grade coal.

Comox Logging had an interest in two timber properties on the north side of the Puntledge River: the "Gwilt claim" (directly across the river from Bevan), which the company bought from lumberman Tom Gwilt in about 1924; and the "Bevan sidehill" (the lower slopes of Mount Becher, Indianhead Mountain, and Mount Drabble), bought in 1928 from the E&N Railway. Showing a clear appreciation of the enduring physical framework of a company town in a staple-driven economy, Comox Logging turned the abandoned mining town of Bevan into a logging camp (the new Camp 2), converted a mining hotel into a logging bunkhouse, and applied to higher elevations the same railway and steam technology it had used on level ground. The Gwilt and E&N properties—and a short-lived summer show at Open Bay, Quadra Island, in 1928—would keep the Bevan Camp 2 busy from 1927 until 1933.

COMOX LOGGING TACKLED the Gwilt claim first. Over the winter of 1925/26, surveyor Hughie Stevens, assistant Wallace Baikie, and Baikie's helper Johnny Duncan (who would later replace Stevens as Comox Logging's chief engineer) surveyed a branch line off the Canadian Collieries main line at Bevan, laid out a railway bridge across the Puntledge, and determined the location of spur lines in the Gwilt claim. In April 1926, bridge foreman Bill Eastman was building the railway bridge across the Puntledge River with a crew that included Joe Cliffe, Dunc Thompson, Lyle McKenzie, and Tim Paterson. In the same month, 35 men were employed on the bridge and grades and felling the settings.

Comox Logging's parent company, the Canadian Western Lumber Company, now took over the empty premises of the mining company. On April 1, 1927, Canadian Western agreed to lease nearly the whole of Bevan for three years. This included the Bevan Hotel for $23.08 a month and, for $8.50 each, "such of the dwelling houses situated in the village of Bevan" as might be needed. The lease mentioned that Comox Logging was "now logging on the lands lying across the Puntledge River near Bevan."

In effect, therefore, in 1927 a logging company took over a mining town. This was not the last time that Comox Logging moved

> The view from the Back Road, Comox, 1945, with the Urquhart farm at left. This is more or less the westward view that Eric Duncan describes in the epigraph to this chapter. The logged-over mountainside at the centre of the photo is what the loggers modestly called the Bevan sidehill, logged by rail by Comox Logging between 1928 and 1933. The snow-covered logging scars, which extend to 3,000 feet, are clearly visible. Above them are the mountains of Forbidden Plateau, all of them about 4,500 feet high. In the foreground the farms occupy the fertile alluvial bottomlands of the lower Tsolum and Courtenay rivers, and Courtenay is in the distance. Comox Lake is nestled in the large mountain valley at upper left.—COURTENAY & DISTRICT MUSEUM & ARCHIVES, 989.96.19

> An aerial view of Bevan, c. 1935, showing the Puntledge River flowing from left (west) to right (east). Comox Lake is just to the left of the photo, and the houses of Bevan are visible near the centre of the photo. Railway lines lead from Bevan to Comox Lake, Cumberland (five miles south), Royston, and the logging show on the Bevan sidehill at top right. The photo shows the aftermath of logging by rail, skidder, and cold decker. On the north side of the Puntledge River are five skidder settings, identifiable by the radial skid roads that lead into the skidder settings at trackside. Bevan is about 400 feet above sea level, but the railway at right leads to switchbacks that extend up as far as 2,700 feet.—COURTENAY & DISTRICT MUSEUM & ARCHIVES, 984.62.7

SILVERTONE PHOTO

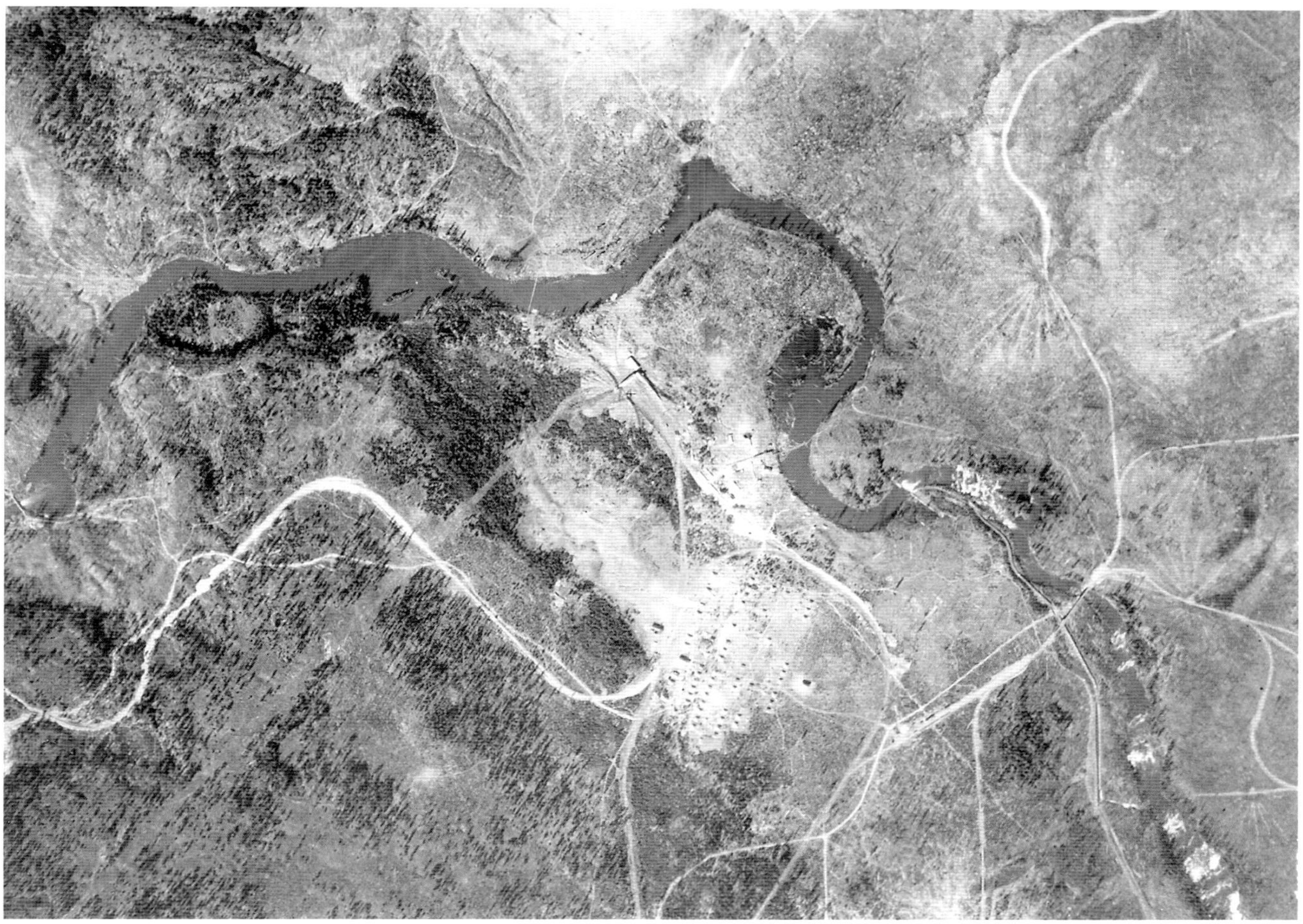

into a mining town upon the exhaustion of coal mining: it would do the same at Ladysmith ten years later.

The bulk of the Gwilt claim was logged in the summer and fall of 1927, and most of the Bevan contingent was then transferred to a temporary camp on an island north of Comox and east of Campbell River. In the fall of 1927, work began at Open Bay on Quadra Island, then known as Valdez Island (pronounced "Valdeez").

"On Open Bay, near Heriot Bay on Valdez Island," noted the *Argus* on October 6, 1927, "a gang of men has been at work putting in grades to take out about twenty million feet of timber that was killed in the fires there two years ago. This will be an operation that will be taken in hand this fall and winter."

Logging began, Wallace Baikie related in *Rolling with the Times*, in about April 1928, after the winter shutdown. "That gave them six months to log the claim, boom up the logs, pick up the steel and ties, move everything back onto scows and get out of there before the southeasters started. This they did, and it was quite an accomplishment. They were working six days a week in those days but they worked the crews ten hours a day and paid them overtime. My brother Harper was head loading and made enough that summer to start buying a De Soto." The fallers were through with their work by May 1928 and returned to Headquarters, leaving the logging crews to yard and load the felled timber.

BILL WALKER PHOTO

The *John Davidson* and *Stormer* at Heriot Bay wharf, Quadra Island, 1929. Coastal tug expert Bob Spearing wrote in a June 2002 letter that "the *Stormer* was not a member of the Canadian Western fleet, but was privately owned at that time." —Gwen Greenwood

BILL WALKER PHOTO

The tugboats *John Davidson* and *Stormer* at Heriot Bay wharf, Quadra Island, 1929. Comox Logging's show at Quadra Island was a northern extension of its regular booming and towing operations.—Gwen Greenwood

WALTER GAGE PHOTO

Landscaping required: The instant company town of Bevan, *c.* 1912. Bevan was built in 1911 to serve Canadian Collieries' No. 7 coal mine, which was closed prematurely in 1922. In 1927 Comox Logging rented the townsite and necessary houses from Canadian Collieries, and effectively relocated Camp 2 to this deserted mining town. Bevan is now abandoned; some of its houses, which were notoriously drafty and cold, were moved to Cumberland and Courtenay.—Babs Hind

Cutting Ties

Vast quantities of railway ties were needed at the Bevan sidehill. They were provided by local men who worked in advance of the steel gangs, as Stan Hanham recalled:

There were three or four tie-makers in Bevan. All they'd do was make ties on contract to Comox Logging. They'd cut them in the bush out of Douglas fir with a broad axe and bring them out of the bush to where the railway would be. They'd pile them up for when they brought the steel along the grade.

Company men, including Austin Blackburn, Bob McQuillan, and Otto Niemi, cut ties in the off-season. In *Rolling with the Times*, Wallace Baikie recalled that, one winter, Bob Filberg gave Harry Sheppard and Harold Grant a contract to cut ties. Filberg instructed Baikie, an assistant surveyor, to count the ties and make sure they were seven inches thick and had a six- or seven-inch face. Baikie described what happened after he culled a few:

Then a month or two later Joe Asselin and his steel gang went to pick up ties and found many undersized ones that they complained about to Filberg. Well, I had slept in one Sunday morning after a Saturday night dance and Bob came storming into our cabin giving me hell over these undersized ties I was half awake and tried to tell him that I had culled a fair amount of them but I guess a fair amount wasn't enough. He said, "Jesus Christ Wally, you should have culled half of them," and turned around and left me wondering if I still had a job.

I never heard another thing about it. Bob was like that, he could raise hell and storm off and be fine next time you met him. He expected respect and never got too friendly with his crews. ■

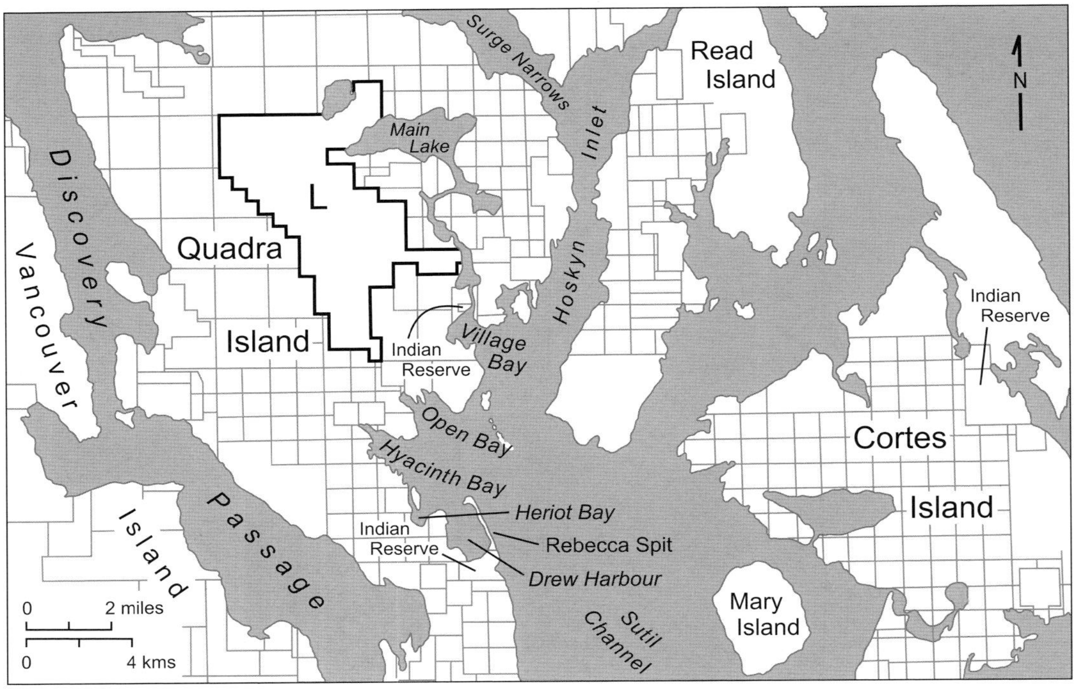

Comox Logging's property (Section L) at Open Bay, Quadra Island, logged by a detachment from Camp 2 in the summer of 1928.

Laying Out the Quadra Show

In *Rolling with the Times*, surveyor Wallace Baikie related that he and Hughie Stevens stayed at the Heriot Bay Hotel while they surveyed Comox Logging's property at Open Bay:

In the spring of 1927 I think it was, Stevens and I had a job to do on Quadra Island. The Comox had this claim of timber behind Open Bay. I guess that the reason it was never logged before was because there was no protection from the southeast wind as it was strictly a summer show. We picked out a suitable place for a dump, ran a line up the valley, laid out five skidder settings, put in the grade stakes and elevations and our job was finished.

Early the next spring [1928], *Bill Eastman and his bridge crew arrived. They first made a slip so they could unload equipment and rolling stock. The grading crew were already grading the road. The steel gang landed next by scow with their loci, steel and ties. They already had established a cookhouse and bunk houses, and Bill Eastman was building the dump. It had to be on pilings so they would have deep water. It was a beautiful stand of old-growth fir. The fallers came next, then when the next scow arrived with the skidder and flatcars all they had to do was to wheel them up to the woods, rig the first tree and start logging. By this time Bill Eastman would be driving the booming ground. Hughie Cliffe the head boomman had a supply of boom sticks already bored. They were logging!* ■

BILL WALKER PHOTO

Comox Logging boom man Bill Walker wrote the following caption: "Drew Harbour 1929. Logs from Open Bay operation Comox Logging & Railway, Quadra Island." The tall figure in the middle is probably Hughie or Harold Cliffe. The boom men are building the log boom between two rows of pilings, and they are standing on large 40-foot Douglas fir logs.—Gwen Greenwood

A high rigger, possibly Jack McKenzie Jr., topping a spar tree at Bevan, *c.* 1930. A large fir was already topped about 20 feet away but was unsound or unsuitable for use as a spar tree. "I can recall they'd top two trees in case one was not good," Arne Anderson remembered. Note the "fell and bucked" timber on the skidder setting in the valley below.—Fred Clifford

High rigger Jack McKenzie Jr. rigging a back spar at Camp 2 or Bevan, late 1920s, above a cold-deck pile and Clyde yarder.—Myrtle Heron

Left in the Dark

Shortly before his death in 2002, Davey Janes described his work as a mat-hooker on a Marion shovel at Bevan:

I went from Camp 3 to Bevan with the grade crew and I was hooking mats on a shovel on the sidehill for Carl Widen [Viden], *who was a Finn or Swede* [Swede]. *He drove a Marion shovel, a gas-electric. The grading foreman was Tommy McQuillan, who was one of the old original shovel operators. He ran one when they first got the shovels. Spit Quinn was also on the grade crew at Bevan.*

I was just a kid at the time—I started when I was 15. I was born in 1912, so this would be 1927 or so. I boarded with my sister Marjorie Blackburn at Bevan. I couldn't board in the grading camp because the pole makers were using it.

To get to work, I had to walk across the swing bridge on the Puntledge River when it was still dark and walk five miles up the mountain up to the eighth switchback every day. There was no crummy at that time. I was mat-hooking with a flashlight in the dark! We put in pretty well the whole grade, right to the top, before they logged it. That is why we had to walk to work—because the grade wasn't in yet. It was all switchback. The steel gang came later to lay the track down.

Janes's memory of crossing the Puntledge River in the dark is supported by an article in the *Comox Argus* on July 1, 1926:

Long before most of us, even the farmers, are astir these days, the woods are alive with men and the "toot-toot" of the donkey engine waking the echoes. Daylight saving has been adopted with a vengeance in the woods this season. Goodness knows when the cooks get up, but they almost meet themselves going to bed, because breakfast is on the table at Camp 3 and Bevan at a quarter past three so that the men can be on the job at four o'clock. But the last toot of the donkey is heard at half past twelve and the loggers have the whole of the rest of the day to themselves.

This change of hours has taken place during the hot weather to prevent operations when there is least moisture in the air and consequently when there is most danger of fire, a long period of records showing that nearly all fires break out in the afternoon between twelve and four o'clock. ■

One skidder, two cold deckers, and a locie—the Five-Spot, a geared Heisler—were brought to Quadra Island via Royston. Most of the men at Open Bay had recently been at Bevan, and shortly before that at Camp 2 at Oyster River, including foreman Len Harding, side-push (assistant foreman) Harold Grant, timekeepers Pete McLoughlin and Alec Tilleard, leverman Harry Sheppard, saw-filer Bob Larson, and blacksmith Lance Berkeley. The boom crew, consisting of Ron Bowen, Harold Cliffe, Hughie Cliffe, Fred Donnelly, and Bill Walker, came from Royston. Powderman Fred White also worked on Quadra. His granddaughter Jackie Gray (*née* Gullett) recalled that he carried explosives in a packsack until Mrs. Filberg, who was visiting one day, came along and said, "No man should have to pack explosives on his back!" After that, Bob Filberg got a packhorse for the purpose.

A limited affair, Open Bay was over by September 1928—a "summer job," according to Comox Logging chronicler Chris Holmes. The *Argus* noted on August 2, 1928, "The Comox Logging Company have ceased their operations at Open Bay, Valdez Island, but they still have a gang of fifteen to twenty men over there, cleaning up. They are taking up the tracking and loading it on scows at Open Bay, whence it will be transported back to the centre of the operation via the [Royston] boom camp." As the show ended, it claimed the life of Russell Goodrich. He was killed at the age of 23 when he was hit by the top of a falling snag at Open Bay in September 1928. The camp at Open Bay was abandoned at the end of that month.

PERSONNEL AND EQUIPMENT from Open Bay returned to Bevan in the fall of 1928 to take advantage of the new E&N property, as the *Argus* noted on October 4:

> A timber deal of considerable importance to the district has just been made whereby the Comox Logging & Railway Company takes over the timber belonging to the E&N Railway to the west of Bevan. There is a very considerable tract of this and the purchase will extend the life of the operations of the company here [at Bevan] for five or six years. There is merchantable timber as high as 3,200 feet, or a little farther than what is known as the First Lookout on the Mount Becher trail. A railway line is now being constructed from the block of land which was logged off by the company last year to cut into the new block of timber. Operations will commence here at once.

The first job was to extend the grade into Comox Logging's large new property. The grade crew worked with fallers and powdermen to fell, buck, and clear the main-line right-of-way. "The grading crew

The Open Bay Social Column

Life and logging were in full swing at Open Bay in early summer 1928, as the *Comox Argus* reported on June 7:

The Boom camp of the Comox Logging Company trimmed the natives at basketball in a very enjoyable game at Gowland Harbour recently. The islanders had to stack up against Harold Cliffe, W. Walker, E. Watt, J. Larson, and F. Donnelly, and they found the going hard. The score was 26–6. It was a very enjoyable game.

The salmon fishing is good and any of the boys who can throw a line or can find a boat are having plenty of fun hauling in the bluebacks. The trouble is there are not enough boats on Open Bay to go around. Picking wild strawberries is also a favourite occupation.

There are 91 men now in the camp at Open Bay, running three skidders, two of which are in actual operation. The logs are boomed up in Drew Harbour as Open Bay is too exposed.

Harry Sheppard, of Bevan, has the only radio in camp and his shack is always well patronized.

Messrs. Tilleard and Pete McLoughlin are now members of the gang at Open Bay. They came down from Headquarters at the week-end and will make the trip every week as long as the operations at Open Bay continue. The work in the woods should be through in August. ■

Bill Lyttback, Bevan Faller

Bill Lyttback (born 1908) recalled how the hiring hall in Vancouver sent him to the Comox Valley:

In 1933 Comox hired me through Black's agency—I always went to Vancouver to get a job. I stayed in Vancouver at the hotels, at the West Hotel on Carrall Street. It is still there. Now it's a rough place. Five falling gangs—fifteen guys—came from Vancouver, and Comox put us up at the Bevan Hotel. Most of us were Finns. We were the only ones staying at the hotel. The hotel was an ordinary two-storey hotel. We ate downstairs in the dining room and kitchen, and we slept at the hotel too. It was a hotel from mining, but no miners were there then.

I started in September 1933 at Bevan. I was never on the hill at Bevan: we fell behind Bevan and right behind Chinatown in Cumberland, right below Chinatown, but then we were all laid off. All the other guys went back to Vancouver, and one guy, [Frank] *Lilquist, went back to Finland, and another guy—Kald—went to the States, and he and his wife started a restaurant there. I never saw him again. These were my partners at Bevan.*

But they gave me a job falling at Comox Lake, where they had been logging for several years at this time, and I moved my family into a four-room mining house at Bevan and commuted to the lake. Viking Johnson [Osterblad] *and I rented the house from Canadian Collieries for $7.50 a month, which included light, water, and everything.*

So I happened to get a job at Bevan and I stayed with Comox Logging ever since! Comox was a good outfit to work for. I lived at Bevan for three and a half years while I worked there and at Comox Lake, and then I moved to Ladysmith in the spring of 1937. The No. 8 mine at Bevan started just before we left, and a group of miners moved down from the Interior and moved into the houses at Bevan. ■

Left to right: Inga, Bill, and Nancy Lyttback at Bevan, c. 1933, part of the town's large Finnish contingent of fallers, buckers, and families. In the background is one of Bevan's mass-produced miners' houses, dating from 1911. "The houses were plain and of frame construction, with siding on the outside and cedar shiplap lining inside," wrote retired miner Bill Johnstone in *Coal Dust in My Blood*. "Each was supplied with a cold-water tap that served a small pantry equipped with a white enamel sink."
—Nancy Hindmarch

Bill and Inga

Finnish immigrant Bill Lyttback reflected on his first working years in British Columbia:

I started to go into bucking contests when I was working for Bloedel, Stewart & Welch around Campbell River. Quite a few Finns were in loggers' sports, like Uno Forsman, Fred Dahl, and Axel Erickson. In 1930 Bloedel's sent Burrill and me to Courtenay to the competition, and I got third place, and they sent Fred Dahl and me to Duncan once. Fred came first, I got second, and Nestor Hagg came third. I never won a bucking contest but only got second, so I got fed up. I said never again!

I met my wife in Vancouver when I was working for Bloedel. Her name was Inga Knutar. She came out with her girlfriend, Linea Holmstrom, whose brothers were in Campbell River. She was two and a half years younger than me. We had lived 25 miles apart in Finland! She hated farming—she used to say, "We would have starved to death if we'd stayed on the farm." She came in 1929 and did housework for families around Vancouver till she got married. She had an uncle in Vancouver who'd been in the States, a logger, so she stayed with him.

I brought Inga to Bevan in the fall of 1933. She died eight years ago [in 1988]. *Goddamn it, sometimes I am so lonely I could throw that sofa out the window.*

Bill Lyttback received good care in his last years and died at a private nursing home in Ladysmith in 2003, aged 95. ■

at Bevan were mostly northern Europeans and Italians," recalled Stan Hanham. "They had horses and a two-handled scoop. They used powder to blow stumps for the right-of-way." One member of the Bevan grade crew was a Croatian, Martin Domovich, who had come to Canada alone in 1927 and who had finally saved enough to buy tickets for his wife, Mary, and their two children; they arrived in Bevan in 1931. Powerful Marion grade shovels were run by Tom McQuillan (grade foreman), Arthur "Spit" Quinn, and Carl Widen (Viden) with 17-year-old Davey Janes as one of the mat-hookers.

Comox Logging, the *Argus* repeated in late October, "is losing no time in putting its tracks into the timber bought from the E&N land company behind Bevan. It will extend about five miles into the foothills of the Island range and it will be in rough country." With the grade and rail in place, the company started logging its new E&N block at Bevan in January 1929, and the following month it started building switchback spur lines to Forbidden Plateau, as the *Argus* reported: "At Bevan this company is constructing extensions from their present railways on the side of the mountain. These railways will go to a point near the present trail to the Forbidden Plateau at an elevation of nearly 3,000 feet. It is probable that when logging has been finished in that vicinity the railway grades can be used as an auto road."

This major new logging show required a good deal of manpower. At least 100 men were required, including fallers, buckers, cold-deck and skidder crews, train crews, section gangs, steel gangs, and the whole varied complement of railway logging. Most of the required men came from the earlier incarnations of Camp 2 at Open Bay and Oyster River.

The Bevan Hotel housed the single men, many of them hired locally at Cumberland and at Bevan itself. In this, Comox Logging continued a local hiring policy begun on the flats earlier in the decade. Wallace Baikie noted in *Rolling with the Times* that Comox Logging took on "quite a few Cumberland boys" when it started logging at Bevan and Comox Lake. In fact, it had already started hiring Cumberland and Bevan men, including Bud Baker, Jim "Slats" Robertson, Charlie Stockand, and Jimmy Weir, on the flats in the 1920s. These men had school or sporting friendships with Courtenay men, or they had married into Homeguard or logging families, or for personal or family reasons they preferred the open air to underground work. The result was that, over time, some coal mining families became logging families, and their paycheques now came from Canadian Western instead of Canadian Collieries.

Spit Quinn at the Bevan Hotel

In *Voices from Bevan*, Betty Annand recorded a conversation with Davey Janes about the legendary Arthur "Spit" Quinn, a grade foreman at Bevan and later at Comox Lake:

Davey didn't have much time to socialize when he lived at Bevan. He does recall one occasion when he visited Skipper Murray's establishment to enjoy a quiet beer. It was nice and peaceful until Art (Spit) Quinn burst into the hotel and high-tailed it up the stairs, taking them three at a time. Right on his heels were six or seven very irate-looking loggers shouting words so colourful that even some of the loggers hadn't heard them before.

The pursuers thought they had him cornered when Spit ran into his room but he fooled them and jumped out of the window. No one was brave enough or foolish enough to follow him since his room happened to be on the second floor. Spit didn't seem any the worse for wear and walked away laughing. Davey never found out what the chase was all about, but knowing how Spit loved a good joke, it wasn't surprising.

Spit was from Courtenay but boarded at the Bevan Hotel for a short time. He was a big powerful man who could pitch a ball with a good deal of speed and accuracy. He played baseball for a Courtenay team and had a habit of spitting before he pitched a ball, thus the nickname "Spit." ■

The Bevan Hotel

Comox Logging took over the old Bevan Hotel and made it into a camp with office, cookhouse, and rooms for the crew, as the *Comox Argus* noted on April 8, 1926:

To accommodate their single men the logging company has taken over the Bevan Hotel which will accommodate a hundred men and they are already in possession. The married men can easily find quarters in the many houses of the Canadian Collieries company that are not now occupied.

Among the residents in 1929 was Merville farm boy Stan Hanham, who recalled that the hotel "was run like a normal logging camp." Pete McLoughlin was timekeeper and ran the commissary from his office in the hotel. The loggers were charged $1.15 a day for room and board, of which 15 cents was a donation to St. Joseph's Hospital. In 2002 Hanham recalled:

I stayed at the Bevan Hotel. It was full right up! The cooks were Chinese. They had a first cook, a second cook, and a flunky, and they prepared the meals morning, noon, and night, and they packed our lunches. We worked six days a week in those days. We'd work a little extra to come out early on Saturday. ■

The Bevan Hotel, 1916. It would be leased by Comox Logging in 1927 as a bunkhouse for loggers working on the Bevan sidehill, shown at the upper right of this photo before a single tree had been cut. One of the hotel residents in 1929 was 18-year-old Stan Hanham, fresh from the family farm at Merville.—Cumberland Museum, c20-19

JOHN TUKHAM PHOTO

"Business is flourishing. Mrs. C.P. MacAulay and myself in Bevan," wrote Comox Logging accountant John Tukham as a caption for this photo in about 1927. Margaret MacAulay, shown here, was the wife of Headquarters car shop foreman Charlie MacAulay.—Dorothy Tukham

As at Merville and elsewhere on the flats, local hiring provided advantages to both men and company. Local men found local work, and the company avoided workers of unknown quality from the Vancouver hiring halls. Among the Bevan and Cumberland men hired after 1926 were Connie (Colin) Aitken, Dave Aitken, Bill Bell, Pete Cameron, Joe Ducca, Dick Sgarbossa, Bobby Watt, Jack Watt, and Wattie (Walter) Whyte. Indeed, this emphasis on local hiring would characterize Comox Logging's operations elsewhere on Vancouver Island. In the 1930s, '40s, and '50s, it would do the same thing at Ladysmith and Nanaimo when the coal mines were closing down.

At first, however, the Bevan foremen were familiar faces. As at Camp 2, the woods foreman was Len Harding, an American from Sedro-Woolley, Washington State, and a personal friend of superintendent Bob Filberg. Harding had come to Comox Logging as a skidder engineer. "He was an original skidder man," recalled Sam Telosky, who worked with him in the 1930s. Harding married Irene Dixon, daughter of fellow American Dick Dixon, known as Pop or Pa, the proprietor of the Riverside Hotel in Courtenay. Harding's side-push was Homeguard and former skidder leverman Harold Grant (born 1897), son of Albert Grant, a farmer from the Dyke Road, Comox. His brother Mack also worked for Comox Logging at Bevan.

The families of Camp 2, who at Oyster River had lived in shacks dotted along the main line, moved into Bevan's spacious but drafty pre-war miners' houses. Bevan School grew with the addition of young families. In May 1933 the teachers, Charlotte Carey and Edna Gear, contributed a procession of children to the May Day Parade in Cumberland. "The children," noted the *Comox Argus*, "were dressed up to represent some part of the operations of the Comox Logging Company." Dorothy Bouleau (*née* Blackburn) recalled "being dressed up like a tree and walking down the street smiling and waving her branches at all the spectators" (Annand, *Voices from Bevan*). School was also an opportunity to assimilate the Finnish families, most of whom spoke little English at home. Buck Niemi recalled how his father, Otto, "showed up at the school with his boys, and the teacher asked him what the trouble was. He said, 'What you teach my boys? When I say "Here!" they don't know whether to come or to listen'" (Annand, *Voices from Bevan*).

As was the case at Comox Logging's other shows, most of the fallers and buckers were Finns and Swedes secured at the Vancouver hiring halls. According to Bob Cliffe, the Finns "were all trained fallers, so they got hired." Bob Filberg—son of a Swedish tailor—also attributed the Finns' and Swedes' success to some previous experience in logging. "I don't think there was any industry

Fred Parkin (born 1928) at Bevan, c. 1933. Fred's father was Norman Parkin (1904–91), who then worked as a loader. "Dad was with Comox Logging then at Forbidden Plateau," Fred Parkin said. "He started as second loader, on steam on the Bevan hill."—FRED PARKIN

Children at Bevan, 1928. Left to right: Norma Sheppard, Madge Sheppard, Myrtle McKenzie, Sonny (Ralph) Gibson, Beverly Gibson. The Sheppard and McKenzie fathers worked for Comox Logging, but Ralph Gibson Sr. was a section foreman for Canadian Collieries. (Even after closing the Bevan mine, Canadian Collieries kept one open at nearby Cumberland, and the company continued to maintain the railway.) "There used to be quite a village at Bevan," recalled Gord Blackburn. "A school, store, post office. The school went to Grade 4: two rooms and two grades in each room." After Grade 4, children went to school in Courtenay.

—MYRTLE HERON

A family company. Visiting a skidder setting at Bevan, *c.* 1931, are (left to right) Hallie Dixon, Mrs. Logie, Norman Parkin, Mrs. Danny Marsh, Mr. Logie. The Marshes were Hallie Dixon's and Norman Parkin's in-laws, and the Logies were friends of the Marshes.—Fred Parkin

Colleen Hughman at her birthday party, Bevan, *c.* 1930, in front of a table of flowers. Colleen's father, Eddy, was then a high rigger at Bevan, and later at Comox Lake and Ladysmith.—Fred Parkin

A chip off the old block. Edna Dixon (*née* Marsh) and her son Jimmy during a visit to a skidder setting at Bevan or Comox Lake, *c.* 1931. Edna was the daughter of Cumberland miner Danny Marsh, and her husband was Comox Logging engineer Hallie Dixon.—Fred Parkin

in [North] America that attracted as good a calibre of labour as logging," he said in a 1960 interview. "The people from northern Europe came here with some experience in logging; they were healthy, husky, ambitious people and they could get jobs quickly and easily, and the pay was better [than in other industries]."

Some Scandinavians married locally or settled in the Comox Valley after bringing their wives from Finland or Sweden; others moved on to Ladysmith after Comox Logging opened its show there in 1936. According to Larry Lehtonen, "quite a few" Finnish loggers and their families moved to Bevan in about 1926. They included Waino Jylha, Matti Kentala, Toivo Koski (Hanhikoski), Frank Lilquist, Bill Lyttback, Joe (Jusso) "Poika" Moelanen, Otto Niemi, Eino Patama, and Harry and Olaf Storr (Storbacka). Among the other fallers were Ivan Wicklund (a Swede), Austin Blackburn (a Merville Homeguard), and Les Marshall (an Englishman), who, according to his daughter, Mona Law, was "really proud" that he could keep up with the Finnish fallers. Although most fallers returned to the hiring halls of Vancouver, all of these listed men and their wives settled in the Comox Valley or Ladysmith.

Toivo Koski and Waino Jylha (pronounced, locally, "Yool-haw") are good examples of those who stayed. According to Erik Piikkila, the grandson of Koski and step-grandson of Jylha, the two men had come together to Canada in 1923, and Koski had found work at Youbou on Lake Cowichan in 1924. By 1928 they were both working as fallers on switchback No. 4 on the Bevan sidehill. Betty Annand wrote in *Voices from Bevan* that they met their wives at the Silver Slipper dance hall in Vancouver, both married in June 1929, and together rented the last available house in Bevan. Bertha Koski and Meini Jylha arrived in Bevan while the men were at work, and they always remembered their reception: "Several Finnish ladies were waiting at the platform to welcome them and take them home for coffee and food." The two couples and a third Finnish friend, Matti Kentala, saved enough money to buy, jointly, a farm on Howard Road at Merville. They built three houses, drew straws to decide who got which house, and left Bevan together in 1932.

LOGGING STARTED IN EARNEST once the trees were down and the rails had been laid by foreman Joe Asselin and his steel gang. Bevan was an ambitious show with two or three skidders and the same number of cold deckers. Each skidder required between 21 and 24 men, and each cold decker another 7 men. Skidder levermen at Bevan included Ray Dawson, Hallie Dixon, Oswald Harmston, and Harry Sheppard. Cold-deck engineers included Dave Aitken, Bill Cessford, Jim Mathers Jr., and Ray Parkin. High riggers at Bevan included Joe Cliffe, Eddy Hughman, Emil "Sailor" Lehtonen,

Otto Niemi

Born in 1898 in Vaasa, Finland, Otto Niemi came to Canada in about 1925. His wife, Maria, came a year later, and in 1928 they moved from Vancouver to Bevan with their son Buck (Armond) Arvo, born 1927. Their son Arfie (Adolf) was born after their arrival in 1928, as Arfie explained:

Dad came first. He worked his way right across Canada. Mother came straight across Canada to marry him. He worked mostly at Comox Lake, but he also worked on the Bevan sidehills. He had never used a crosscut saw before. He got together with another Finn guy who was a faller and who told him to get on the other end of the saw!

We lived there—in the first house on the right at Bevan. The loggers would walk from the train to the hotel, and to the family houses in Bevan. We kids would wait for them to get off work to get cake and cookies from them left over from their lunches. Later we lived right on the Puntledge River.

Dad was a typical Finlander. He built his own clinker-built boat: he steamed the planks in his woodshed, and made the boat in the woodshed too. He also made homemade skis for me from alder. Dad was good with an axe and a broad axe. His axes were like razors! He made railway ties on Little Mountain on the way up Forbidden Plateau, for Tom Carlyle, who sold them to Canadian Collieries. They were eight feet long and all Douglas fir. The trees were only a foot and a half thick, and he could get five or six ties off a tree. He worked by himself and he was paid so much a tie. First you would chip the tree with a sharp axe and then stand on it and shape it with a broad axe. My brother and I were only 12 or 13 at the time but we skidded the ties for Tom Carlyle with a horse.

Otto Niemi died in October 1972. ■

Above: Comox Logging faller Otto Niemi (right) and his partner toppling a big Douglas fir, Bevan or Comox Lake, late 1920s or early 1930s. This photo was probably taken for the benefit of family back home in Finland. Normally, fallers got right out of the way of a falling tree in case of "widow-makers" (falling tree limbs) and other overhead debris.—Arfie Niemi

It goes with the territory. Finnish faller Otto Niemi in the early 1930s. The fallers had strong legs and superb balance from standing all day on springboards, and were often accomplished dancers. Buck Pedersen recalled that his father, a Danish faller named Al (Axel) Pedersen, took part in the tug-of-war team at the July 1 loggers' sports in the 1930s: "They had fallers and rigging men on the tug-of-war. The rigging men always beat the fallers because the rigging men were used to pulling line all the time. They had strong hands, which was the key. The fallers just had rhythm."
—Arfie Niemi

Jack McKenzie Jr., and Bob McQuillan. Loaders included Harper Baikie, Emil Dosen (later Dawson), T. Eng, Ali Helm, and Norman Parkin. One hooker was Bob Aitken, brother of Dave and Connie. Others on the large logging crews included Bill Bell, Jack Berkeley, Bill Burgess, Pete Cameron, Bill Chalmers, Reg Churchill, Stan Hanham, I.J. "Lefty" Larson, Lyle McKenzie, Alvin Parkin, Dick Sgarbossa, Bob Strain, and Albert Trotter.

Train crews included engineers Alec Chalmers, Harry Grieve, and Bill Rourke; brakemen Martin "Pidgy" Davison, Dick Delin, and Jack Mitchell; and fireman Tommy Cessford.

Other Bevan employees included Lance Berkeley (blacksmith), Pete McLoughlin (timekeeper), Bill Shevlock (section foreman), Bill Westwood (shop crew), and Wattie Whyte (watchman and hostler).

The steep grade of the Bevan sidehill put a strain on railway logging technology, and Comox Logging required geared locies, switchbacks, tail tracks (short extensions at the ends of the switchbacks), and high stakes on flatcars. "It was quite an engineering feat," recalled Harry Trotter. Geared locies were essential because ponderous main-line locies could not pull or push the empty flatcars up the sidehill. In the spring of 1929 Comox Logging benefited from the closure of its sister company, the Columbia River Lumber Company, when three geared Heisler locies were shipped from the Selkirk Mountains in southeastern British Columbia to the Vancouver Island Mountains to be used by Comox Logging. In 1980 Joe Cliffe recounted to Evelyn Henderson the challenges of railway logging on the Bevan sidehill:

> You can put a railroad anywhere, if necessary, but my goodness to try to get your logs down! And worse still is to try to get your cars back up those mountains. We logged up where the Plateau ski lift is, right next to the first building they put up was our home tree. We logged up to 2,700 feet there with railroad. In order to get back two miles to Bevan, we had 12 miles of track, what you call switchbacks, along the side of the mountain, back and forth. And that meant you could only run a small operation because you couldn't put another locomotive on there. There was no place for sidings or pass places, a 5 per cent grade all the time. We had geared engines, you see; they had come into favour then. The old rod engines would never work there ... They would probably go themselves but they wouldn't take any load.

A particular problem at the higher elevations was snow, which meant longer winter shutdowns. On March 3, 1932, for example, the *Argus* announced that logging would commence after the shutdown with two skidders working at Bevan. But a week later, three

Fallers Austin Blackburn (left) and William Beckman working out of Camp 2 or Bevan, 1920s, and shown here with a large Douglas fir. Blackburn, son of a Merville farmer, served in the First World War with the Royal Canadian Engineers and started work as a faller for Comox Logging in the early 1920s. In the 1930s he ran the crew boat *Joan* on Comox Lake. Beckman was a Swede or Swede-Finn whose brother, Axel, was the third member of this falling gang. Their tools include a sledgehammer for pounding wedges, a seven-foot crosscut saw, and a double-bitted axe. Hanging next to Blackburn's arm is a bottle holding a saw lubricant, such as coal oil.—Donna Edwards

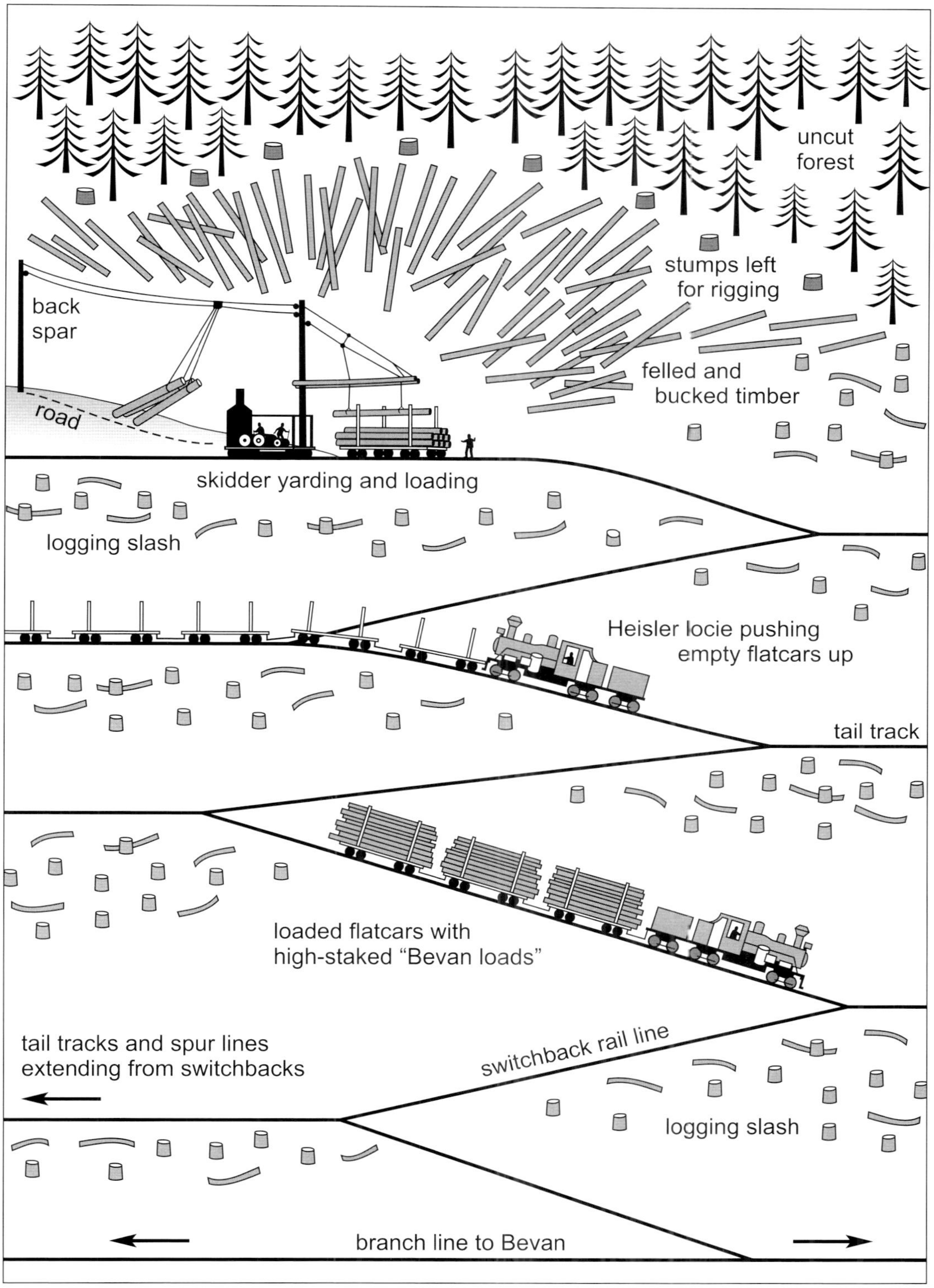

Mountain logging: Comox Logging's railway and skidder show at the Bevan sidehill, *c.* 1930. This was the company's only attempt to log a high mountainside by rail. The venture required powerful geared Heisler locies, switchbacks with long tail tracks, and high stakes on the flatcars to hold the smaller trees of high elevations.

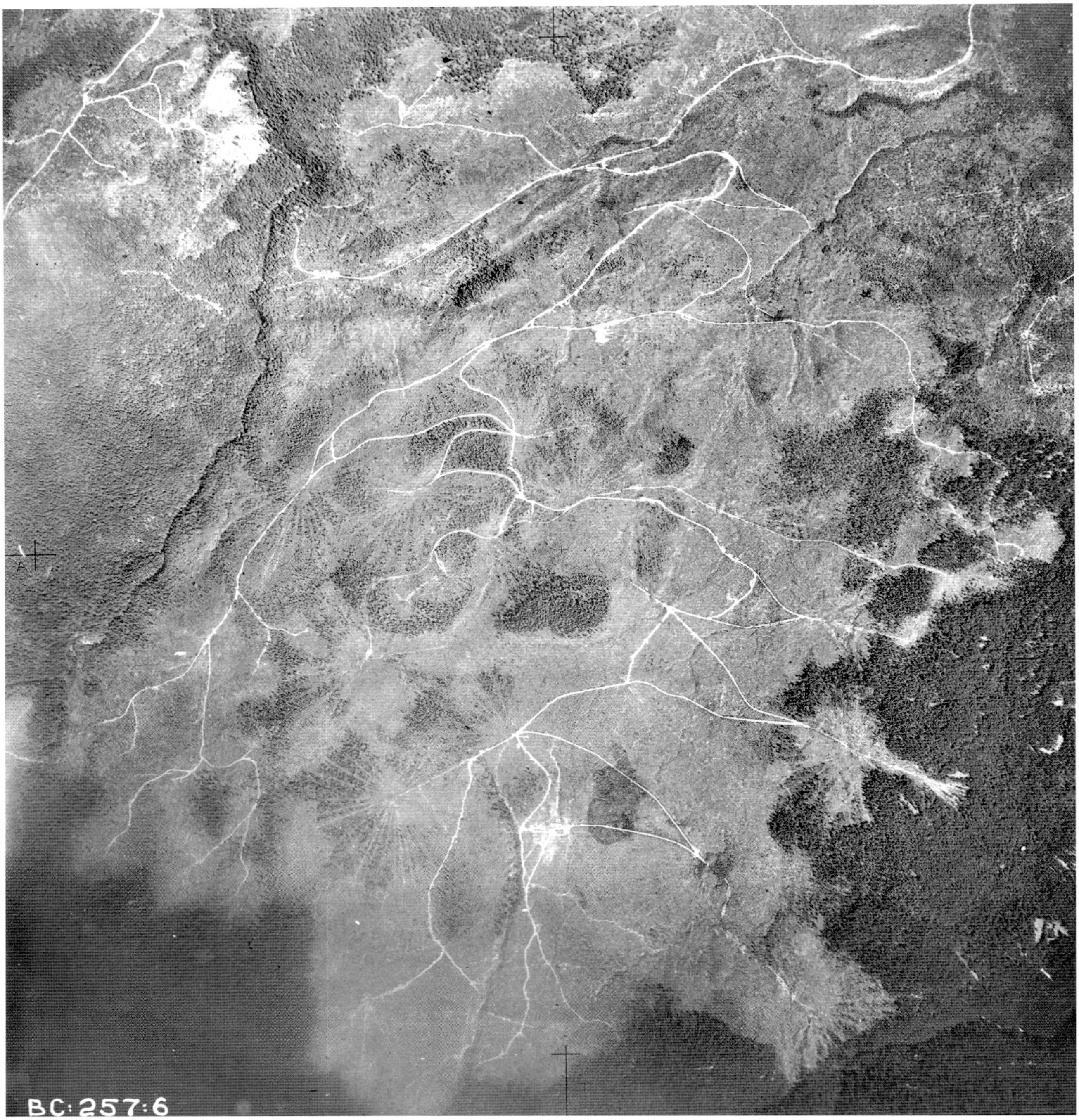

An aerial view of the Bevan sidehill from an elevation of 16,000 feet, July 20, 1946, looking east. The result of logging between 1928 and 1933, this large scar occupies the lower slopes of Mt. Becher, Indianhead Mountain, and Mt. Drabble. Despite the passage of time, Comox Logging's skidder settings are easily identifiable by the radial skid roads leading into the skidder landings. The company's network of switchbacks and tail tracks is also visible. Browns River is at left, a large bend in the Puntledge River is at top right, and the Bevan townsite is just out of the photo at top right. Forbidden Plateau Lodge is visible in the cleared area near the bottom of the photograph; from it the Mt. Becher trail leads up to the mountains and lakes of Forbidden Plateau. The difference in elevation between the top and bottom of the photo is about 2,300 feet: the Puntledge River is 400 feet above sea level and the logged-out settings at the bottom of the photo are at about 2,700 feet. An "auto road" replaced the railway when logging on the Bevan sidehill ended in the spring of 1933.
—British Columbia Air Survey Film Record, Roll No. 257, Frame 6, Air Photo Warehouse, Victoria, BC

Ivan Wicklund

Betty Anne Hemstad recalled her father's early life on the British Columbia coast:

My father, Ivan Wicklund, started on Burrard Inlet. He immigrated to Canada from Sweden when he was about 22, and being a "true Swede," it was only natural that he would go to work in the woods. While working in the woods my dad was always a faller. I recall him mentioning that he could hardly believe the size of the huge trees here in comparison to the woods back in his homeland. I believe that in 1933 he was working in Franklin River for Bloedel, Stewart & Welch. My mother came from the old country, from Norway, when she was 17 or 18. They met in Vancouver. She was a waitress; she couldn't speak English! They lived at Bevan when they were first married, in 1935, and he continued logging at Comox Lake. Logging in those days was not easy, and at least for some of that time while at Bevan, the family lived in a tent. This was a real problem during the heavy snows of winter in this area.

Dad was always an ardent hunter and fisherman. While he lived at Franklin River and at Bevan, his hunting and fishing exploits helped feed many a family in the logging camps. My father was an ambitious man and trying to better himself, so while at Bevan, in his spare time, he drove a taxi. Some of his trips took him from Courtenay to Nanaimo. My father and family moved to Ladysmith in 1936, and [he] *continued working for Comox Logging until 1945, when he went into full-time turkey farming. In 1942 my parents had bought a house and 20 acres of land in the Diamond District of Ladysmith, and this later became known as the Wicklund turkey farm.* ■

A three-man falling gang, *c.* 1935. Left to right: Uno Forsman, Fred Erickson, and Fred Dahl. Fred (Sigfrid) Erickson joined Comox Logging as a faller in 1927 and worked at Headquarters, Camp 3, Bevan, and Comox Lake before becoming a sleighmaker at Headquarters in the 1940s. Erickson and Forsman were Swede-Finns—Finns of Swedish background—and Dahl may have been too.—Harry Erickson

The Sauna from Hell

In *Voices from Bevan,* historian Betty Annand recorded some Finnish sauna stories from Bevan:

Emil "Sailor" and Saime Lehtonen: *Most Finnish families built their own saunas, and the Lehtonens were no exception. Having the river situated so close to the house was a bonus. They could come out of the hot sauna and jump into the cold river to cool off.*

Otto and Maria Niemi: *One of the first things that Otto did was to build a sauna right beside the river. It became very popular, not only with the Niemis but with many of their new friends.*

Toivo and Bertha Koski: *An unpopular guest drank too much and passed out. Not knowing what to do with him, they decided to put him in the sauna. He later told someone that when he came to, he was so hot, he was soaked with sweat. Then, when he saw the red-hot cinders, he was sure the devil had come and taken him to hell.* ■

and a half feet of snow had fallen where the skidders were working, and logging had to be stopped, and in late March one skidder was still buried under more than four feet of snow. Both skidders were finally at work in early April, in more than three feet of snow.

The smaller trees at higher elevations created other problems. On the flats north of Courtenay, the company had used cheese blocks to support loads on flatcars, with large trees piled strategically between the blocks. (As Walt Edwards explained, "A cheese block was a wedge, one on each side with a chain to hold it in place. You'd knock it out and the load fell off.") On the flats, most loads contained between three and eight 40-foot logs. The use of cheese blocks was, in fact, a sign of extraordinary natural abundance: forests that had been allowed to mature for almost 300 years. However, on the sidehill the company found that cheese blocks could buttress only a small pile of logs; anything more would roll off. To make it economically worthwhile to log the Bevan sidehill, Comox Logging copied a method used by the pole companies: instead of using cheese blocks, they fitted Douglas fir stakes into the "pockets" on the sides of flatcars and then stacked the logs vertically between the stakes. "The timber was not big on the sidehill at Bevan, so they started putting high stakes on so they wouldn't have to use so many flatcars," brakeman Pidgy Davison told his son, Frank. As a result, "Bevan loads" are readily recognizable in photographs. The use of stakes, however, meant additional time and expense.

Log-laden trains made their way from the sidehill to Bevan. "When Comox Logging logged Forbidden Plateau, they came down the mountains and crossed the Puntledge River on a high trestle into Bevan," recalled Bryson Whyte. The company's own locies hauled logs off the sidehill, across the Puntledge bridge, and along a railway spur to Canadian Collieries' tracks, depositing their loads at a siding behind the Bevan School.

Canadian Collieries then took over and hauled the logs to Royston, usually with their large Twenty-Spot engine. A contract stipulated that Collieries locies would pick up the loaded cars at Bevan, haul them to Royston, and bring back the empty cars. "The tracks of the colliery company run close to the scene of operations and it was of mutual benefit to conclude the agreement," the *Argus* observed in April 1926. As a result of such deals, the town of Bevan was almost entirely surrounded by railways. "I can remember the trains chugging through Bevan, taking coal [from Cumberland] to Union Bay and logs to Royston," recalled Fred Blackburn. "They had a little switch to determine where the train went."

Comox Logging benefited from the demise of its sister company, the Columbia River Lumber Company of Golden, B.C. That company collapsed, in part, because of the expense of railway logging on mountainsides, and three geared Heislers came to Vancouver Island from Columbia River Lumber in 1929 in time for the Bevan show. "When Comox Logging needed a geared locomotive, [James] McCormack [Canadian Western Lumber Company vice-president] would get one from Golden," recalled Howard McQuinn. "He knew where to get them. They needed geared locomotives for mountain logging." This is the builder's photo of a locie built in 1911 for Columbia River Lumber. In 1929 it became Comox Logging's Eight-Spot.
—Elwood White

Builder's photo of the Columbia River Lumber Company's No. 3. Built in 1912, it became Comox Logging's Nine-Spot in 1929. Geared Heislers were ideal for the steep grades at Bevan. "They only used Heislers up there," noted Frank Davison. Comox Logging got two decades of use from the Eight-Spot and Nine-Spot before they were scrapped in 1949 and 1950, at the start of the truck-logging era.
—Elwood White

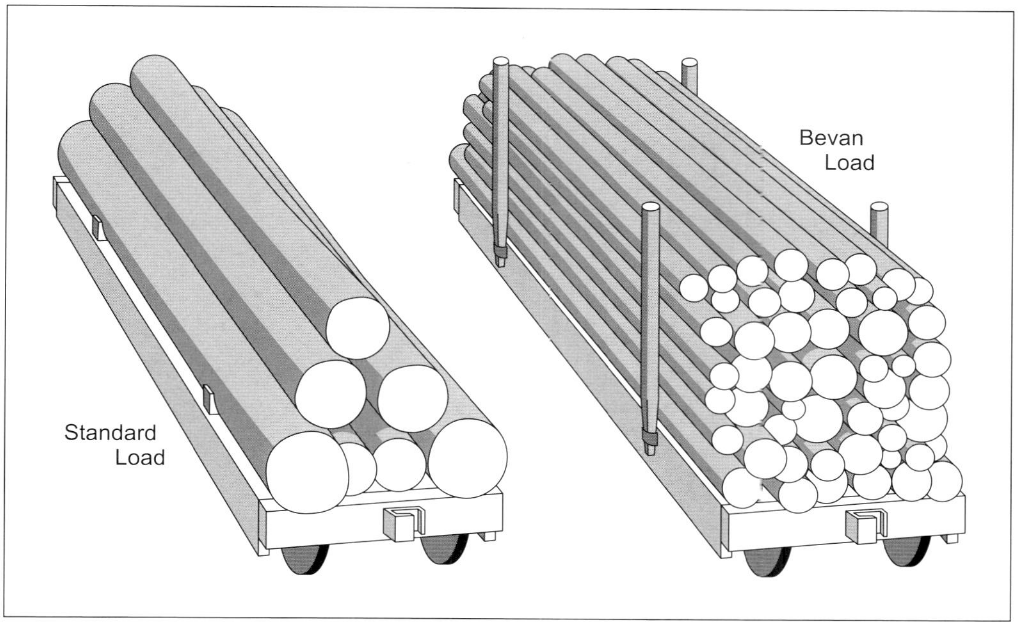

Two kinds of flatcar loads. In normal lowland conditions Comox Logging loaded logs on flatcars between cheese blocks; on the Bevan sidehill, where trees were smaller, the company devised the "Bevan load," in which logs were piled between high stakes on flatcars.

Canadian Collieries No. 20 at Bevan, *c.* 1930. This locie hauled Comox Logging's logs from Bevan to Royston and took the empty flatcars back to Bevan. "Canadian Collieries took all the logs to Royston," recalled Stan Hanham. "Comox had their own locies on the Bevan side, then they left them on the siding at Bevan, and Canadian Collieries took them to the boom camp." After the Bevan show ended, No. 20 continued to haul Comox Logging's loads from the Comox Lake load-out to Royston until rail hauling ended in 1953.—Elwood White

Two of the K & K Pole Company's Percherons, probably Barney and Bruce, at Tsolum, *c.* 1930. The pole companies came in before Comox Logging's main falling crews and removed cedar suitable for telephone poles. The pole companies used horses for yarding and loading poles until the late 1930s, when they switched to small Caterpillar tractors. During poling, these horses "hardly ever saw a field," according to Buck Pedersen, but during the winter and fire seasons they were kept in local farmers' fields.—Buck Pedersen

The Pole Company

The K & K Pole Company moved in to log the small cedar as soon as Comox Logging had completed the railway grades. "That whole hillside," wrote Wallace Baikie in *Rolling with the Times*, "from where the Forbidden Lodge is now down to Bevan, although having a good stand of medium sized fir, also had an excellent crop of cedar poles." Davey Janes recalled, "The grade crew went in first. They were followed by the steel gang, and the pole crew came next. Usually they laid steel before the pole company came in." Comox Logging's falling crews came next. In October 1928 the K & K started to build bunkhouses at Bevan. Davey Janes recalled:

They built the pole camp above us, at the tenth switchback, and at Oyster River they had another camp. Pretty near all the cutters and peelers were French-Canadian. Canadian Western at Maillardville hired from Quebec too. They took the small cedars for cedar poles. They used horses and cold decked [the poles] *into great big piles next to the railroad tracks; then they'd spot a flatcar next to them. They had a little gin pole or tree rigged, then they loaded them onto the flatcars. The flatcars had stakes on them. The teamster would back up slowly and the log would drop between the stakes.*

The old teamsters had to get up an hour ahead of anyone else to get grub for the horses. They had one horse that could cross a trestle! Up to a couple of dozen men worked for them, including the peelers and the teamsters. They always used three little dogs on the cedar logs. Dogs were little spike things on a piece of chain. They used to hook one end to the horse, the other to the log. They needed these three little dogs so they wouldn't hang up.

They took only cedar. They dumped them at Royston, where they had their own little booming grounds. They tell me a lot of cedar poles went to New York. ■

Pole men loading a flatcar, Comox Valley, *c.* 1930. The poles are being loaded by a combination of gin pole and horse team. Most poles taken out of the Comox Valley were destined for the American market.—Courtenay & District Museum & Archives, 986.15.10

Comox Logging's locies hauled the K & K Pole Company's products out of the woods for them. Here, the Eight-Spot Heisler waits beside a pile of sniped poles, possibly on the Bevan sidehill, 1930s. The standing timber is still intact: horses have removed only the small cedar, and the locie is preparing to haul the poles to Royston.—Elwood White

BY 1932 THE OLD CAMP 2 personnel were tired of the small trees and the laborious travel by switchback, and they eagerly anticipated the move to the large timber of the upper Puntledge and Cruickshank river valleys. Jack Berkeley recalled, in a 1995 interview with his son Don, the skidder crews' frustration with the small timber on the Bevan sidehill:

> I got a job on the Bevan hill [in 1932]. That was a railroad show and they had 12 switchbacks to get up the mountain, back and forth, and I worked mostly on the top of the eighth switchback, but it went up to 12. That's where the present ski lodge is. I worked there for a couple of years, and then that camp moved up to the lake. There's another little incident about that set-up there: we were hearing rumours all the time about the big timber up at Comox Lake, and we just couldn't get off that mountain soon enough to get up the lake to get into this big timber, you know!

In November 1932 the *Argus* announced that Comox Logging was getting to the end of its work at Bevan. In late April 1933 one skidder was still "cleaning up" and loading from two cold-deck piles. This work was to be finished in ten days, noted the paper, and then the tracks would be torn up. By late June 1933 the steel gang had taken up all the track and the Bevan show was over for good. The railway grade was, indeed, converted into an "auto road" that formed the basis of the modern switchback highway up to Forbidden Plateau. Logging, a resource industry of the early 20th century, almost literally paved the way for the Comox Valley's modern recreational industry. ■

Yarding on the Bevan sidehill: a turn of logs approaches the landing pile at a skidder setting, *c.* 1930. From the landing pile the logs would be loaded by hayrack boom onto the flatcars in the foreground.—FRED PARKIN

CHARLES SILLENCE PHOTO

> Four flatcars loaded with logs from Bevan and shown at Royston, *c.* 1930. Gone are the three-log loads from the flats. "They started using stakes at Bevan because the timber was small," explained Frank Davison. Left to right: unknown, Len Harding (Bevan foreman), Bob (Robert) Filberg (superintendent), Dick Delin (brakeman).—BOB BERKELEY

Turning Grey on the Switchbacks

Shuttling loaded flatcars down a mountainside was harrowing work for the train crews. "They say that Bill Rourke, the engineer, got grey hairs on the 13 switchbacks at Bevan," remarked Frank Davison. Stan Hanham recalled that the "switch" or "tail track" at the end of every switchback had enough room for seven or eight cars; therefore, only that many flatcars could be brought up or down at a time. Davey Janes thought that "six or seven empty flatcars was all the locies could handle" on the sidehill, and both Harry Trotter and Walt Edwards remembered a maximum of seven flatcars on the tail tracks. "Seven was the Bevan hill number," said Edwards.

Some of this activity was visible from below, in Courtenay. "As kids we could see the train working back in the mountains," said Trotter, but Eddy Berkeley recalled that the trains were invisible from the Berkeley family farm on the edge of Courtenay. "My brother, Jack, once said that from below you couldn't see the locies on the switchbacks, just their steam showing."

Bryson Whyte (born 1924) remembered visiting his grandfather Wattie (Walter) Whyte at Bevan when he was a boy:

Wattie used to be watchman on the sidehill at Bevan. As a matter of fact, my brother, James, and I went up on a logging train to visit Wattie! Harry Grieve was engineer, Tommy Cessford was fireman, and brakie was Jack Mitchell. We went up on the switchbacks. That was up to the timber line, which was as far as they went. It was like a big yellow-cedar forest at a certain level.

As on the flats, families liked to harvest berries on logged-over land. "My mother and my sisters used to go up there in the morning and pick blackberries," recalled Eddy Berkeley. ■

Canadian Collieries' No. 20 at Royston boom camp, 1937. Just in front of the locie the trestle extends out into Comox Bay. "Canadian Collieries hauled all logs from Bevan and from the siding at Comox Lake," said Frank Davison. "Collieries' No. 20, their big engine, did the hauling. It was like Comox Logging's No. 3 but bigger, but it was much the same."—Elwood White

Never Stand in the Bight

In a 1995 interview with his son, Jack Berkeley (born 1913) recalled his early days as a whistlepunk at Bevan:

By this time I was 16, so I started working in the woods as a logger. The first job, I was a whistlepunk, otherwise [known as] *a signalman, in the woods ... My uncle* [Oswald Harmston] *was woods foreman at that time, so that's why I got the job, actually.*

I didn't know one end of a log from another! Absolutely green! And one little incident: they used to coach me, you know. One thing I really remember. I didn't have enough whistle wire, and they were yarding maybe 1,200 feet. This was an old steam skidder, the old steam skidder. They were a long way out and I had maybe 800 feet of whistle wire, something like that, and I had to climb a tree to get up so I could maybe see the crew, you know!

And I remember one time I was up on the sidehill and we were just started back on the road, and the receding block—there's two blocks out on the back end of the claim that the haulback line runs through. In the case of the steam skidder, they call that the receding line, same thing, the block was here and I was down in here, in other words in the bight of the line, which was a no-no, you know! I found that out later on. I was there I think all one morning and early afternoon, and as we worked back on the road I got past that, beyond that, and further up the sidehill, and then, oh, I guess along late in the afternoon, they had a real heavy turn on the skidder, and he had to hold the receding line, in other words tight-line the whole turn up so it wouldn't hang up, and by doing that he yanked this stump out, right where I'd been standing all morning and part of the afternoon, so that would've ended my career kind of abruptly! I never forgot that. So they told me after: you should never stand in the bight. ■

WALTER MONTGOMERY PHOTO

Skidder 4 at Camp 2 or Bevan, late 1920s. Top row, left to right: Albert Trotter, unknown, Bill Chalmers, Harper Baikie, three unknown, Bill Burgess. Ray Dawson is the engineer at bottom right, and Alvin Parkin is also in the photo. Harper Baikie left Comox Logging in 1928 after the deaths of his friends Arthur "Skinky" Boyd and Percy Anderton, but there would be no more deaths among Comox Logging crews until 1935.—Olive Burgess Donation, Filberg Heritage Lodge & Park Association

Loading on the Bevan sidehill: loaders pose with a log suspended by tongs from a hayrack boom, *c.* 1930. The view is to the east, looking out over Browns River towards Merville and the Strait of Georgia. Comox Logging had spent the previous 20 years logging the flats in the distance.
—Jean Berkeley

Rigging crew at Bevan posing during the lunch break, *c.* 1931. They have attached their choker lines to a log and instructed the engineer to lift the lines up towards the skyline and carriage. One of the crew is chokerman Peter Cameron. Only the snags and saplings have been left standing, and this photograph captures the desolation of a logged-over skidder setting.
—Karl Cameron

The crew of Skidder 3 at Camp 2 or Bevan, late 1920s. Skidders 3 and 4 were transferred to Bevan in the late 1920s, while Skidders 5 and 6 remained at Camp 3. Top row, left to right: Harry Sheppard (leverman), Oswald Harmston (leverman), unknown, Jack McKenzie Jr. (high rigger). Standing in the middle row, third from right, is foreman Harold Grant, whose dog, Cap, is in front. Among the other men is Bob Strain.—Myrtle Heron Donation, Filberg Heritage Lodge & Park Association

Skid Marks

Smaller trees at higher elevations made it possible to challenge and beat logging records set among the giant trees on the flats. In November 1929 a Bevan crew set a record:

The Bevan camp of the Comox Logging & Railway Company established a record of loading yesterday which, it thinks, will be hard to beat. Len Harding's crew at No. 2 skidder loading on the hill behind Bevan, broke all previous records when they loaded in 8 hours, 618 logs on 25 cars. Those mainly responsible were Hally Dixon, leverman; Ally Helm, head loader; second loaders, T. Eng and E. Dosen [Emil Dawson]*; Ray Dawson, skidding engineer; Joe Cliffe, unhooker; Bob Aitken, head hooker; and Bob McQuillan, head rigger. Pete McLoughlin held the watch and counted the logs.* (*Comox Argus*, November 28, 1929)

And the following spring:

Side No. 2, at Bevan, during March, yarded and loaded no less than 8,131 logs. This is an average of 313 logs each day for every working day of the month which works out at the rate of one log loaded every minute and a half during the time the logging machine was operating, for an entire month ... Only one thing makes records of this kind possible. The equipment used is not the important part, the personnel of the men doing the work and the complete co-operation of every man in the logging crew is what did the trick. (*Comox Argus*, April 3, 1930) ■

Clyde cold decker, crew, cold-deck pile, and back spar at Camp 2 or Bevan, late 1920s.
—Myrtle Heron

***Detail of photo above*: The crew of a Clyde cold decker with the edge of the cold-deck pile at right, Camp 2 or Bevan, late 1920s. Left to right: Bill Cessford, two unknown, Len Harding, unknown, Jack McKenzie Jr.**
—Myrtle Heron

Should Have Stayed on the Farm

In 1929 Stan Hanham (1911–2003) was transferred with Comox Logging to Bevan:

I worked up there on the Bevan sidehill! I was blowing whistles for a cold decker to start, then I was setting chokers for $4.00 a day on the cold decker. I worked on the fourth switchback. The skidders could only go out so far, so cold deckers brought the logs to where the skidders could pick them up. We made piles of several hundred logs.

We went to work on a railway crummy. One of the cold-deck engineers was Ray Parkin, who went to Ladysmith later, and Eddy Hughman was there. Harry Sheppard was skidder engineer, and Len Harding was head of the whole camp at Bevan.

It was now the spring of 1930 and the Depression was coming on. Wages had dropped down to $2.50 a day. Besides, I didn't care for the woods. My stepfather, John Blackburn, said, "I'll give you $25.00 a month and board if you come back to the farm," so I went back to the farm at Grantham in the spring of 1930 and took it over after Blackburn's death in 1937. There was a lot of demand for farm produce. Mining was very strong at Cumberland in the Depression, there were lots of coal ships at Union Bay, and I provided eggs, potatoes, hogs, and chickens to stores at Union Bay and Cumberland. I was never short of money. ■

Three cold deckers and crews on the Bevan sidehill, *c.* 1930. The two cold deckers at left are on a platform made of log cribbing. The logging railway is visible at lower left. The Clyde cold decker (at right) is above the railway track. "The rigging's up," said retired rigger Walt Edwards. "Maybe the one [cold decker] at right is getting ready to be loaded onto the tracks. They're getting ready to load it out." The middle of the three men seated at front is Hallie Dixon. The tall man standing fourth from right is Bill Bell. "Bill Bell was always a foot taller than anyone else! I spent a lot of time with him," said Sam Telosky. The short man to the right of Bell is Dick Sgarbossa. Second from right is Joe Ducca, and next to him (seated) is Mack Grant. Sam Telosky also identified the light-haired man fourth from right, back row, as donkey engineer Dave Aitken. "Dave Aitken sitting there on a donkey! He definitely worked at Bevan hill."—Laurie Mathers

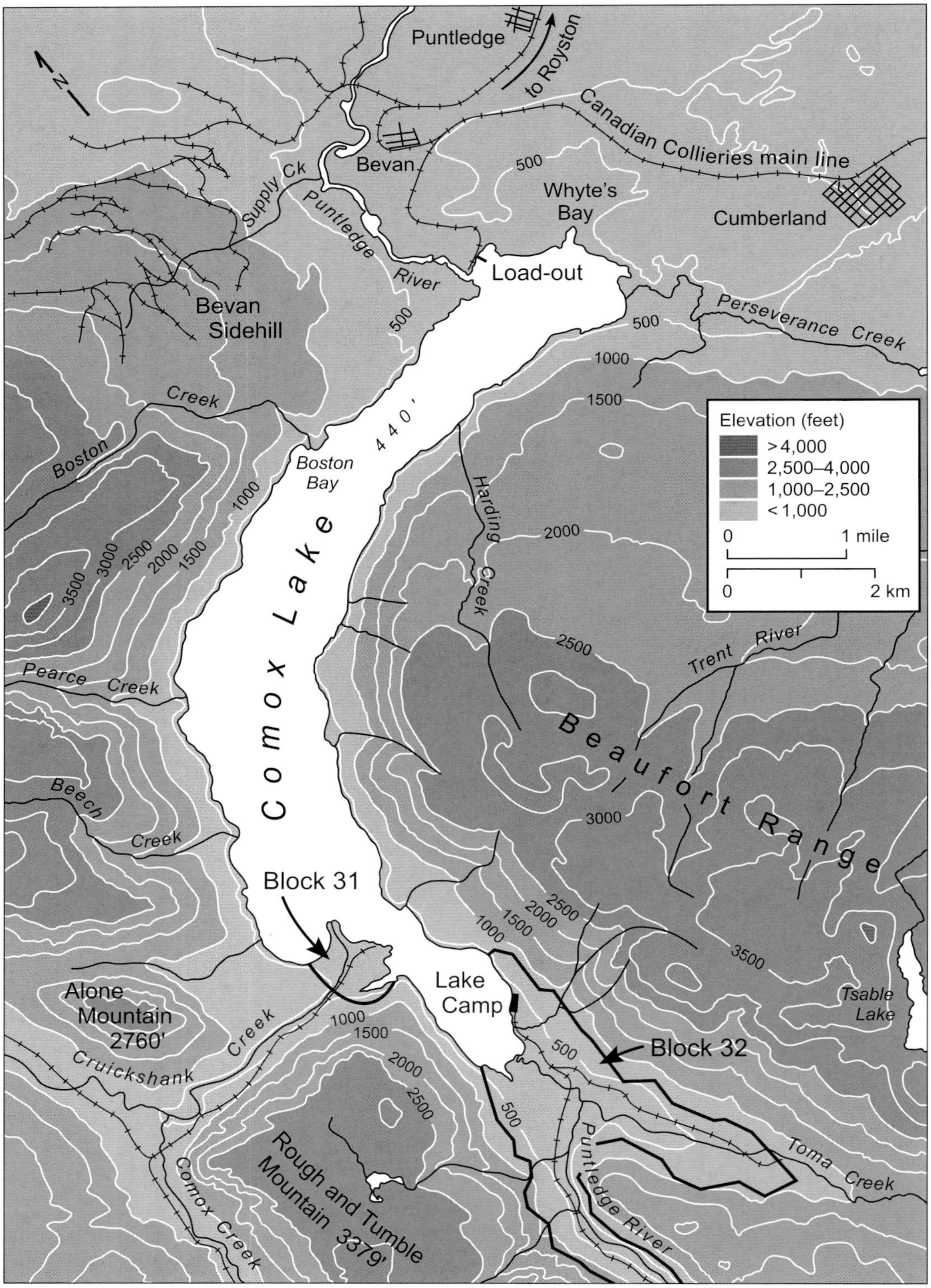

Comox Lake and vicinity

The operations of the Comox Logging Company are now not entirely on the bottom lands of the Comox Valley, where logging was comparatively easy, but on the benches back of Bevan, and on the steep sides of Comox Lake.

—*COMOX ARGUS*, JANUARY 23, 1930

4

LOGGING BY RAFT

THE BEVAN SIDEHILL WAS ONLY A PRELUDE TO LOGGING AT COMOX Lake. In 1888–89 the Chemainus Company (Victoria Lumber & Manufacturing Company) had bought timber properties at the top end of Comox Lake (Blocks 31 and 32, Nelson Land District) from the Dunsmuir land grant. According to timber cruiser Eustace Smith, these blocks, purchased for a dollar an acre, were cruised by Joe Hooker. "In all cases where these big blocks were secured, the areas selected were picked for heavy stands per acre, with a preponderance of fir, which was the only timber considered of value," Smith observed (quoted by Wellwood, "The Smith Saga"). Blocks 31 and 32 included the lower valleys of Toma Creek and the Puntledge and Cruickshank rivers, where they entered the lake. These areas were, the *Argus* noted in October 1927, "extensive limits in some very fine timber." Victoria Lumber's original plan had been to float logs down to the bottom (east) end of Comox Lake and then down the Puntledge River to Comox Harbour. But the company's 1906 experiment in river driving on the lower Puntledge and Tsolum rivers had proved a disaster, owing to the shallow water and huge logs. As a result, the rich Comox Lake properties had remained untouched while the company logged the more accessible timber of the flats adjacent to the Strait of Georgia. In about 1910, Victoria Lumber sold its Comox Valley holdings, including Blocks 31 and 32, to Canadian Western.

These two blocks at the top end of the lake, however, represented only a portion of the lakeshore timber, and Canadian Western was anxious to obtain more of it. Surveyor Wallace Baikie wrote in his 1985 memoir that in the mid-1920s, in anticipation of the move to the lake, "Comox Logging had or was acquiring all of the timber around Comox Lake." At various times, the company hired timber cruisers Harry Baxter and Eustace Smith to survey what Baikie called its "numerous patches of timber" at the lake. Therefore, the whole lake was essentially open to the company's ambitions.

Little could be done until a branch line was extended from Bevan to Comox Lake, and until the load-out works had been built. In April 1927 Comox Logging leased enough land from Canadian Collieries to build both a railway right-of way from Bevan and what the legal documents called a "receiving and loading plant" at the lake. Early in 1928 surveyors Hughie Stevens, Alec Tilleard, and Wallace Baikie laid out the branch line from Bevan and surveyed the trestle and load-out works, which were built in the same year by Bill Eastman and his bridge-building crew. In August 1928 the grade crew and fallers were making the Comox Lake extension; in September a gang under Harold Grant was laying steel to the lake; and by October the railway from Bevan to the lake, as well as the load-out, had been completed.

The sides of Comox Lake were so steep that railways extending right to the company's lakeside timber were completely out of the question. Conditions were ideal for A-frame logging, a coastal logging technology that had been perfected in the first decade of the 20th century on the steep inlets of mainland British Columbia. In A-frame logging, a donkey engine mounted on a raft yarded logs from the steep lakeside, or from the side of a coastal inlet, down into the lake or sea. This method required a substantial and buoyant multi-layer raft and a high A-frame, which was, in effect, a floating spar tree consisting of two large logs lashed together in the shape of an A. Combined, raft and A-frame gave the steam donkey the stability and the highlead (that is, the lift) needed to yard heavy logs down into the water. Of necessity, cold decking was dispensed with altogether, and skylines were strung to a stump on the shore rather than to a back spar. A-frames could be used for cherry picking selected logs, as was done in the wartime spruce programs in the Queen Charlotte Islands, or for removing everything of commercial value, as at Comox Lake. A-frames were also employed in the late 1920s by Bloedel, Stewart & Welch on the shorelines of Sproat and Great Central lakes, the other large interior lakes of central Vancouver Island; indeed, the same A-frame was used at both lakes, which are separated by only a few miles at one end.

In its essentials Comox Lake was much like Carter & Allen's donkey show at Knight Inlet, described 20 years earlier by Martin

> An aerial view of Comox Lake (looking to the west), July 18, 1923. Comox Lake separates the Beaufort Range in the south from Forbidden Plateau in the north. In the distance are Comox Glacier and the mountains of Rees Ridge, all of them in Strathcona Provincial Park and reaching about 6,000 feet in elevation. Comox Logging would log the lower shores of Comox Lake, which is nine miles from end to end, with an A-frame between 1929 and 1934. By the late 1930s the company owned most of the land around this lake and along the lower portions of the major rivers flowing into it—the Cruickshank and Puntledge.—Courtenay & District Museum & Archives, 984.62.8

> Looking north from the top end of Comox Lake, *c.* 1935. "I went to Comox Lake once," recalled retired faller Gunnar Jonsson. "They had a fire there and took everyone up to fight it. It was very steep; I thought the hill would never end! We had much flatter country at Headquarters and Camp 3." At left is Castle Crag (later renamed Castlecrag Mountain; 5,700 feet), and the smaller hill in front of it is Alone Mountain (2,780 feet). To the left of Castle Crag is the Cruickshank River valley, where Comox Logging would log by rail after 1937, and to the right are the mountains of Forbidden Plateau, including Mt. Becher (4,540 feet). On the lakeshore in the centre of the photo is the delta of the Cruickshank River, and just to the left of it is Mosquito Bay. In the foreground is the slashing caused by A-frame logging, and at bottom left is the beginning of a boom, with boomsticks.—John Holmes

CHRIS HOLMES PHOTO

Grainger in his logging classic *Woodsmen of the West* (1908). The lake resembled Knight Inlet in its mountainous setting, in the dramatic vertical drop of timber, and in the steam technology employed there: both locales had steep hillsides, lacked roads, and employed a shoreline engine and A-frame mounted on a raft to log the handy timber. The beauty of A-frame logging was that it did not require an expensive railway or logging road, but merely a tugboat to tow timber down the lake or inlet. Once the logs were in the water, the tug played the part in transport that the main-line locie played on dry land. The much cheaper tugboat transport made this the ideal Depression-era logging show.

The company's property at the top of the lake was known for its exceptionally tall Douglas fir, and in the summer of 1928 Harold Grant built a temporary camp at the head of the lake to find suitable timber for the A-frame that was to be built at the bottom end of the lake. For her book *Voices from Bevan*, Betty Annand talked to Stan Hanham, a whistlepunk on a yarder, who recalled his winter at this camp:

> In November 1928 the company sent the yarder and the crew to work at the top end of Comox Lake. The accommodations consisted of two small cabins by a creek. One was the cook house and living quarters for the cook and the other one housed the foreman Harold Grant, Ed Hughman, and the

Between 1929 and 1934 Comox Logging ran this A-frame on the shores of Comox Lake. The skidder engine on the raft dragged the logs down to the lake, where chasers removed the chokers and then boom men pushed the logs into bag booms for transport down the lake. A camp complete with cookhouse, bunkhouses, and first aid shack was also attached to the large float.

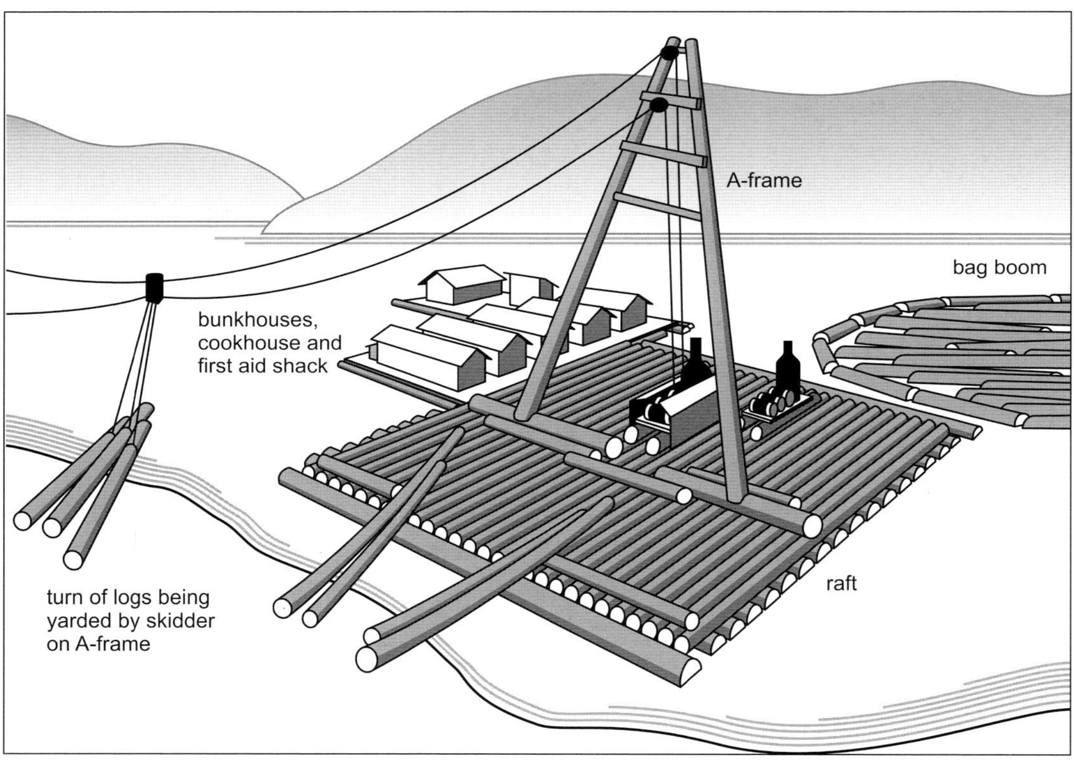

An aerial view of the central portion of Comox Lake, July 20, 1946, looking east from an elevation of 16,000 feet. Between 1929 and 1934, Comox Logging extracted patches of timber from the lower slopes around the lake by means of an A-frame, which could yard logs from about 1,200 feet up the steep lakesides. Some 15 years later the logging scars are still visible. The A-frame was moored occasionally to the shore, and logs were yarded down to it. Two major creeks are visible on the north side of Comox Lake: Pearce Creek comes down in the middle of the photo to a fan-shaped delta, where the characteristic radial scars of an A-frame setting can still be seen, and Beech Creek flows down to a small log-cluttered bay near the bottom of the photo. Tied up to the shore near the top of the photo, on the south shore of the lake, is a round bag boom resembling a shiny teardrop. Opposite the bag boom is a large logging slash caused by Cat logging, the roads of which resemble white branches reaching up the mountainside toward the uncut lower slopes of the Beaufort Range, which in this photo extend to about 2,000 feet. Comox Lake itself is 440 feet above sea level, and in places the mountain slopes on the north side of the lake reach as high as 4,000 feet.—British Columbia Air Survey Film Record, Roll No. 256, Frame 29, Air Photo Warehouse, Victoria, BC

engineer. A tent was put up for Stan and the six other men in the crew, two of whom were fallers. The tent had a wooden floor and was boarded up four feet on all sides. A big pot-bellied stove kept the men warm and cozy even on the coldest winter nights. The crew's job was to fell and prepare timbers for an A-frame that would be used in logging the shores of the lake. The timbers had to be 150' long and 14" at the top. By March of 1929 the timbers were ready to tow down to the Bevan end of the lake where the A-frame was to be built.

One of the timbers was 220 feet long and 18 inches in diameter at the top end. "The tree was perfectly straight and if cut off at an eight-inch top would probably have been 250 ft. long," the *Comox Argus* noted in February 1929.

High rigger Joe Cliffe was also sent to the head of the lake, under orders to find a Douglas fir flagpole for the Australian government. He found one that measured an awesome 272 feet. The A-frame and flagpole logs were floated down to the bottom end of the lake, where construction of the A-frame and floating camp began early in 1929, again under Harold Grant's supervision. In February the *Argus* announced that Comox Logging had started to build a floating camp "to be used in developing the timber in the vicinity of Comox Lake ... The camps are all to be on large rafts made of logs ... There will be five floats 60 by 150 feet, besides a number of smaller floats. The camp will have accommodation for about 200 men." Built by the Royston Lumber Company, the first bunkhouses were brought to the lake by flatcar.

THE COMPANY TREATED COMOX LAKE like a coastal inlet. Early in 1929, on a raft made of logs obtained by Harold Grant's crew, Comox Logging erected a towering A-frame and fitted it with a

Distance by Evinrude

In *Rolling with the Times*, Wallace Baikie described a novel survey method employed at Comox Lake in about 1925 by Harry Baxter:

The Comox [Logging Company] *had or was acquiring all of the timber around Comox Lake, so they brought in a cruiser by the name of Harry Baxter to get a cruise on the numerous patches of timber. Harry had to plot these areas of timber on the map and devised a scheme to measure these distances without struggling through the brush around points and bays. He measured off a line in a straight strip of beach, and with his small boat and outboard timed himself over a measured mile. In this manner he was able to plot these different patches on his map, estimate the distance the timber ran up the hillside, get his distance across the patch by timing himself with his boat, estimate the stand per acre and he would have his cruise. It was remarkable how close his cruises came out, and by having this information the company could log the larger areas with their skidder* [A-frame] *and contract the smaller lots out to gypos.* ■

The A-frame camp under construction, c. 1929, with the view towards the bottom (east) end of Comox Lake. Beyond the row of bunkhouses are the water tank and the long roofline of the cookhouse. The bunkhouses have been separated to lessen the fire hazard, and platforms built between them. These bunkhouses would remain in use on Comox Lake until the 1950s.—Dorothy Tukham

The A-frame under construction at Comox Lake, *c.* 1929. In the foreground are long logs intended for the raft, which consisted of three layers of 150-foot logs pulled into place and bound with line with the help of the donkey engine in the middle of the photo. The piledriver at right was also helping with float construction. At left is a separate raft holding seven bunkhouses, which would later be tethered to the A-frame float. A bunkhouse is under construction in the centre of the image. In the background are the Bevan hills leading up to Forbidden Plateau.—Myrtle Heron

The Comox Lake load-out and A-frame under construction, *c.* 1929. At right is the trestle for loading logging trains for the trip to Royston. A spar tree and hayrack boom were eventually erected on the trestle to lift logs out of the water and place them on flatcars. The long logs in the foreground—some of them 200 feet long—were destined for the massive A-frame float. At left are a piledriver and donkey engine. The railway track turned to the right, and from the load-out it was a mile into Bevan.
—Myrtle Heron

A-frame Whistlepunk

Born in 1911, Stan Hanham came from England with his mother in 1917 and grew up at Grantham at the farm of his stepfather, John Blackburn. Before his death in 2003 he told the story of his brief career in the woods.

I went to work in the woods in 1928, when I was almost 17. All the young fellows were quitting school to go into the woods for $3.20 to $4.00 a day. I started blowing whistles on a yarder at Headquarters, back up towards Constitution Hill. Eddy Hughman was rigger.

Only six weeks later, the yarder and crew of seven were transferred to Bevan with the specific job of getting the timbers out for the A-frame at Comox Lake. They were sent to the top end of Comox Lake, where they spent four months getting the long timbers out for the A-frame. "We stayed in a tent all the winter of 1928/29," recalled Stan. His boss at the head of the lake was Harold Grant. The yarder, mounted on a raft, pulled the A-frame timbers into the lake:

They needed a whistlepunk. I worked on the lakeshore, and the engineer on the raft gave the signals to me. The engineer then pulled the trees to the water. The A-frame logs were 160 feet long and the top end was 14 or 15 inches across. We cut over one million feet of timber for the A-frame. ■

powerful skidder engine to yard logs off the lakeside. As the *Argus* reported on February 28:

> There is considerable merchantable timber growing around Comox Lake and the system the Comox Logging & Railway Company will use in the logging of that timber is to place a logging machine on a raft and drag the logs directly into the lake. They will then be rafted and towed to the east end of the lake, where they will be loaded on flatcars and hauled by the Canadian Collieries Company over that company's railway to the booming grounds near Royston.
>
> The raft on which the logging engine will be placed for logging the timber which can be reached from the shore of Comox Lake, is a very large one. It will be 150 feet wide and 200 feet long. There will be three tiers of large logs, which will give the raft a depth of about 12 feet. There will be over one million feet of timber used in the raft. On the raft an "A" Frame 150 feet high will be used. In making up the raft and lashing the logs together several miles of steel cable will be used. For making this raft full length trees 200 feet long were carefully selected and put in the lake by gasoline logging engines.

A-frame logging started months before the onset of the Great Depression and continued despite interruptions caused by the flat markets. "In 1929 started to log at Comox Lake—around the

The A-frame show at Comox Lake, 1929–34, showing the stump rigging at the back of the show, the sometimes abrupt descent of logs from the lakesides, and the float and boom at the lake.

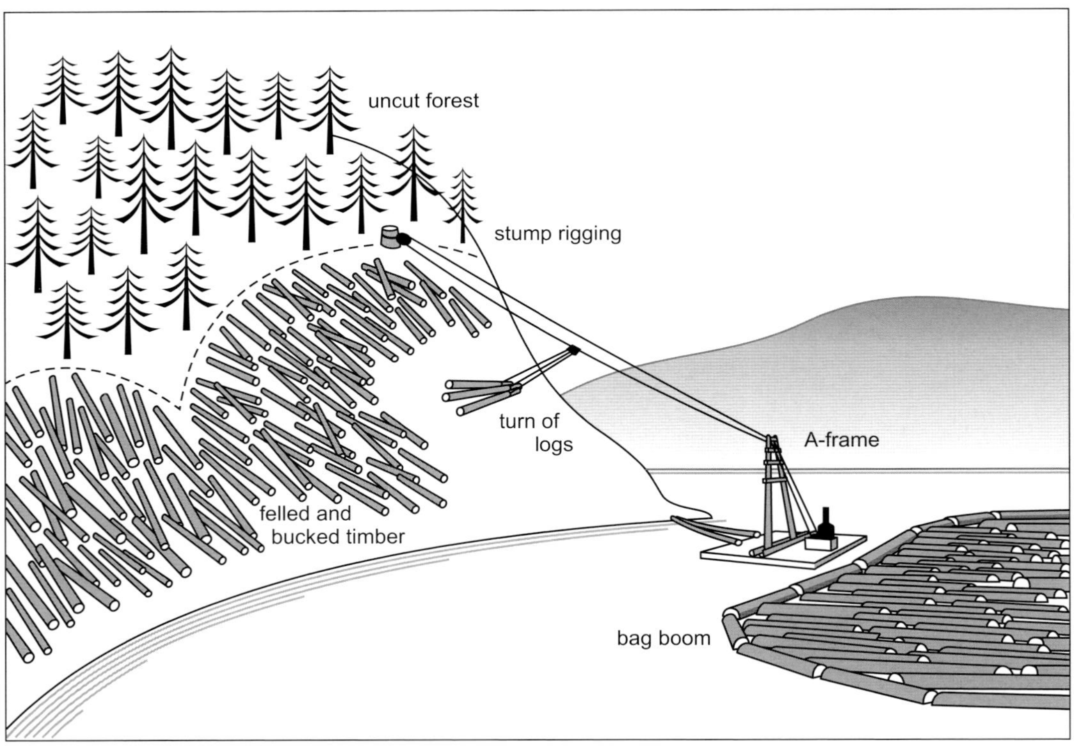

The A-frame and the tug *Eva* tied up at the load-out at the bottom of Comox Lake, *c.* 1930 (view to the south). Notches have been cut every four or five feet in the A-frame logs for lashing them together with wire rope. Behind the tug are two engines with vertical boilers. At left is a Lidgerwood skidder for yarding logs down to the lake, and at right is a skid-mounted donkey used for moving the raft along the lakeshore. In the distance is the bunkhouse float, which would later be moored to the A-frame.—RAY POJE

Detail of above: The tug *Eva* tied to the A-frame float at Comox Lake, *c.* 1930, with a skidder at left and donkey engine on skids at right. The tug's name and home port, "Eva New Westminster B.C.," are on the stern of the vessel. Note that the lakeshore timber in the distance has not yet been cut; the lake is still pristine and park ike. Built at New Westminster in 1891, the *Eva* had a varied career on the Fraser River and Strait of Georgia before being bought by Comox Logging in 1927 for use on the lake. According to John Holmes, the *Eva* was a sister ship to Comox Logging's boom tug *Joyful*. Comox Logging's Arthur Hilton said in 1954 that the *Eva* was originally owned by a tugboat captain named Kopiezinski. For her first years at the lake she was attached, literally, to the A-frame float, under her captains Alex Brackett and Bert Radford. Her job was twofold: to tow booms down to the bottom end of the lake, where the logs were loaded onto flatcars for shipment to Royston, and to move the floating camp down the shore of the lake as needed. The *Eva* was destroyed in about 1936 when, while she was beached at the bottom end of the lake awaiting new planking, some boys set her on fire. —RAY POJE

NORMAN SCHWARZE PHOTO, SCHWARZE PHOTOGRAPHERS, NANAIMO

The A-frame working along the shore of Comox Lake, *c.* 1930. "It was an A-frame skidder, about 22 men," wrote Mauno Pelto in a caption to a similar photo (in Babs Pelto's collection). "It was the most efficient way. They had the same idea at Alberni [Inlet] and also at Great Central [Lake] and elsewhere." Two boom men walk along the boomsticks in the foreground. These long logs show in all the A-frame photographs. They helped anchor the A-frame to the shore and served as pathways for the boom men and logging crews. The opposite side of the lake has not yet been logged. —LORNA CLIFFE COLLECTION, COURTENAY & DISTRICT MUSEUM & ARCHIVES, 2000.158.55

NORMAN SCHWARZE PHOTO, SCHWARZE PHOTOGRAPHERS, NANAIMO

Detail of photo at left: Two boom men, balancing their 16-foot pike poles, return to the A-frame float.—LORNA CLIFFE COLLECTION, COURTENAY & DISTRICT MUSEUM & ARCHIVES, 2000.158.55

> *Above*: The A-frame at Comox Lake, *c.* 1930. Sitting at left is boom man Andy Radford, son of veteran Comox Valley logger Arthur Radford and Florence Cliffe. Andy Radford worked at the booms at Royston, Comox Lake, and Ladysmith before taking over the family farm in Comox in about 1939. Like many of the Cliffe clan, he was a noted athlete who took part in loggers' sports competitions in the 1920s. Among his first cousins were loggers Joe and Harold Cliffe and Big Jack and Bill McKenzie.—PHYLLIS CURRIE

> *Below*: The load-out at the bottom end of Comox Lake, early 1930s (view to the south). A spar tree and hayrack boom have been erected at the load-out, where the tug *Eva* is moored; its funnel is visible below the hayrack boom. Moored in the distance is the bunkhouse float.—ELEANOR KELLER

> The A-frame moored on the south shore of "Mesquiti" (Mosquito) Bay on Comox Lake, just south of the mouth of the Cruickshank River, 1931. On the float itself are the skidder (in front) for yarding logs off the lakeside, and a donkey on skids (behind) for moving the raft along the lake. The skidder has been walled in, presumably to protect the crew from the wind. To the left of the skidder is the first aid shack, with a single window, and to the left of that is a row of six bunkhouses. Note the man standing on tongs suspended from the skyline carriage, demonstrating the method that brought logs down to the lake. At top is the company's boat *Mary*, and at far left is the floating office of Chris Holmes, timekeeper and first aid man. "Dad used to go up and down the lake in the *Mary*," recalled John Holmes. "*Mary* was a company runabout. Joe Rickaby brought her up there from Nanaimo. Dad used to bring visitors up the lake in her and she also served as the crew boat, but usually the crew went up and down in the barge." The A-frame has already logged the lower reaches of the far side of the lake.—Babs Pelto

WALTER MONTGOMERY PHOTO

WALTER MONTGOMERY PHOTO

The A-frame crew on the raft at Comox Lake, 1931 (view to the northwest); this photo was taken at the same time as the one above. The raft is moored on the south shore of "Mesquiti" (Mosquito) Bay. The mountain behind is Castle Crag, and the hill in front of it is Alone Mountain, known to the loggers as "All Alone" or "Little Joe." At far right is the edge of the skidder frame; the vertical timber is one of the A-frame spars. Standing, back row, left to right: Mike Willis, Harold Grant, Joe Ducca, Bob Strain, Mike Koserich, Dick Monks, Rod McAllister, Danny Stant, Joe Super, Peter Cameron, Joe Picketti, Harry Sheppard, Bert Radford, Chris Holmes, Tommy Cessford. Sitting, front row, left to right: Mike Kosac, Andy Radford, John Picketti, Ed Lefley, Bobby Watt, Dick Downey. Standing at the post is Murdie (Murdoch) McAllister, who was killed while rigging behind Bevan (not for Comox Logging) in 1937, aged 29.—Phyllis Currie

Peter Cameron notching a stump at Comox Lake, *c.* 1930. This is the only known photo of A-frame logging on the sides of Comox Lake. The A-frame and floating camp are just visible in the background. Cameron had earlier worked for Comox Logging at Bevan; his 1912 birth certificate states that he was "born at No. 7 mine," Bevan.
—Karl Cameron

Springboards and Vertigo

In a 2007 correspondence, Erik Piikkila, himself a forester, described the Comox Lake work of his Finnish grandfather Toivo Koski and Koski's falling partner, Waino Jylha:

Using a monstrous A-frame, every accessible piece of ground around Comox Lake was logged. As operations were winding down on the Bevan hillside, Toivo Koski and Waino Jylha became part of the Comox Lake operations which were just down the road from Bevan.

The terrain around Comox Lake is steep, broken, rocky ground with many ledges, bluffs and rocky outcrops. Falling and bucking operations in big timber on fairly even ground can be challenging and dangerous enough, but steep sidehills add increased complexity to an already dangerous situation.

Every day Koski and Jylha had to pack up and down the hill two six-foot-long springboards, two doubleheaded axes, a 12-foot handsaw, several heavy steel falling wedges, a big sledgehammer, a bottle or two of coal oil for the saw to cut through pitchy trees, a gallon of water per day for each man, and a lunch.

After determining the lean and the desired direction to fall the tree, they chopped notches into the stump to place the springboards. The fallers placed their springboards roughly 45 to 90 degrees (depending on the size of the tree) to the intended direction the tree was to fall. Trees were often felled parallel to the hillside, so one of the fallers might have to stand on a springboard suspended over a cliff edge. My grandfather and Jylha described the ground at Comox Lake as very steep. When my grandfather first started as a faller on steep ground in the Capilano River in North Vancouver, the old-timer who was training him cut down several saplings as a mat to create a visual barrier that appeared to stop the steep slope at that point and help prevent vertigo. I am sure this technique might have been used once or twice on the steep slopes above Comox Lake. My grandfather said that, even as a boy, he did not have a great head for heights.

Moving a springboard that was attached to a tree while standing on it was an art as well. It involved hooking one foot below the springboard, then kneeling with the opposite leg on the springboard with that leg hanging over and sort of bunny-hopping the springboard in an arc to another position.

If [the men] *worked as buckers, which they often did, springboards were not required but an 8-foot measuring stick was needed. This was many years before retractable, metal, Spencer logging measuring tapes became the standard.* ■

lake with 'A' frame," wrote logging chronicler Chris Holmes in his Comox Lake scrapbook. Holmes's son, John, recalled:

> I was born in 1924, and when I was a kid I went up Comox Lake in a boat and they were A-framing on the side of the lake. My dad was timekeeper and first aid man on the A-frame. A-frames were a common sight in the early days. They were all up and down the coast. It was a tried-and-true technology. The A-frame went along the lake and set up in places where there was a bit of timber and where there were no rock bluffs so they could bring the timber safely down to the lake. They worked their way up the lake and moored wherever they could get any timber off with the A-frame. I think the A-frame could reach about 1,500 feet, but they didn't have a donkey up there [at the back end] because it was so hard to move, and it was so steep. It would have been uneconomical too. They'd drag the logs down to the water and take the chokers off in the water, and they held the logs off the beach with boomsticks. It was a low-cost operation. The lakeshore logs were the cheapest logs you could get! You needed no railway, no road, to get them.

Bridge foreman Bill Eastman and his crew built the A-frame. "They put a skidder engine on it," recalled Stan Hanham. "Then they went all around the lake with the A-frame; they went along the shoreline and logged as far back as possible, about 1,200 to 1,500 feet, hauled the logs into the lake, and boomed them with a little boat." High rigger Walt Edwards, who started working at the lake in 1937, agreed. "They had an A-frame at Comox Lake and got all the easy stuff with it. They had a skyline on the A-frame." The whole camp was moored repeatedly in places suitable for lakeside logging. Fallers, buckers, riggers, and chokermen had to be unusually fit and "catty" (cat-like or agile) to negotiate the slopes and overhead hazards at these sites.

Logs yarded by the A-frame were kept in a boom attached to the float itself before being towed by the *Eva* to the bottom end of the lake, where they were stored until needed at Fraser Mills. Forester Bill Grange recalled that the steep lakesides contained timber much in demand at the mill. "Another interesting item," he wrote by e-mail in 2005, "was the high-grade tight-grained Douglas fir that grew along the slopes of Comox Lake. It was favoured for edge-grain fir flooring which Fraser Mills cut for special orders. It was so tight-grained it was as hard as oak." The lake was ideal for log storage owing to the fresh water and absence of wood-boring teredo worms. When needed, logs were transferred onto flatcars for shipment to Royston. The lake was 440 feet above sea level, so it was a gentle downhill for the nine miles to Royston.

WARD PIERCY PHOTO

The *Eva* tows the A-frame down the east shore of Comox Lake, *c.* 1931, with the Beaufort Range at right and Mt. Becher in the distance. The delta of the Cruickshank River is at far left. Ward Piercy provided this caption: "The A frame Comox Lake just when it [has] been towed down the lake it is working down by where we took the boat, it goes along the shore and hauls the logs off of the Mountains and into the lake. The frame is one hundred and seventy feet high. It was raining when I took this picture."—DONNA EDWARDS

BILL WALKER PHOTO

< Harold Cliffe on the A-frame float at Comox Lake, *c.* 1932. Born in Comox in 1909, Cliffe was the youngest child of Lu (Lucius) and Alice Cliffe. "We were always close to the water and the woods," Harold said, "and I guess it's natural that several members of my family would enter the forest industry" (*Crown Zellerbach News*, June 1969). His uncle Tom Cliffe was captain of a Canadian Pacific Railway ship, and his brother Joe and uncle Hughie were well-known loggers. Harold started working for Comox Logging as a boom man at Quadra Island in 1927, when he was a student at the University of British Columbia. "He was the youngest of the Cliffes of his generation," recalled his niece Myrna McTaggart, "and he was the only one to get a university education. The others subsidized it. They could see that he was clever. My father, Joe, helped him out especially." However, he quit UBC after three years when Bob Filberg offered him a permanent job with Comox Logging, and he was working at Comox Lake in 1934 when he met Mabel Olson, daughter of the shovel mechanic at Headquarters. They married in 1937. Cliffe worked for the company as a scaler, bull bucker, surveyor, statistician, and rail construction foreman, but he kept going back to booming jobs, and in 1936 he was sent from Comox Lake to Ladysmith to help open Comox Logging's new booming grounds there. Like many of the Cliffes, Harold was an excellent athlete, playing rugby and football for UBC and climbing in the mountains around Comox Lake. In the 1950s he succeeded his uncle Hughie as manager of Comox Logging's sister company, the Canadian Tugboat Company.—Gwen Greenwood

This logging show became a community of sorts. "The A-frame was a *camp*," asserted John Holmes. A floating camp with office, commissary, cookhouse, bunkhouses, and blacksmith was attached to the A-frame raft. The camp lacked married men's quarters; women were absent. A large crew of about 50 men was needed, including crews for falling, bucking, choking, yarding, and booming.

Comox Logging's local hiring policy continued at Comox Lake. Bevan men were transferred to the A-frame and new men hired from Cumberland. Convivial timekeeper Chris Holmes was transferred from Headquarters, where he had worked as a scaler. "He also ran the boat on the A-frame. Lots of tourists seemed to go up to the lake and he hosted many of them," stated his son, John.

The workforce was dominated by men from old Homeguard and farming families from the Comox Valley, and supplemented by refugees from the Cumberland and Bevan coal mines. Ray Dawson, Ed Lefley, Jim Mathers Jr., Ward Piercy, and Harry Sheppard were on the A-frame as engineers. Crews attached to the raft, many of them Cumberland boys, included Pete Cameron, Dick Downey, Joe Ducca, Mike Kosac, Mike Koserich, Murdie (Murdoch) McAllister, Rod McAllister, Dick Monks, Joe Picketti, John Picketti, Danny Stant, Bob Strain, Joe Super, Bobby Watt, and Mike Willis. Another Cumberland boy was Clive Banks, injured when he fell 30 feet at the load-out in November 1929. Boom men included Harold Cliffe, Andy Radford, and Bill Walker. The *Eva* was manned by Alex Brackett and Bert Radford (captains), Paul Hellan (engineer), and Tommy Cessford (crew member). Hallie Dixon was leverman at the load-out.

In 1932, even with the A-frame still active, Comox Logging built a large camp at the top end of the lake, where the Puntledge River entered it. Here, the company moved ashore with cold deckers, skidders, railways, and all the noise and expertise of early highlead logging. In January 1934 the *Argus* reported that the A-frame crew was taking out the last few blocks left on the shoreline. The A-frame was dismantled in 1935, and Lake Camp took over as the base for logging the level river valleys at the top of the lake. ■

WARD PIERCY PHOTO

Skidder leverman Harry Sheppard in the woodyard of the A-frame float, *c.* 1930, holding a six-foot peavey and standing in front of a round of Douglas fir. His colleague Ward Piercy wrote the following caption for the photo: "This is Harry Sheppard my partner by a block of wood I sawed at camp. It is over six feet through." The little boom above swung these two-foot rounds from the woodyard onto the bunkhouse floats. Later they would be split into many cords of firewood needed for the bunkhouse and cookhouse stoves.
—Donna Edwards

From Sloane Square to Comox Lake

Born in 1887 in Hillingdon, Middlesex, England, Harry (Herbert Henry) Sheppard was a boy when he went to work as a servant in Sloane Square, London, in the house of a British army general. As he told his daughter, Norma Oven, when he was 17 he set sail for Canada and "rode the rails" to Vancouver, where he found work as a cook in a logging camp. In 1912 he joined Comox Logging's sister company, Columbia River Lumber Company, at Golden, British Columbia, and got into the logging end of camp life. He was at Camp 5 at Williams Beach just before the First World War, and during the war he logged spruce in the Queen Charlotte Islands.

Sheppard runs through five decades in the history of the company, and he worked with hundreds of men at several different camps. In 1919 he married Rosalie Williams, daughter of Charlie and Margaret Williams of Williams Beach, and the couple had daughters Norma and Madge. By 1921 Harry Sheppard—known as "Shep"—was a skidder leverman, and in that year he left logged-out Camp 1 at Miracle Beach for Camp 2 at Oyster River. He remained at Camp 2 until 1926, when he was sent to Quadra Island, Bevan, and Comox Lake; he was skidder engineer on the A-frame when it logged the steep lakesides. The *Comox Argus* announced in January 1930 that "H.H. Sheppard, working at the Comox Lake camp of the Comox Logging Co., established a record. As a skidding leverman, he has been yarding a distance of about 1,500 feet and he has only broken his slack line once in six months. A breakage of four or five times a week is not a bad average." The fact that he was yarding downhill might have made his job easier.

Sheppard moved with the company to Cruickshanks Camp in about 1933, and in 1941–42 was leverman with Elmer Orr on Skidder 3 on the South Fork of the Cruickshank River (Comox Creek), right at the end of the skidder era. In 1947 "Old Shep" was running donkey at Comox Lake, and soon afterwards he worked the bundling machine at Royston. He retired in 1955, when he was 67, and died in Vancouver in 1980. ■

Beat That Tacoma Pole

Joe Cliffe was high rigger at Comox Lake in 1928 when British Columbia's premier, Simon Tolmie, asked Bob Filberg to procure a tall Douglas fir for the government of Australia as a flagpole. Tolmie wanted the pole to be taller than the 150-foot Douglas fir pole that was then one of the attractions at Britain's Kew Gardens and, if possible, to be taller than the world's tallest, a 250-foot pole at Tacoma, Washington. Filberg told the crew, "Beat that Tacoma pole!" Filberg later recalled:

We were logging at the head of Comox Lake at the time. It was in a narrow valley with high mountains; the trees reached high to the sun. Len Harding was foreman of the operation. I told him exactly what I wanted. Len and the crew, which I think included Joe Cliffe, "snubbed" down [lowered by line] *three fine trees. The men worked hard to snub the trees down. It was cleverly done. It was impossible to tell which of the trees was the best until they were put into Comox Lake. Once in the water they were able to detect the best and straightest of the three giant Douglas firs. Finally one was selected and towed down the lake, loaded on a logging rail car, and taken to a platform behind the Native Sons Hall in Courtenay. The tree was barked, levelled, and left to season. It measured 272 feet.*

Arrangements were under way for the shipment of the tree when the Depression came in 1929. Money was scarcer than 272-foot trees, and the pole presentation to the Australian government had to be shelved. The pole sat on the platform until the middle of the 1930s and proved to be a conversation piece, but otherwise an eyesore. Reluctantly it was cut into short logs. Thus was the dismal end of what could have been a monarch among flag poles. It is not likely that it will be equalled again. (Crown Zellerbach News, September 1961 and January 1964) ■

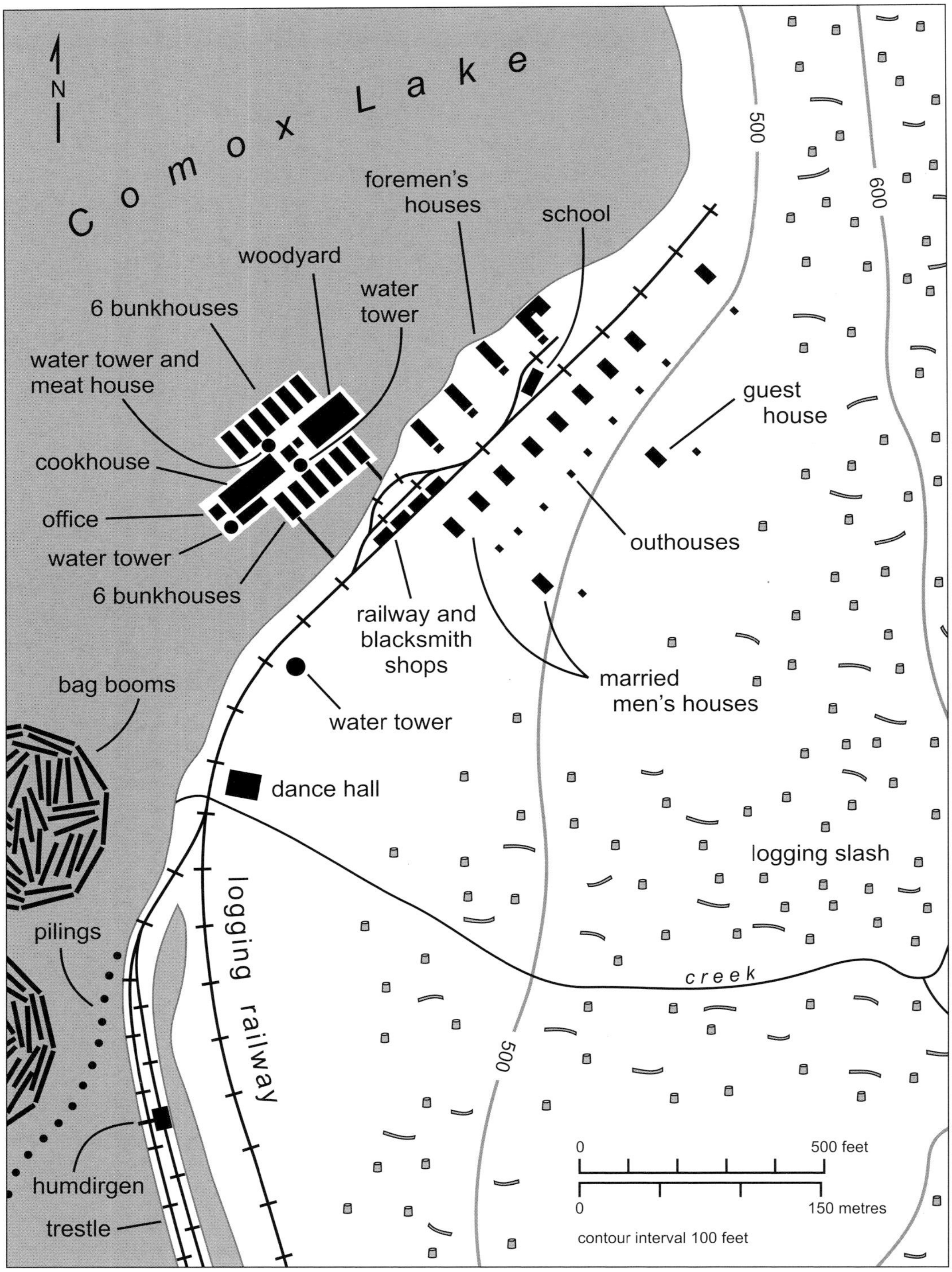

Lake Camp. Between 1932 and 1937 this was the base camp and industrial hub of Comox Logging's railway show at the western end of Comox Lake. Married men lived in houses on land while single men stayed in the bunkhouses on the floating camp.

There is no comfortable covering of the soil to walk upon; moss and huge trees alike grow on the very rock, sustained by the heavy winter rainfall upon a scanty pretense of soil.

—MARTIN GRAINGER, *WOODSMEN OF THE WEST*, 1908

5

LAKE CAMP

IN 1932 COMOX LOGGING WENT ASHORE FROM THE A-FRAME AT the top end of Comox Lake and built a major railway camp that remained open until 1937. Situated close to where the Puntledge River and the old Comox–Alberni trail met the lake, it was known as Lake Camp, though really it remained Camp 2. Lake Camp was the staging point for a major railway-and-skidder logging show in the heart of the Vancouver Island Mountains. This camp was the necessary transport base between the watery expanses of the lake and the "long sides" in the narrow river valleys to the west. An impressive photographic record has survived of this austere Depression-era camp, which was marked by an appropriate minimalist simplicity and almost puritanical absence of frills or ornaments. Everything at the camp was movable, as *The Timberman* noted in September 1935: "Lake Camp is a typically British Columbia institution, with all the [main] camp buildings on floats, which may be towed to any desired location along the lake shore."

Lake Camp was also part of an even older Canadian tradition of temporary, male-dominated workplaces in the wilderness. There have been such places, perched at the far end of transport systems in remote parts of New France and then British North America, since the 16th century. They led back, over many miles, to faraway mills, markets, and cities. Like these earlier camps, Lake Camp resulted from a sudden rush of capital and labour into an unpopulated, remote, and resource-rich area. It was typical of a resource-based work camp in its purpose, in its isolation, in its setting at the "edge of the world economy," and ultimately in its very transiency.

THE OPENING OF LAKE CAMP, first contemplated in 1926, was postponed repeatedly until 1932 for at least two reasons. The purchase of additional timberland from the E&N at Bevan in 1928 kept crews active there for several more years; and the use of the A-frame, which began in 1929, enabled the company to log the lakeshore methodically and economically during the first years of the Depression.

Comox Logging finally started construction of the new camp at the head of Comox Lake in October 1932, "near where the floating camp now is," the *Comox Argus* noted in November. Work was coming to an end at Bevan, the paper announced, and men and families would be moved from there to the lake. The camp would enable loggers to live "on the job" and avoid the long commute from Bevan or Courtenay. Initial plans called for 50 houses for married employees; each three-room house would measure 22 by 24 feet. Other buildings necessary for a woods camp would also be built. The bunkhouses, cookhouse, office, and a few related buildings would remain on floats, while the married men's houses, railway facilities, and other industrial facilities would be built on shore.

The *Argus* noted that the company's extensive timber limits at Comox Lake would "keep them occupied for many years to come" and added, far too optimistically, that the camp would be "of a permanent nature." "One of the blessings of the camp is a plenitude of spring water," the paper continued. "Hugh Stevens found an excellent spring on the side of a hill not far from camp, and it has been piped down to camp where it provides 65 pounds pressure by gravity flow."

The camp, of course, was geared to purely industrial ends: to permit the production of sawlogs and their transport from the remote confines of the Vancouver Island Mountains to the Strait of Georgia. Additional floating bunkhouses were required immediately because, unlike the A-frame shows, logging on land involved cold deckers, skidders, railways, and their crews and maintenance. A great many men came and went at the bunkhouses, including skidder crew members Tommy Gray and Fred Balmer, sawfiler Bob Larson, faller Bill Lyttback, and riggers Jack Berkeley and Sam Telosky.

In April 1933, Courtenay carpenters led by Malcolm Burnett and Walter Rigler were building the married men's quarters. Men and families arrived at the camp as soon as the houses were ready: levermen Oswald Harmston and Norman Parkin and sawfiler J. Marty moved their families to the lake in April. They were

> *Above*: Comox Lake, with the view towards the west and the Vancouver Island Mountains. Great Central Lake and Sproat Lake are in the valleys in the distance, and beyond them (not visible) are Tofino, Ucluelet, and Bamfield on the Pacific Ocean. This photo was taken by a surveyor in 1934 from the "Comox Overlook" (elevation 4,054 feet) near the Mt. Becher trail. Lake Camp, with its whitewashed buildings, is visible near the top of the lake, as are Comox Logging's lakeshore A-frame slashes and the logging railways extending up Toma Creek and the Puntledge River at the end of the lake. At centre right is the Cruickshank River valley, where logging would begin in 1937. The photo gives a good indication of the vertical limits of railway logging.—BRITISH COLUMBIA ARCHIVES, I-64242

WARD PIERCY PHOTO

Looking east from Rough and Tumble Mountain towards the Comox Lake camp, *c.* 1933. The two rows of married men's houses are clearly visible. Camps like this resulted from sudden incursions of capital and labour into remote but resource-rich locations. This large camp would, in fact, stay in this location for only five years. Lake Camp provides a vivid demonstration of a transient work site in an austere and isolated mountain setting.
—DONNA EDWARDS

N.C. STEWART PHOTO

WARD PIERCY PHOTO

Detail of small photo on page 164: **Looking east from Rough and Tumble Mountain towards the Comox Lake camp, *c.* 1933, with the Beaufort Range behind. Lake Camp was set on a strip of land gouged by the A-frame from the old-growth forest. Ward Piercy wrote in his caption, "Before I went to Mud Bay we went acros the Lake and clum a Mountain a little way and took some pictures. See the bunk house the second one from the left it is where I sleep. The camp looks like it was on flat ground but there is a boom of logs tied up the camp."** —Donna Edwards

joined in May by blacksmith Hughie McKenzie and by Fred Carlson and H. Wayne. "They had a big blacksmith shop at Comox Lake where they made tongs, peaveys, chokers," recalled Darrell McQuillan. In early June the *Argus* noted, "There are now eight families living at the head of Comox Lake." By the end of June the number had jumped to 14, including those of engineers Alvin Parkin and Eino Patama, and in the next month 17 families inhabited the camp. In late July 1933 the *Argus* reported:

> The camp is growing fast in size and permanence. It's coming ashore. The original buildings at the south end of Comox Lake were all afloat as was only natural as the headquarters of the "A" frame gangs wandering round the lake for timber to haul down the steep slopes. Now the "A" frames have finished and the crews have all come ashore. Gradually the camp is settling down.
>
> Seventeen families have moved up the latest settlers being R. Strain, Fred Erickson, V. Carlstrand, S. Carlson and S. Dosen and more will come as soon as houses are built. The company is providing the framework of the houses leaving the tenants to finish them off according to their own fancy and inclinations ... Three weeks ago Mrs. Pattema [*sic*] was in Bevan; today she has a garden with bulbs she brought from Bevan, a sink and running water in her kitchen, the beginnings of a basement and one or two rooms sealed in. Before the fall rains come it will be as snug a little house as you could wish to hang your hat in. Mrs. Pattema's house was the only one that the writer had the opportunity of looking through but no doubt many of the others were expanding just as quickly.
>
> Another advantage on a hot day is the wind blows up the lake. Marty who files the big saws can work in comfort no matter how hot the day is and all the buildings are as cool. Fire insurance examiners, who recently visited the camp, declare it is the finest camp site they ever saw.

Houses were reserved for the married men and their families. Former teacher Pat Williams (*née* Holmes) recalled that the houses on the waterfront during her era (1935–37) were occupied by Len Harding (camp foreman), Oswald Harmston (assistant foreman), Hughie McKenzie, Norman Parkin, and their families. As foreman, Harding had succeeded Harold Grant, who died of pneumonia in June 1935 at the age of 37. Later, Bob Strain lived in one of the foremen's houses. Other married men at this time included locie and humdirgen engineers Harry Grieve and Alvin Parkin, rigger Bill Parkin, Finnish faller Eino Patama, boom man and skipper Bert Radford, and Swedish faller Gus Rundquist.

Chris Holmes, *c.* 1940. "He was a fine gentleman all around—talk about integrity!" recalled Malcolm Carwithen. "You couldn't meet anyone better." Holmes performed the white-collar functions on the A-frame and then at Lake Camp. Genial and presentable, he was also the camp's tour guide, taking groups of visiting nurses, sailors, journalists, and photographers around the camp. The son of an English engineer in Burma, Cedric Chrishop Holmes was born in 1891. He came to Canada straight from his English boarding school in about 1909 and worked as a carpenter for Comox Logging. After serving in the First World War, where he was badly gassed, he farmed at the Soldier Settlement at Merville before rejoining Comox Logging as a scaler in the 1920s. Holmes then became timekeeper, storekeeper, and first aid man at the A-frame and at Lake Camp. In 1937 Holmes was transferred to Headquarters, where he spent the rest of his working years as office manager. He died in 1974.—John Holmes

Hughie McKenzie (left) and Hallie Dixon at the McKenzie house on Comox Lake, *c.* 1934. McKenzie, son of Courtenay blacksmith and mayor John McKenzie, had worked as a blacksmith at Camp 2 and Headquarters before moving to Comox Lake in 1933 with his wife, Birdie, and daughters, Babe (Joy) and Toots (Doris). He was, according to Bob McPhee, "a really good blacksmith," famous for his tongs. "He was one of Comox Logging's top blacksmiths," agreed Sam Telosky. "He made good tongs—some blacksmiths' weren't good. Head loaders and second loaders liked them—well, everybody liked Hughie McKenzie's tongs. He made the best tongs!" After the war he was fired for selling his famous tongs on the side, and by 1948 he ran a little machine shop in Courtenay. Hallie Dixon, born in about 1902, grew up at Headquarters and at his family's Riverside Hotel in Courtenay. By the late 1920s Dixon was a skidder leverman at Bevan. In the following decade he also ran the humdirgen at the Comox Lake load-out. "Hallie was a hell of a good loading engineer," recalled Sam Telosky. Hallie's sister, Irene, married foreman Len Harding.—Fred Parkin

WARD PIERCY PHOTO

Looking southeast from Rough and Tumble Mountain into the valley of Toma Creek, a tributary of the Puntledge River, *c.* 1933. Comox Logging's Alberni line ran up the Toma Creek valley for several miles, towards Alberni Pass. Ward Piercy's original caption: "This is looking [toward] Alberni and there is a good trail all the way. You can't see the camp it is at the left but you can see where they dump the logs in to the lake at the left. To the right is the end of the lake." At left is Tsable Mountain (4,856 feet). Comox Logging's trestle and booming grounds are visible at bottom left. —Donna Edwards

The floating camp at the top of Comox Lake, 1932. The steep hillside looming at left is Rough and Tumble Mountain, and in the distance is the distinctive Castlecrag Mountain. The tug *Eva* is towing in four bunkhouses on a raft to add to those already moored to the A-frame. The large building at right is the cookhouse. Pat Williams (*née* Holmes), who lived at the camp from 1935 to 1937, recalled that "lots of waterfalls came down the mountain into the lake. You could look from the married quarters to a small waterfall."—Donna Edwards

WILMER GOLD PHOTO

In 1934 photographer Wilmer Gold climbed a little hill and took this photo of Lake Camp, eight miles up Comox Lake and built within the A-frame's shoreline slash. This camp was the centre of the company's Comox Lake operations from 1932 until 1937, when the camp was towed two miles to the mouth of the Cruickshank River. The large building in the distance is the dance hall. The eight houses in a row are for married men and their families. Note the row of identical outhouses behind them. The house at far left is the guest house, where timekeeper Chris Holmes and his daughter, Pat, lived over the winter of 1934/35. "Dad and I shared one of the company houses," she wrote in an October 1996 letter. "The houses were pretty basic to say the least, outdoor 'plumbing' of course, but we did have water piped in—one tap, cold, but it sure beat carrying buckets from a well." The large building next to the railway track at lower right is the school, completed in September 1933. The foremen's houses are along the lakeshore to the right. The floating part of the camp includes office, woodyard, boiler house, cookhouse, and bunkhouses for single men. In the distance are the booming grounds and, at top left, the trestle with humdirgen log unloader. Experimental Caterpillar tractor logging has taken place on the far shore *(see Chapter 6)*. "They started Cat logging in the distance," said Fred Parkin. "That area had been logged by Cats in about 1933."—Wilmer Gold Collection, United Steelworkers Local 1-80, Duncan, BC

> Detail of the Comox Lake camp showing the trestle, humdirgen, log dump, and booming grounds at Comox Lake, 1934. Logs from the Puntledge River and Toma Creek valleys were dumped and boomed here for the eight-mile trip to the load-out at the bottom end of the lake. "Harry Grieve and Uncle Alvin [Parkin] ran the humdirgen," recalled Fred Parkin. The dance hall, completed in May 1934 by volunteer labour, is at right. To its right is the water tower for the steam locies. In the foreground are snags, stumps, slash, and outhouses. In the distance is the mouth of the Puntledge River. "There was a bunch of scrub bush where the river entered the lake, and we'd row there sometimes. They kept a rowboat at camp and we'd row up that way," recalled Pat Williams (*née* Holmes).—Fred Blackburn

WILMER GOLD PHOTO

Another detail of the Comox Lake camp, 1934. Floating on the lake are, top to bottom, the office, cookhouse, meat house, bunkhouses, boiler house, and woodyard. The two rows of bunkhouses are for single men, and between them is Mike Kisisky's floating woodyard with donkey engine, ramp, and boom for lifting logs from the water to fuel the donkey engine. The donkey engine, in turn, generates power to run Kisisky's saw, which resembles a "Wee McGregor." Austin Blackburn's boat the *Joan* is docked on the shore below the bunkhouses. The row of four buildings at left includes the car shop, machine shop, and blacksmith shop; there was also a filing shed at the camp. Outside the buildings is a wooden railway platform. Scattered around the camp are three railside tanks for water or crude oil. The school is the large building at bottom. "The foremen lived along the shore, in this order from top to bottom: Oswald Harmston, Norman Parkin, Hughie McKenzie, Len Harding," recalled Pat Williams (*née* Holmes). "Only Len Harding's house was specially made. The rest were converted bunkhouses which had been winched onto the land. They all had a kitchen-living room in the centre and a bedroom at each end. I can't recall there being any gardens at Comox Lake."—Fred Blackburn

WILMER GOLD PHOTO

Babies, toddlers, and school-aged children also called the camp home, and a school was completed in October 1933. The first teacher, Mrs. Agnes Henning, stayed from 1933 until July 1935. She was replaced by the timekeeper's daughter, Patricia Holmes. In June 1935 Harold Grant's daughter, Nan (later McPhee; born 1928), was sent to Lake Camp to be with her aunt and uncle Bessie and Eddy Lefley. Nan recalled that Mrs. Henning's school was built alongside the railway tracks and that the children enjoyed watching the logging trains go by every day.

Twelve bunkhouses at Lake Camp were home to some 110 single men. With its many bachelors and families, Lake Camp needed a community centre of some sort. The Comox Lake dance hall was completed in May 1934, with the proceeds from the opening dance going to St. Joseph's Hospital, Comox—the loggers' hospital. Pat Williams has vivid memories of the dance hall:

> That was quite a building, put up by the employees and made from logs either in the round or split, as in the top of the porch railings, or squared for the steps. The hall was beautiful! I don't know who designed it—it was built before I lived there—but it was a "spare-time" camp project, so perhaps it was the result of a group plan. Every night after work people worked on that hall. A lot of the little finishing touches were the result of individual ideas, things the men saw in the bush and brought back, like the strangely shaped limbs used for handrails and the table made of a large burl cut through the middle horizontally with a section of trunk left at each end to act as legs ... We were very proud of the fact that the hall had a spring floor, not too common at that time. There was usually a dance on pay nights.
>
> It was a fine hall. There was a raised bandstand in the front window corner and a stage at the rear where Joy McKenzie, later Woodrow, gave dancing lessons and where we had a dance recital every year. We also had the school Christmas concerts there. Joy's mother, Birdie, played the piano. I had only about ten children at the school but they all took part very enthusiastically.

FOR MANY PEOPLE in this elemental setting and during this stark Depression decade, the cookhouse and its abundant meals were the most important part of the camp. "I lived in the bunkhouses at Comox Lake once in a while," recalled faller Bill Lyttback. "Before my wife moved up [in 1933], I stayed at the Comox Lake camp for a month. They had good food there—Chinese cooks." The cookhouse staff consisted of five or six Chinese men: a head cook, two assistants, and two or three flunkies who prepared vegetables, set and cleared the tables, and washed dishes. Pat Williams (*née* Holmes) recalled that "all except one were Leungs. Leung Sing was

Deep in the Heart of the Island

Journalist Bruce Hutchison visited Comox Lake in the spring of 1933:

But in the deep woods winter lingers on. There are no buds, no flowers, no statesmen. At the end of a long, narrow lake, lying between sheer walls of rock, we board a launch with the lumberjacks going to work early in the morning. While we watch the snow-covered mountains rising straight up on either side, and the flat white cap of a glacier behind them, like a frosting on a cake, the lumberjacks beguile the long trip up the lake by playing cribbage. It is surprising to what a degree of skill and endurance they have developed cribbage in the lonely places of the island, where men are men, and have never heard of Culbertson and jig-saw puzzles.

At the other end of the lake, near a neat village of white houses, all floating in the water, we board a flat car in front of a tiny, chuffing engine and with a farewell toot of the whistle, push up the narrow valley which runs deep into the heart of the island. (From *Victoria Times*, reprinted in *Comox Argus*, April 20, 1933) ■

PATRICIA HOLMES PHOTO

Betty Cleland, of Merville, on the steps of the dance hall at Comox Lake, *c.* 1934, photographed by her friend Pat Holmes (later Williams). "The company provided the wood," John Holmes recalled, "and the boys built it. It had a spring dance floor: two big logs underneath ran the length of the hall and acted as springs. They would take bargeloads of people up from Courtenay for the Saturday-night dances. It was quite a famous place." The *Comox Argus* reported on May 17, 1934, "Loggers from as far as Vancouver flocked to the dance at the new hall at Comox Lake. The hall is made of logs, built by the men in their own time. 50 × 30 feet, fine spring floor holding 250 people and featuring the Imperial orchestra, Comox Lake's Own—two accordions and two violins—Ole Beline, Oscar Berkenstock, Dick Woodrow, and Alvin Parkin. The crowd left in boats at 4 a.m." Norman Parkin and Bill Lyttback were among the builders; Bill Wilson was among those who attended the opening party. The company moved the hall to Cruickshanks Camp in about 1937 and turned it into a blacksmith shop.

—Patricia Williams

Out of the Woodwork

Born in Cumberland in 1915, Babe (Marion Joy) McKenzie grew up at Camp 2, Camp 3, and Headquarters, and moved to Comox Lake with her parents, Hughie and Birdie, in 1933. By this time she was already an accomplished dancer. At Comox Lake she met and married logger and musician Dick Woodrow, and in her adult life Joy Woodrow became Courtenay's best-known dance teacher. Debra Martin of the *Comox District Free Press* wrote about her on April 1, 1977:

Mrs. Woodrow began dancing as a small girl in logging camps where her father, Hugh McKenzie, worked as a blacksmith for the Comox Logging Company. She and her sister [Toots or Tootie (Doris)] *were always organizing little shows.*

Up at Camp 2 at Oyster River there were many abandoned houses which Woodrow and her sister used as theatres for their shows. All the kids in the camp took part and charged the loggers five cents admission. Blankets carried by a couple of boys served as curtains and anybody could get up and perform. Some of the performances were hilarious and Woodrow recalled one older girl who gave a rather awful performance of "Sally in our Alley." The entire audience was doubled over trying not to laugh out loud and the singer walked off the makeshift stage in disgust. Woodrow and her sister immediately went on and sang "[I Wonder] *What Became of Sally."*

All through her childhood Woodrow was hungry for dancing knowledge and wanted to learn every type of dancing. She "picked up lessons everywhere" until she had a good "smattering" of skills that made her in big demand at local talent contests, she said. Her favourite dance in those days was the Sailor's Hornpipe because it had "lots of zip and pep." She would perform this and other dances at the open-air dance pavilion on the shore at Royston. "It was gorgeous there," she said, with all the girls in long formal dress and music by Mart Kenny's big band from Vancouver. Local talent contests were held there too and Woodrow said people were always asking her to do a dance. Then requests for lessons started coming in, so Joy started teaching what she knew in the dance hall at the [Comox Lake] *logging camp for 25 cents a lesson.*

Later Woodrow studied with qualified teachers in Victoria and Vancouver, gave shows and recitals, and had a family of four children. She offered lessons in all styles of dancing: tap, ballet, jazz, Highland, and national dances. She also served on the Scottish Official Board of Highland Dancing and was qualified to judge international competitions. With her colleague Beryl Regier, she produced shows for many years, including *Hello, Dolly!* and *The Nutcracker.* Marion Joy Woodrow died of cancer in 1977 aged 62, mourned by hundreds of friends and former students. ■

head cook, and I also recall the name Leung Chung Jik, and one was named Leung Chow. They would be from south China—most Chinese were, in those days. There were never Chinese families at the lake. Chinese women were practically barred."

Although without families of their own, the Comox Lake cooks made Chinese cookies, kept a variety of pies on display, and generally had a soft spot for the camp children. "The cooks used to slap a great big piece of pie into our hands," recalled Fred Blackburn. The cooks gave Mary Holmes bolts of silk cloth and hand-painted silk handkerchiefs, and every year on his birthday, her son, John Holmes, got a three-tiered cake from the cooks. "You needed a chisel to crack the icing!" John remarked. "The Chinese were very generous, but I treated *them* with respect. I learned respect when I was seven or eight—by living in the camp with its different nationalities."

John's sister, Pat, who visited the camp frequently between 1932 and 1935, before she started teaching there, recalled the splendid breakfasts:

> We arrived on the crew boat on Monday mornings just as the men were finishing breakfast. By the time Dad and I came to breakfast the men were all leaving. It was regular breakfast food for hungry people. They served porridge, which I didn't bother with, but they also had bacon and eggs and sausages and pancakes, and lots of it! And probably hash browns. That was a bonus. We'd sit near the kitchen end because it was less distance for the kitchen crew to carry stuff. Visitors also sat near the kitchen end.

Although Lake Camp was remote (ten miles by boat and rail from the nearest town, Bevan), the young families there rose to the challenges of isolation. Pat Williams (*née* Holmes) described the spartan existence at the work camp on Comox Lake during her years as a teacher there (1935–37):

> At that time people worked six eight-hour days, which didn't leave much time or energy for a social life. Also, there were none of the household gadgets we take for granted. Each house had a cold water tap—period! So for the women, cooking on a wood stove, heating water for scrubbing and washing clothes, often with a tub and scrubbing board, recreation was a neighbour dropping in for a cup of coffee. There may have been some social gatherings on Saturday night, but I was always at Mother's [in Courtenay], except for the pay-night dances when she often came up.
>
> We didn't have electricity in the houses, except for the four down by the lake which got it from the floating camp, and I think that was turned off around 9 p.m., and after that it was

> *Top, left*: Cookhouse assistant, Jocko the dog, and head cook Leung Sing at the Comox Lake camp, *c.* 1934. "The guy at left we called Shorty," recalled Sam Telosky. "He was a flunky. And that's Jocko, who rode the rail! And that's Sing." According to Ruth Masters, one of the Chinese cooks at Comox Lake had a favourite expression. He would say, "Long time dead!" when a high rigger did a foolish or dangerous stunt. This photo was taken looking towards the east, with the dark treed delta of the Cruickshank River below and Mt. Becher, Mt. Drabble, and Indianhead Mountain above. To the right of Sing is an empty box labelled "Sultanas."—John Holmes

> *Top, right*: The original caption by the unknown photographer was "Ah Chow in kitchen" (*c.* 1934). "I think one of the cooks was named Ah Choo, but I won't swear to it!" said Pat Williams (*née* Holmes).
—John Holmes

> *Bottom*: Inside the Comox Lake cookhouse, *c.* 1935. "In those days camp food was great," wrote Bill "Bronco" Moncrief in an August 2002 letter. "I still don't know why loggers bitch about the food." Of his coastal woods work in the 1930s, journalist and ex-logger Arthur Mayse wrote in 1985, "The largest frame building would be the cookhouse-dining hall, where perhaps 400 men satisfied their gargantuan appetites twice a day. Lunch was eaten on the job and washed down with strong-as-sin coffee brewed in a donkey-engine boiler." Note the lunch boxes left out, the solid-white ironstone vitrified china, the row of enamel jugs on the far shelves, the cook toiling in the kitchen, and the two large wood stoves in the centre aisle. Electricity for the overhead lights came from Kisisky's boiler house. John Holmes recalled that breakfast and dinner were the main meals of the day at this establishment. Lunch, with most of the men in the woods, "was basically only one table." One of Lake Camp's barrel-shaped "Lumberman" stoves, made by McLennan, McFeely & Company of Vancouver, is now in the Museum at Campbell River.
—John Holmes

CHRIS HOLMES PHOTO

Leung's Menagerie

Leung Yong, a Chinese cook at Comox Logging's A-frame on Comox Lake, indulged his parental instincts by hatching a little zoo there, as the *Comox Argus* reported in two articles:

Leong [sic] *Yong, head cook at the Comox Lake camp of the Comox Logging Company, has a brood of six ducklings which he hatched out on the top of his hot water boiler. The queer part of it is that during the day the top of the boiler is very hot while during the night, it is almost stone cold. How did the eggs come to hatch?* (*Comox Argus*, August 1, 1929)

A visitor to the camp of the Comox Logging Co., at the Lake Cumberland [Comox Lake], *was quite taken with the family of pets the cook of the camp has. Some couple of months ago visitors were amused at the clutch of duck eggs hatching out on the hot water boiler. There are some 23 handsome white ducks swimming around the camp now, and in addition a hen and a half grown brood of little yellow peeps. Also there are six lively, frisky, young half setter and half Cocker spaniel pups and their mother, and a tame blue jay. The whole family are on the best terms and were noticed sharing their common supper dish.* (*Comox Argus*, October 24, 1929) ■

CHRIS HOLMES PHOTO

coal oil lamps which the rest of us used all the time. We didn't even have radios except for the odd battery one, and the batteries didn't last. Eating was a big part of our lives. We all had piles of canned goods, and if anyone dropped by after supper it was an excuse to whip up a "snack."

The second winter I was there [1936/37], the snow really piled up. The phone line went out, and the snow was so deep that the lake could not be reached from the Courtenay end. Just as we got cut off, we all started to come down with the flu and all the women had to start taking turns while they were still well or had recovered to care for the ill ones and their families. The whole darned camp got the flu! I was one of the first to get it, and Mrs. Strain came to the schoolhouse and looked after me and brought me food, and the school closed down because the kids were all sick; and then Mrs. Strain came down with the flu and I had to look after her. I had to cook for her and her husband and baby daughter.

Fortunately, no one became critically ill, and food-wise we were okay, but the camp was completely cut off with no phone or any contact. The single men had left, so it was just the married couples, the families, left at camp. Even the cookhouse was shut. There was one man looking after the meat shed, and he said, "For goodness' sake, come and get the meat, Filberg won't want us to go hungry." Nobody went hungry. We just had to ask for what we wanted.

Transport was provided both by Bert Radford, who skippered the company's boats *Mary* and *Eva*, and by former faller Austin Blackburn, whose two boats, the *Evelyn* and the *Joan*, carried men and families up and down the lake. The *Evelyn*, Fred Blackburn recalled, had a round bottom, and when it snowed, "Dad would snowshoe up to the lake [from Bevan] to shovel off the boat so it didn't capsize." The *Joan*, launched in 1933, was more seaworthy and remained on the lake until about 1945. Blackburn charged company employees and their families 25 cents a ride in the *Joan* from the bottom end of the lake up to the camp; if you weren't with the company, the fee was 35 cents. The *Joan* was equipped with a powerful headlight. "I ran the boat for a time with Dad," recalled Gord Blackburn, "and we would aim the light 150 to 200 feet in front to watch out for deadheads. The light was so powerful that it could pick out a chunk of bark 150 feet ahead."

The *Joan* also carried milk from valley farms, and groceries from Courtenay merchants, up to families at Lake Camp. Bill "Bronco" Moncrief recalled that these deliveries earned *Joan* the nickname "Milk Boat"—highlighting the contrast between the pinched and rocky settlement on Comox Lake and the lush and pastoral Comox Valley, 15 miles away.

> Comox Lake School, 1935. Front row, left to right: Ruby Rundquist, Bernice Rundquist, Fred Parkin, Donna Mae Coleman, Ethel Harmston, Avril Grieve. Back row, left to right: John Grieve, Eva Rundquist, Bernice Lefley, Pauline Harmston, George Grieve, Charlie Strain, Patricia Holmes (teacher). This photo nicely illustrates the settlement history of Vancouver Island. The Parkins had been on the Island since the Hudson's Bay Company era (1854), the Harmstons since 1862, and the Grieves since 1874. All were Homeguard farming families who made the transition to woods work in the early 20th century. Pat Holmes was the daughter of English immigrants, Bernice Lefley was the granddaughter of a Scottish-born Bevan coal miner, and the Rundquists had arrived in the 1920s from Sweden.—Wilmer Gold Collection, United Steelworkers Local 1-80, Duncan, BC

The Rundquists of Comox Lake

Gus Rundquist was a Swedish faller who arrived with his family in the 1920s, as Bill Lyttback recalled:

Gus Rundquist was from Sweden. He ran a taxi on weekends from Bevan to Courtenay. His wife was from Sweden before they came over. His whole family moved here; he had three daughters. The youngest daughter married a Swede and moved back to Sweden. Gus Rundquist was later in Ladysmith. He quit falling and went fishing, that's all I know.

Former teacher Pat Williams (*née* Holmes) also recalled the Rundquists:

[Gus's daughter] *Eva Rundquist was crippled, but she had a beautiful singing voice. She was crippled in the hip. She was sent for an operation, but it didn't work and she was always lame. She had a gorgeous contralto voice, and when Babe McKenzie put on the concerts she didn't do anything in the dancing line but she certainly made up for it in singing!* ■

WILMER GOLD PHOTO

The Reluctant Teacher

Born in England in 1917, Pat Williams (*née* Holmes) grew up at the Merville Soldier Settlement and, after 1920, at a more fertile farm on nearby Headquarters Road not far from Courtenay:

I left Tsolum School in 1933 when I was 16. I did not want to be a teacher, but the school inspector talked Mother into it, and Mother talked me into it! I trained to be a teacher at Victoria Normal School. It was a year-long course. I boarded in Victoria over the winter 1934/35, and then I went right to Comox Lake School. The school was a well-built place, divided by a lengthwise wall with about two thirds of the width on one side for the schoolroom and the other third divided into a common entrance hall and a kitchen-living room and a bedroom for the teacher. It was very cozy. There was a big drum heater in the schoolroom, and if it was stoked up at night, it was still warm in the morning and just needed some smaller wood and a bit of draft to come to life. Of course all the cooking ranges were wood burning. There also wasn't a school board—the company took care of the school, any repairs, wood for the stoves, library books, etc. Dad handled that too, along with his timekeeper, storekeeper, and first aid work. At the end of that school year [1937], a lot of the families were transferred to Cruickshanks or Ladysmith, and the school closed.

I never went back to teaching after my two years at Comox Lake. Instead, I took a physical education course, and then I got a job in Vancouver. I was in the book department in Spencer's store! I was selling books. People would come in and say they wanted a book for such and such a person, and I would suggest an appropriate book. I loved it. This was still before the war.

After the war Pat Williams moved to the Interior of British Columbia. Ninety years old in 2007, she has a garden not far from a mountainside behind Falkland, and occasionally coyotes and bears come down to the settlement. ■

CHRIS HOLMES PHOTO

Keep the home fires burning. Chris Holmes's caption for this *c.* 1934 photo of Lake Camp was simply "Bunkhouses at top end." "I stayed in that float camp when I worked in Willemar!" exclaimed Sam Telosky. Jack Berkeley told his son in 1995, "It was all railroad and steam skidders. I stayed right at camp, at the float camp on the lake. There was a million feet of cedar logs that it took to build that float camp, the float itself! Cookhouse, bunkhouses, shops, that kind of stuff—everything on a great big float." The tug *Eva* is docked at left, and Austin Blackburn's crew boat, the *Joan*, is moored to the walkway. —John Holmes Collection, Courtenay & District Museum & Archives, 990.67.21

The *Joan* moored at Lake Camp, spring 1937. "Mr. Austin Blackburn," noted the *Comox Argus* on March 2, 1933, "who has been employed by the Comox Logging & Railway Company for a number of years, has now purchased a boat, which will be put on Comox Lake in the near future. He will carry passengers and freight to and from any part of the lake. This service should be especially advantageous to Courtenay and Cumberland merchants when married quarters are established at the new camp." The horizontal plank at right is a diving board, and the vertical stick at right is a water meter, at the top of which is written "Feb 1935" (indicating the level of the Comox Lake flood of that month). —Jean Berkeley

Diving and swimming at Comox Lake, *c.* 1937. According to Fred Blackburn, the little stovepipe at the back of the boat was attached to a "little miniature wood stove in the boat that warmed up passengers on their way up and down the lake. It's a wonder it never caught fire—the guys would be throwing diesel oil into it!" The scar left by the A-frame follows the shore behind.—Jean Berkeley

Tommy Gray and Ethel Harmston (daughter of foreman Oswald Harmston), at the top end of Comox Lake, *c.* 1934. "Oh, Christ, Grayhound!" said Sam Telosky when he saw this photo of his old friend. "That's a good picture. He was always good for a laugh! Always joking, always a smile." Tommy "Grayhound" Gray grew up on a farm at Dove Creek and joined Comox Logging in 1932. He worked as a loader on the skidders at Comox Lake and later became head loader at Tsolum. His brother Eric had joined the logging company in 1926. Their father, Arthur Gray, a veteran of both the Boer War and the Great War, had joined the Empire Settlement at Dove Creek.—Fred Parkin

The *Joan* at Lake Camp, *c.* 1937. This boat was built by Austin Blackburn and Paul Hellan. Hellan, engineer of the company's tug *Eva*, was a well-known boat builder in the Comox Valley. Among the other boats built at his yard on the Courtenay River was the *Pauline J* (1918).—Bryson Whyte

Coming down Comox Lake on the *Joan*, 1937. The men have bathed, changed, and left their work clothes behind. It is probably a Saturday afternoon, and they are on their way to Courtenay for their brief weekend.—Jean Berkeley

CHRIS HOLMES PHOTO

Mountain timbers. The bunkhouses at Comox Lake camp, *c.* 1935, seen from within the timber frame of the water tower. "When you're waiting for the launch, you'd look around that point," recalled Pat Williams (*née* Holmes).—John Holmes

Pat Williams recalled the transport arrangements at Comox Lake between 1935 and 1937:

> The launch *Joan* provided transportation for the men who commuted from Courtenay, and every day it brought up the mail and milk. On Saturday evening the boom tug *Eva* went down with a small scow with benches, taking any of the residents and bunkhouse workers who wanted to go to Courtenay. It left right after supper and came back Sunday evening. Dad [Chris Holmes] and I went down that way and Mother met us at the foot of the lake. The trip took about an hour, I think, and that's where I learned to play crib.
>
> There wasn't a doctor, but Dad knew some first aid, and when the Industrial First Aid program started he was one of the first to get his ticket. If there was an accident the person involved would be taken down the lake by boat and from there to the hospital. People in the camp used to leave their cars near the dock.

Many people remember the flood of February 1935, which raised the lake level by two feet. The flood followed a January blizzard that dumped three and a half feet of snow on Courtenay. "The flood of 1935 took out all the houses on the lake," recalled Fred Parkin. "The flood came right up to the school! It snowed and then melted. There was six feet of snow on the bank, and then it rained for two weeks. Dad [Norman Parkin] and I took the *Eva* down the lake, but the snow was like ice in the water and we couldn't make it to the bottom end." Fortunately, the buoyant wooden houses were retrieved and winched back to their proper moorings on land.

LAKE CAMP WAS FAMOUS for its pet raccoons. Pat Williams recalled the details of Josie the raccoon's early life:

> One spring a tree was felled that turned out to contain a coon's nest. The mother was seen carrying one of the two babies away, but she did not return for the other, so the next day the scaler [Harold Cliffe] put it in his pocket and took it back to camp. He and Dad had to use an eyedropper at first to feed it milk, but it grew into a real pet. Visitors to the camp were really impressed!

In July 1933 the *Comox Argus* told Josie's story to the world:

> No logging camp is complete without its mascot, and Josie, the little coon, is the pet of the Comox Lake camp of the Comox Logging Company. She was a very wee thing when the axe brought down a big fir but she rolled herself into a ball and took the bump without hurt. She travelled to the camp in Harold Cliffe's pocket. As much care has been given her rearing as if she were a silver fox and worth hundreds

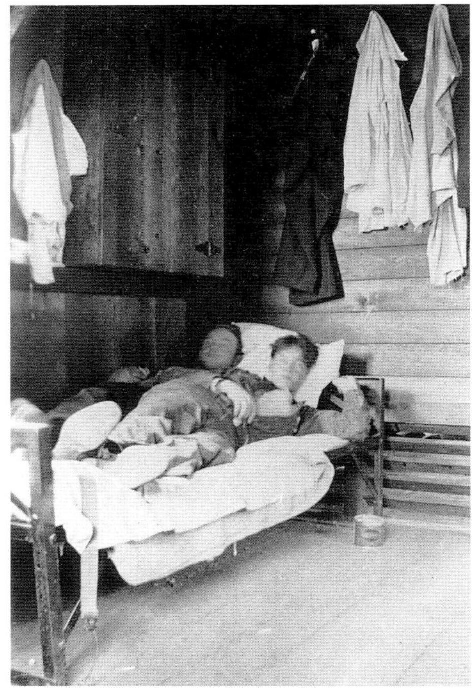

Shove over! Two unidentified young loggers in a cot at the Comox Lake camp, 1937. One of them has thrown his braces onto the foot of the bed. Note the spartan furnishings: metal cot, tobacco tin for ashtray, nails for hanging clothes, simple cupboard for personal belongings, rough and unvarnished lumber, and low wooden vent leading into the other half of the bunkhouse. In 1985 journalist and ex-logger Arthur Mayse described the coastal logging bunkhouses of the 1930s: "The bunkhouse interior was stark. Its plank floor, well-scarred by boot calks, was carpetless, its windows bare of curtains. Down each side were six orphanage-style metal cots. Two more across the far end brought the total of beds to 14. The area around each cot was the individual logger's domain. A wooden box nailed to the wall held his toilet articles and small gear. His packsack and suitcase were shoved under his bed and his town suit, wrapped in brown paper, dangled from a wall spike. The heart and centre of any bunkhouse was the converted oil drum heater mounted in a crib of sand or stones midway down its length."—JEAN BERKELEY

Lake Camp, *c.* 1937, just before it was moved to the mouth of the Cruickshank River. The view is to the south, into Toma Creek valley. The bunkhouses are at left, and the long cookhouse is at right.—FRED BLACKBURN

The camp at the top of Comox Lake (view to the north), *c.* 1935. The tug *Eva* is moored to the float, beside the water tower that kept the meat house cool. "In those days," wrote Wallace Baikie in *Rolling with the Times*, "a meat house consisted of an eight by eight shed with a roof over it with the walls boarded up half way and netting on the top part to let the air blow through. They usually had water sprinkling on the roof to keep it cool." Water was pumped into the water tower and constantly drained down on four sides of the meat safe, keeping the meat cool in all weather. Just visible at far right is the railway grade and track that connected the camp to the logging operations in the mountains. At upper right are the scars left by the A-frame, which reached timber within about 1,500 feet of the lakeshore. The higher elevations would remain intact until the truck logging era of the late 1940s and later.—JEAN BERKELEY

of dollars. For weeks she was fed milk out of an eyedropper and has only just learned to lap it up herself. She is the most affectionate little thing you ever saw and clean as a cat. Her guardians were naturally afraid that she would be snapped up by some stray dog and were very pleased when the big retriever of the camp boss constituted himself her protector and growled off all other intruders. She can and does climb anywhere and everywhere and is as inquisitive and gentle as a mongoose.

Some of the men at the camp were less impressed by Josie, as rigger Jack Berkeley told his son Don in 1995:

> Somebody brought in a coon from the woods when it was small. Found it stranded or whatever. So they tamed this thing and it used to wander around the camp. It had the free run of the camp. It got to be a bloody nuisance, you know! You'd be lying in your bunk at night, and this coon would climb up on your chest and be digging in your pocket to see what he could find! It's amazing how they could work those front feet, just like hands, you know. But it got a little frisky around the meat safe—at that time they used to have coolers with wire around them, this was the days before the fridge—and the Chinaman had the meat hanging up in there, different kinds, and Josie—they called her Josie, this coon—she used to sneak in there and help herself the odd time.
>
> And this went on for quite a while. So one day Josie turned out missing. You didn't have to guess too many times to figure out what happened to Josie. The cook with the cleaver, he fixed Josie! Guys used to kid him about having coon stew there for a while!

This was also cougar country. Pat Williams (then Holmes) sensed a cougar once when she was walking on the track just past the dance hall, where the railway track went through a cut. "Anything could have been on top of that cut!" Shovel swamper Bill Wilson was walking from the float to one of the family houses to play bridge when a cougar screamed near him. "It scared the daylights out of me! It's the most blood-chilling thing you've ever heard."

Pat Williams recalled an alarming incident when the children, picking berries, assumed that a bear was the camp dog. "Once, Georgie and Johnny Grieve and other kids, probably all the kids in camp, were picking berries up the tracks from the camp ... There were lots of berries there: they were always best where they'd logged it off. It was there that they saw that black dog Jocko ... but he wouldn't come. They went a little closer, and it was a big black bear!"

CHRIS HOLMES PHOTO

Raccoon City. "The old gentleman with the coons," explained John Holmes, "is Mike Kisisky, who cut wood in the camp. You see, a coon called Rosie [not to be confused with Josie] lived in Dad's office. She would sleep on his bed and in the morning pry his eyes open! But people fed the coons, and others came and soon they were crawling along the float and they became a nuisance. A year or two later they were inundated with coons." Pat Williams (*née* Holmes) said, "Rosie used to sleep on Dad's bed. Rosie would get on the bunk and hiss at the men, and she wouldn't let them get in their beds! They'd have to call for Dad and get him to take her away." Carpenter Malcolm Burnett's daughter, Babs Pelto, said, "One morning, Walter Rigler was asleep, and a raccoon came and jumped up on the bed and lifted up his eyelid and looked at his eye!"—John Holmes Collection, Courtenay & District Museum & Archives, 990.67.2

CHRIS HOLMES PHOTO

Mike Kisisky at Lake Camp, c. 1934. Kisisky had been a member of the Headquarters steel gang in 1921–22. "A real nice fellow," according to John Holmes, Kisisky ran the woodyard and steam plant at Comox Lake. With a little donkey and boom on the floating camp, Kisisky lifted logs and two-foot blocks of firewood from the water, pulled them up a ramp, sawed them into rounds with a steam-powered saw, and split firewood for the heaters in the bunkhouses and for the cookhouse stoves. His steam donkey also provided a hot water supply for the camp's bathhouse and cookhouse. Note the whitewashed wall behind and the two shackles at his feet for securing lines. Kisisky later married a widow named Hannah Brown, of Cumberland.
—John Holmes

Hot Shower in a Cold Camp

The presence of Mike Kisisky's boiler house meant that hot showers were available at Lake Camp, in the bathhouse next to the cookhouse. Women, though, had to take extra precautions because of the single men's bunkhouses next door. Pat Williams (*née* Holmes) recalled the bathing arrangements:

The boiler house was for getting pressure up for electricity and hot water for camp. They used an awful lot of hot water at the float camp. The bathhouse, or steamhouse, was in the middle. It was the only place where they could get hot water. And they'd need it in the cookhouse too. There were showers but no baths. They must have had clothes-washing facilities, but I can't recall. The men would come in from the bush and head right for the bathhouse! They'd be covered in pitch and sap.

Women used the bathhouse in the daytime. Two or three women would go and put a sign on the door saying, "Keep out. Women inside!" so that they could take a shower. Babe [Joy] *McKenzie and I went there together. We'd pile firewood on the inside of the door in case someone was absent-minded ... The showers were in cubicles. The door was at right, and then there was a row of showers. When the men started coming back from work, no women hung around that part of camp!* ■

Give Me Some Goddamn Cornflakes

Children were thought to suffer from their exposure to loggers and their rough language, as surveyor's wife Kathleen Crickmay recalled:

Our daughter Lois had picked up the most appalling swearing from the loggers. I shall always remember going to Comox to have tea with Jim's relatives, and sort of holding my breath. It was at the Elk Hotel, quite an English place, where they brought out the hot water in a silver jug. Our daughter reached over and, being curious, lifted up the lid to see what was in there. In a high piping voice that you could hear from the far corners of the room, she said "Jesus, God, that's hot!" I was rather glad to leave. (Baptie, *First Growth*)

In *Rolling with the Times*, Wallace Baikie told a similar story:

Two logger kids had got into the habit of swearing all the time and their mother asked the husband how to stop them and was told to give them a good cuff on the ear every time she heard them swearing. So the next morning as they were sitting down for breakfast she asked the first one what he wanted for breakfast. He said, "Give me some goddamn cornflakes." His mother gave him a good clout across the head. She asked the second one what he wanted and he said, "You can bet your sweet ass it's not going to be cornflakes." ■

> **Betty Cleland, of Merville, on a visit to Comox Lake. Betty Cleland was a guest of her school friend Pat Holmes, who identified this photo as follows: "Trestle over [Puntledge] river flowing from the Little Lakes to Comox Lake. In the evenings we could borrow the hand-car and go up to the Little Lakes to fish. We were just visiting Comox Lake. It would be the summer holidays before I was teaching there, probably 1934."**
—Patricia Williams

PATRICIA HOLMES PHOTO

LAKE CAMP WAS MORE than a generic work camp in the far reaches of a logging company's extensive property. Socially, it was a microcosm of Vancouver Island. Comox Logging brought its working crews, with their typical ethnic mix, to this remote place. They included the versatile Homeguards of the Comox Valley: the local descendants of farming families that arrived in the 1860s and 1870s. Among them were Bill, Victor, Alvin, Norman, and Fred Parkin, whose family origins were in 1850s Nanaimo. The Parkins also, through the Malpass family, gave the camp a direct connection to Edward Gibbon Wakefield's colonial settlement theories of the late 1840s. Joe Cliffe and Harold Cliffe were here, with their family's Cariboo gold rush origins, as were their cousins Oswald Harmston, Bert Radford, and Jack Berkeley. (Berkeley found himself related to practically everyone in the Comox Valley because one grandmother was a Cliffe, the other was a Piercy, and his mother was a Harmston.) The interconnected families of Harry Grieve, Ward Piercy, and Wallace Piercy had come from New Brunswick in the 1870s. Alec Chalmers of Denman Island was here; his mother was also a Piercy. Blacksmith Hughie McKenzie's father had arrived in 1879; Bob McQuillan was a son of Ulster Protestants who arrived in the 1880s; Harold Grant's family had a productive farm on the Dyke Road. In short, the camp included many men who were tied together in the web of relationships Jack Berkeley referred to as "the old 42nd cousin thing."

Newer parts of the Comox Valley were also represented at Lake Camp. Wattie (Walter) Whyte and Herb Webber came from the Canadian Collieries works at Cumberland and Bevan. Arthur "Spit" Quinn, Allan and Dick Woodrow, and Eddy Boyd were from Courtenay and elsewhere in the valley.

The camp also included more recent immigrants. Engineer Harry Sheppard was Lake Camp's representative of what might be called the Great Migration—or more accurately the "CPR Migration"—from the British Isles to British Columbia between 1886 and 1914, when hundreds of thousands of British immigrants took advantage of the newly constructed Canadian Pacific Railway

PATRICIA HOLMES PHOTO

Timber trestle at the top end of Comox Lake, *c.* 1934. Pat Holmes (later Williams), who took this photo, captioned it "Trestle across river and wet land not very far from camp. That was just past the dance hall, where the track went through a cut. It was in this area that a very large and much-photographed cedar was found." Note that planks for walking have been laid down the centre of the trestle. They are pockmarked by the spikes from caulk boots.
—Patricia Williams

Fred Parkin and his father, Norman, at Comox Lake camp, *c.* 1934. Fred was born a fifth-generation Vancouver Islander.—FRED PARKIN

A PRINCE OF A GUY

Son of Ike (Isaac) and Margaret Parkin of Dove Creek, Norman Parkin was a fourth-generation Vancouver Islander whose father's family was part of the Hudson's Bay Company era at Nanaimo.

Born in 1904, Norman Parkin grew up at Headquarters and started driving spikes on the steel gang in 1917, when he was 13. The next summer, mechanic Dave Stafford hired him at the Headquarters shop to chip out a boiler with a chisel and hammer. "They built a tin sleeve to slide me in and out, and I've had claustrophobia ever since," Parkin told Martin Hagarty in 1984. In 1921 he was still working seasonally at the Headquarters machine shop.

Meanwhile logging work beckoned. "I started in the woods when I was 15 years old, and moved up to rigging crew and then to running machines," Parkin told the *Crown Zellerbach News* in May 1980. He started as a whistlepunk at Dove Creek, and then he went chasing, scaling, and second loading at Bevan, eventually head loading on a skidder. "I was too light for heavy work and too heavy for light work," he recalled. By the early 1930s Parkin was a loading leverman at Comox Lake, and he also ran some of the early Cats there. In the Second World War he was a second lieutenant in the Pacific Coast Militia Rangers.

In the 1940s Norman Parkin was woods foreman at Bare Rocks (Comox Lake), and in the following two decades was logging foreman for Comox Logging's successor, Crown Zellerbach, at camps on the mainland coast and Queen Charlotte Islands. He lamented the transition from steam to internal combustion engine. When asked what he thought of modern logging equipment compared to the old steam pots, Parkin replied, "I don't know why we ever left them. Steam was efficient and powerful, all we needed was some wood and water" (*Crown Zellerbach News,* May 1980). He retired in 1969 to his hobbies of fishing, wood carving, and rock collecting. "He plans to live in Courtenay, purchase a pick up truck and camper unit, and travel around the province to hunt and fish."

"Norman Parkin was a fine man and a real logger," recalled Bob Muir, who worked with him at Bare Rocks. "He wasn't excitable like some of the machine operators. And he always had the time of day for you. He was a prince of a guy." Parkin died in 1991. ■

to make their way west. The Soldier Settlement at Merville was represented by timekeeper Chris Holmes; loader Tommy Gray came from a family of Empire Settlers of the 1920s. A slightly later English arrival was Nellie Booth (later Piercy) from Lancashire, who answered an ad for domestic help.

And there were other nationalities as well. Three American employees at Lake Camp were Dick Delin, Len Harding, and Bob Larson. The camp also included the Chinese cookhouse staff and pervasive Swedes and Finns, mainly fallers and buckers, including Viking Johnson (Osterblad), John Jungaro, Emil "Sailor" Lehtonen, Bill Lyttback, Joe (Jusso) Moelanen, Otto Niemi, Vic Olson, Eino Patama, Nels Peterson, Eric Sundberg, Carl Sundin, Ivan Wicklund, and Walter Widen (Viden). The grade crew included Croatian Martin Domovich. All of them gratefully received work here in the dark years of the Great Depression. ■

> Wilmer Gold, self-photographed in front of the enormous western red cedar at Comox Lake, *c.* 1935. A short walk from the Comox Lake main line, it had a diameter of 15 feet at the base and a maximum circumference of 47 feet. "A near record-sized Western Red Cedar," wrote Gold in *Logging as It Was*, "estimated age of over 2,000 years, located in a swamp-like area at the western end of Comox Lake." John Holmes recalled, "The big tree was to the right of the log trestle. It was down in the same area as the trestle, half a mile from the camp at Comox Lake. When they raised the lake, they flooded the area where the tree was."—Wilmer Gold Collection, United Steelworkers Local 1-80, Duncan, BC

WILMER GOLD PHOTO

< Comox Logging timekeeper Chris Holmes with the great cedar at Comox Lake, 1935. "I remember even seeing a picture of it in an English magazine with Dad standing in front, arms outstretched," recalled his daughter, Pat Williams. This tree was much photographed before the Annand brothers—Alex, Art, and Gordon—of Comox Logging cut it down in the 1940s. The tree was so big that Gordon Annand "had to get inside the tree to finish falling it," wrote Betty Annand in a 2002 correspondence. "Art does remember that the tree was very rotten and fell apart when it hit the ground. Also it had been home for a lot of raccoons. He remembers seeing them running in every direction after the tree hit the ground."—Wilmer Gold Collection, United Steelworkers Local 1-80, Duncan, BC

WILMER GOLD PHOTO

An aerial photo taken from a height of 16,000 feet on July 20, 1946, at the west end of Comox Lake, looking east. All the logged land in this photo was cut between 1932 and 1937 by Comox Logging, mostly within Block 32. The blanched remains of Lake Camp are visible on the lakeshore near the top left side of the photo. Lake Camp was the base of a railway logging show that extended into Toma Creek (above) and the Puntledge River (bottom), which here are separated by the unlogged lower flank of Mt. Hennessy. Willemar Lake is just out of sight at the far bottom right. Several skidder settings, with their distinctive radial skid roads, are visible. The forested land at the top of the photo is the lower fringe of the Beaufort Range, while the uncut upper reaches of Rough and Tumble Mountain (elevation about 3,400 feet) occupy the bottom left side of the photo. The swampy zone at the end of Comox Lake was home to the famous and much-photographed "great cedar," a western red cedar of great size and antiquity, shown on the previous two pages.—British Columbia Air Survey Film Record, Roll No. 255, Frame 62, Air Photo Warehouse, Victoria, BC

Great things are done when men and mountains meet; This is not done by jostling in the street.

—WILLIAM BLAKE, *GNOMIC VERSES*, 1793

6

LOGGING THE HIGH VALLEYS

LAKE CAMP WAS A WORK SITE WITH CLEAR INDUSTRIAL objectives. It was headquarters for Comox Logging's assault on Block 32, which contained the rich, valley-bottom timber of the Puntledge River and Toma Creek at the top of Comox Lake. For three or four miles from the head of the lake, the valley bottoms climbed less than 200 feet. This was a gentle ascent, ideal for railway logging; but the valleys were narrow, between a quarter and a half mile wide, and spur lines were steep. The valleys were known as "long sides"—logging shows that took place in long and narrow valley-bottom holdings. Exceptionally isolated, Block 32 could be reached only from the remote work station known as Lake Camp. Reaching Block 32 required train journeys from Royston and Bevan followed by an eight-mile boat ride, which was followed by another train ride. A place of innovation, the west end of the lake would become known for the introduction of logging by Cat and arch in 1933. This development would be tied directly to Comox Logging's revolutionary Cat, arch, and truck show at Ladysmith in 1936.

Comox Logging had first nibbled the timber of Block 32 over the winter of 1928/29 to find logs for its A-frame float and the world's tallest flagpole (for the Australian government). In the summer of 1928 a crew under Harold Grant put up two small cabins and cleared about a mile of right-of-way for a railway grade up the Puntledge River valley. A pattern of planning, postponement, and delay then set in, much of it caused by the Great Depression. In February 1929 the *Comox Argus* wrote that railways would soon be constructed at the west end of the lake, and Joe Cliffe recalled (in a 1980 interview) that in about 1930 the company started to prepare the railway grades. But in July 1930 the *Argus* noted that logging operations had been suspended. The Depression delayed the start of this show until 1932. By this time, the A-frame had taken care of most of the lakeshore timber, and Lake Camp was built over the winter of 1932/33.

Lake Camp was home base for a railway show with branch lines heading into the two major valleys to the west. The first branch, known as both the "Trout Lake branch" and the "Willemar line," went up the Puntledge River to Willemar Lake (an expansion of the Puntledge River), then also known confusingly as both Little Lake and Trout Lake. (Willemar Lake and Forbush Lake together were, and still are, sometimes known as Little Lakes or Trout Lakes.) These lakes had been surveyed early in 1933 by Johnny Duncan's crew, which included Bill Bell, Art Constable, and Bill Walker. In April 1933 the *Argus* noted that the company had completed the track to within a mile of Trout Lake, and a track machine, skidder, and cold decker were working on the Trout Lake branch.

At the same time the company started construction of a second branch, the "Alberni line." It followed an established trail up the Toma Creek valley for about three miles, towards Alberni Pass. A switch in the track about half a mile from Lake Camp separated the lines. In late July 1933 the *Argus* reported on progress:

> So far there's been no logging Alberni-wards but track has been graded for over two miles and the steel gang has just arrived from Bevan and track laying will soon start. All the logging at present is towards the Little Lakes to the west, along the narrow valley of the Puntledge. The logs are brought out to the booming ground at the mouth of the Puntledge—here a small mountain stream—and boomed there with the help of Tommy Tame's [Tame Jr.'s] humdirgen.

Here, in these two remote and narrow mountain valleys west of Comox Lake, accessible only by rail and boat, Comox Logging again deployed the big skidders in a virtuoso display of early high-lead logging methods. The company moved back to the land with an entire cast of fallers, grade crews, steel gangs, cold deckers, skidders, cherry pickers, locies, and all the expertise of highlead and railway logging. The narrow mountain valleys required adaptations to the usual skidder-setting layout; for example, back spars were dispensed with in favour of stump rigging. A striking set of photographs of this remote railway logging show has survived.

The *Argus* reports make it possible to recreate the main logging advance. In January 1934 a single skidder was working behind Lake Camp itself. From there, activity moved steadily west, up

> The west end of Comox Lake, where Comox Logging extended its railway show between 1932 and 1937. From its base at Lake Camp, the company built branch lines up Toma Creek, deep in the Vancouver Island Mountains, and up the Puntledge River to Willemar Lake, some 20 miles inland from the log dump at Royston.

> Full steam: yarding and loading on Skidder 3, Comox Lake, c. 1935. Left to right: Martin McLoughlin (fireman), Harry Sheppard (yarding leverman), Eino Patama (loading leverman). "That's Harry in the middle," recalled Sam Telosky. "Shep we used to call him. Shep was a good engineer." This photo clearly shows what loader Davey Janes called the "little wee tiny spur" upon which the skidder sat. The flatcars with cheese blocks and landing crew are at far right.—John Holmes Collection, Courtenay & District Museum & Archives, 990.67.27

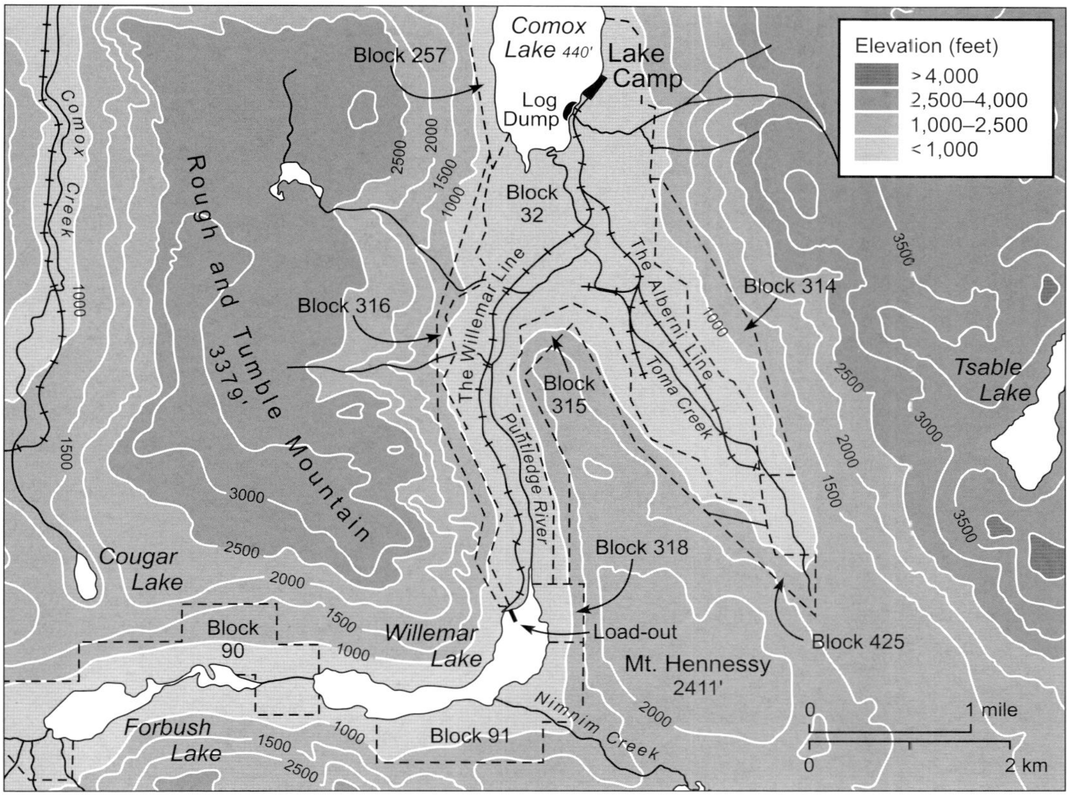

CHRIS HOLMES PHOTO

Toma Creek on the Alberni line and up the Puntledge River on the Willemar line. Several cold deckers and a single large unit—Skidder 3—were kept busy on these lines from 1934 until about 1937.

The usual complement of company personnel went to work here. Most men lived at Lake Camp, but some commuted daily by boat from Courtenay or Cumberland. One of Lake Camp's bunkhouse men was cold-deck rigger Jack Berkeley, who told his son in 1995, "We worked regular shifts up there—six days—and the train picked us up. The track ran right close to camp, just along the lakeshore, and we just walked out 100 feet to the track and climbed on and went out into the woods. It was all railroading and skidders and cold deckers. And they were steam—steam cold deckers at that time."

Len Harding took over as foreman after the untimely death of Harold Grant in June 1935, with Bob McQuillan as his side-push. McQuillan's hiring was consistent with two local patterns: he was a Homeguard promoted through the ranks, and he was a high rigger turned foreman, a common trajectory. In 1926 both Harding and McQuillan had left the old Camp 2 on Oyster River, and although this new logging show contained other remnants of Camp 2, many new recruits were employed, including men the company had picked up in the intervening years from Bevan and Cumberland. Personnel included the usual mix of Comox Valley Homeguards, Courtenay boys, coal mine refugees, and new immigrants from Scandinavia and eastern Europe; the new immigrants were especially prominent in the falling, grade, and steel gangs. Shovel operator and grade foreman was Tom McQuillan, whose swamper, Bill Wilson, recalled that they "built the last railway grade up there [west of Comox Lake] in 1935."

This was a railway show, and among the locies at the top of the lake were the Eight-Spot Heisler (engineer Harry Grieve) and the Ten-Spot Heisler (engineers Alec Chalmers and Martin "Pidgy" Davison). Frank Davison (born 1923) recalled that when he was a boy he'd visit his father and find him using the Ten-Spot to pick up the steel from old spur lines and push gravel cars into new spurs. Train crews included Alvin Parkin and Dick Delin.

Fallers included Bill Lyttback, who worked at the lake from 1933 to 1937. In 2002 he remembered that Viking Johnson (Osterblad), John Jungaro, Vic Olson, Nels Peterson, Gus Rundquist, Eric Sundberg, and Carl Sundin also worked on the falling gangs.

High riggers at the top of Comox Lake included Joe Cliffe, Eddy Hughman, and Emil "Sailor" Lehtonen; second and back riggers included Jack Berkeley, Bill Braes, Dick Harding, Vic Parkin, Ross Scott, and Walter Widen (Viden).

Big Jack's Pranks

In a 1995 interview by his son Don, Jack Berkeley (1913–96) recalled his work as second rigger for Jack McKenzie Jr. in the 1930s. McKenzie was a well-known prankster and put Berkeley through a high-elevation initiation:

I'll tell you another little thing: I forget how this came about now, but I got pulling rigging for Big Jack McKenzie one day ... We were in the middle of rigging the tree. So Big Jack went up and tied on, and then I went up. The rigger ties on first, you see, and then the second rigger goes up, so you always pass your rope around: you reach around and give him your rope, and he takes it around and hands it to you and then you tie your cat's paw [knot], *you see. You have a little loop on your rope in your belt, and you tie a cat's paw in it, then you adjust your rope so you're standing at the angle you want to be.*

So I got the rope—he hands me the end of the rope, and I get the cat's paw tied, and I get her adjusted to the angle I want it to be. He says, "Okay, John," McKenzie says, so I say, "Okay." I holler, "Haul off!" you know, I want to slack the pass line. But McKenzie gave me about two feet of slack and I sailed off! I stood straight out from the tree! And I grabbed the rope! I was just about riding off the tree! "You all right over there?" he says. He was cackling away on the other side of the tree. He pulled that trick on more guys!

He was just like a cub bear on a tree when he climbed; he just put one step after the other. Usually you throw your rope and take two or three steps, then throw your rope and take two or three more, you know, but he just stepped one step after another, just like walking up a tree, you know! Steady as hell. God! Oh, he was a good guy to work for. Always playing tricks. You never knew what the hell he's going to come up with! He was a good rigger, though. ▪

Comox Lake boss Len Harding (left) and his "side-rod" (side-push) Bob "Rocky Mountain" McQuillan, 1938. These two men were in charge of logging at Lake Camp and Cruikshanks from 1935 to the early 1940s. "That's Rocky Mountain at a skidder setting! Sure as shit! He was side-rod when I was up there," said Bill Woods, who started at Comox Lake in 1941. Harding and McQuillan are leaning on an empty flatcar above a skidder. "You parked your empties there, and when you wanted them, you'd pull them down," said Woods.
—Filberg Heritage Lodge & Park Association

Rocky Mountain

Bob McQuillan (1894–1976) was a son of Harry McQuillan, a native of County Derry, Ireland, one of three brothers who came to Canada in the late 1880s and crossed the country on the new CPR. One brother owned a livery stable in Courtenay, but Harry and his wife, Mary, farmed at Dove Creek.

In 1909 Bob McQuillan delivered milk from the family farm to Comox Logging's first camp, but like many other Homeguards he got caught in the logging company's wide net, as his son Darrell recalled:

Father stayed on the farm until about 1920, when he joined Comox Logging. Before that they'd sold the bulk of their produce to Cumberland, and before that they took it to the Comox wharf. Dad was the last one to stay on the farm. He and his brothers left the farm: Dad, Tom, and Stewart all ended up working for Comox Logging. At one time five McQuillans worked for Comox Logging: my father, Bob; myself; my brother Edwin; my uncles Tom and Stewart.

By 1923 Bob McQuillan was head rigger on a skidder at Camp 2 at Oyster River. He later worked at Bevan, Comox Lake, Cruickshanks, and Tsolum. During the Depression, in the off-seasons, he also hewed ties and timbers with a broad axe for Canadian Collieries. By the late 1930s he was side-push to foreman Len Harding at Lake Camp and Cruickshanks. In the 1950s he became Crown Zellerbach's foreman on Quadra Island.

At Comox Lake he earned the nicknames "Rocky" and "Rocky Mountain." "Dad was known as Rocky because at Comox Lake he used to go up to rocky and bare knolls to plan his settings," said Darrell. He would sit on the rock bluffs and plan the company's progress up the mountain valleys.

Bob McQuillan married a granddaughter (Dorothy) of the Anglican missionary John Booth Good and retired in 1958. "My father never raised his voice. Maybe in the logging camps now it's different, but they were really nice men. Loggers were real gentlemen," said Darrell.

Bob McQuillan's cousin Harry McQuillan, also a logger, was the Progressive Conservative member of Parliament for the Comox-Alberni constituency during the Diefenbaker era. ■

Don't Fiddle the Fallers

Bill Lyttback (1908–2003) recalled his work at Comox Lake:

I started at Comox Lake in the fall of 1933. I bucked first, and then the head faller died and I went falling. The [new] *head faller was Fred Swanson, and he had problems with his wife. I lived in Bevan right beside Austin Blackburn, who was a scaler then before he ran the crew boat. I also commuted to the lake with Blackburn, and for a month I lived in the bunkhouses at the end of Comox Lake.*

There was nice timber at Comox Lake! This is when they were working at the far end of the lake. I took the train and speeder to work, and I came back on the speeder too. We had the speeder there all day. I worked towards Port Alberni, on the line between Port Alberni timber and Comox timber. We had to fall them on the Comox side of the line. I liked being a faller. I had no boss except the bull bucker and scaler. The scaler followed behind measuring logs. He scaled your logs once every day. Oh yes, the scalers were honest! Once [bull bucker] *Eric Flinton asked a Swede faller named Ollie Graft to become a scaler, and Graft questioned him, and Flinton said he'd never fiddle the fallers.*

My buddy at Comox Lake was a Finnish guy named Joe [Jusso] *Moelanen. We also called him Poika, which means "Boy." Before Comox Lake he had worked in the hills at Bevan; he was at Comox Lake when I came there. He's gone too. When we needed a new partner I went to Vancouver and got Viking Johnson* [Osterblad], *who had come on the boat with me from Finland.*

In the hungry '30s the price was down to 38 cents a thousand [board feet]*: this was the hand faller's wage. We were paid by the thousand. This was at Bloedel, but at Comox I never got below 45 a thousand. I moved to Comox because the prices were better. In 1934 I made $1,500 in the whole year! It was a lot of money, and a dollar bought you something. I left Comox Lake for the new show at Ladysmith in 1937.* ■

Levermen at work on Skidder 3, Comox Lake, c. 1935. "You can see the iron wheels!" said Art Beech, who worked as a chokerman on a skidder crew in the 1930s. "Those wheels turned sideways. You had to turn them and block them." The wheels are reminders that skidders were mobile units hauled by rail from setting to setting. Skidders were secured by four steam jacks, one at each corner, and sat on a cribbing of railway ties. "They'd jack it up and block it up with ties," said Davey Janes. Chaser Dick Bailey recalled, "The guys knew every tie."—Laurie Mathers

A rigging-up goat at Mt. Hennessy, Alberni Pass, near Comox Lake, 1934. Left to right: unknown, Jack Berkeley, unknown, Joe Cliffe (rigger), Emil "Sailor" Lehtonen, Wallace Piercy. "That looks like a 10 × 12 rig-up goat," said Sam Telosky. "Often they were Empire—built in Vancouver." The rigging-up goat and crew raised and rigged head spars and back spars ahead of logging crews.—Wilmer Gold Collection, United Steelworkers Local 1-80, Duncan, BC

High Hazards

Joe Cliffe (1897–1985) kept on rigging despite losing the toes of his right foot in 1930, when the hammer of a piledriver crushed them. He began working for Comox Logging as a high rigger in 1921 and worked at Comox Lake from the late 1920s until 1936. He recalled in a 1980 interview:

In 1921, that's when I started to climb trees. I thought I was going to die climbing trees! I think I climbed trees 15 years. That was quite a long time especially with skidders—you are up in the air so much. It wasn't like the old high lead where you just go up a tree about twice a month.

Oh, there was more to do with skidders. Quite often you were moving the lines around a bit so the lines won't cut the tree, and all that sort of thing, so many lines on a tree and so many blocks on a tree. Later on I had to rig all the trees and that's all I did was rig, for many years, just sort of up in the air every day for hours and hours ... With six skidders it took a lot of rigging. I also had to take it all down! Of course it didn't take too long to take it down, not near as long as to put it up.

In September 1961 he told the *Crown Zellerbach News* about an incident at Comox Lake in about 1930:

I worked up there off and on quite a bit, especially rigging the trees. While a high rigger I had several narrow escapes. Once the wind suddenly changed and blew the top of a tree I was topping back towards me, and I swayed to and fro about 80 feet above the ground. I just managed to claw my way around to the other side of the trunk as the top swished by.

Cliffe was transferred to Ladysmith as assistant foreman in 1936, and in 1961 he retired from his job as coastal logging manager for Crown Zellerbach Canada, one of Comox Logging's successors. ■

WILMER GOLD PHOTO

The crew of Skidder 3, Comox Lake, 1934. Early highlead logging was labour intensive, as Bill "Bronco" Moncrief recalled: "Skidders, as you are probably aware, had large crews. For example, yarding engineer, loading engineer, fireman, hooker, second hooker, four chokermen, whistlepunk, a head loader, two second loaders, a third loader, and a chaser. Plus a bunkmaker. Also head rigger, second rigger, two or three on the back rigging, with a small donkey or back rigging goat." Back row, left to right: Wattie (Walter) Whyte (fireman), Bob Phye (loading leverman), Roy Morrison, Ron Beaumont, George Colwell, Bill Stant, Oswald Harmston (yarding leverman), Herb Webber, Tommy Gray, Norman Parkin, Dudie Coombs, Eino Patama, Lyle Coleman, Harold Hames. Front row, left to right: Bob McQuillan (head rigger), Walter Widen (Viden) (second rigger), Oscar Berkenstock (head loader), Ed Lefley, Bill Parkin, Bob Dawson, unknown. "I'm guessing," said Darrell McQuillan, "but I think this was taken near Willemar Lake in the Comox Lake area. My father [Bob] was foreman of this crew." Recalled Sam Telosky, "Bill Stant got hurt. He pretty near tore his hand off. He got compensation for life out of that."—Wilmer Gold Collection, United Steelworkers Local 1-80, Duncan, BC

CHRIS HOLMES PHOTO

Engineer Eddy Boyd on his Clyde cold decker, yarding logs into the cold-deck pile at right, Comox Lake, *c.* 1935. Later, the skidder will yard the contents of the cold-deck pile trackside. "The yarders went around the perimeter of the setting, and they yarded logs into the perimeter, and then the skidder yarded them into the landing and they were loaded," explained Bill Wilson. John Holmes noted, "This is a three-drum machine in operation. The three drums held the strawline—which was a light line—the haulback line, and the mainline."—John Holmes

Swinging the tongs, also known poetically as "casting the crooked steel." Second loader Pete Telosky is at a skidder setting near the Willemar Lake load-out, *c.* 1936. A few years later Telosky would break his neck while second loading at Cruickshanks (and, miraculously, later return to work). —Sam Telosky

Shadow, skidder, and steam: loading via hayrack boom near Willemar Lake, *c.* 1936. The silhouetted figure is probably second loader Pete Telosky.—Sam Telosky

Where the Money Was

From Cumberland's Italian community, Chuna Tobacco worked as a faller at Comox Lake in the 1930s and 1940s. He recalled his work in the coal mines at Cumberland and in the woods at Comox Lake:

I was born in Cumberland in 1909 and worked in the coal mine: in No. 4, No. 5, and No. 8. I could have stayed as a miner, but I went into logging because that's where the money was! I switched to logging to make more money, not because it [coal mining] *was dangerous. It's where the money was. I worked for Comox Logging for about 15 years. I was a faller almost the whole time. I worked at Comox Lake on the rigging for a while, on a skidder, and then I was falling.*

[What training did you need to be a faller?] *You'd watch them do it and then try it out! We used a double-bladed axe to put the undercut in, and then we used crosscut saws—that was before the power saws came in. We used springboards if the tree was too big. You'd have to get up high. You packed your saws in on your shoulder, and you brought them in every night to be sharpened. They'd file the saws and sharpen them up at Comox Lake. They had a man who did that for us. They had a number, not your name, on the saws. You had a special rack in the filing shop, and your saws were kept in one place.*

We had an oil bottle for the saw. Sometimes we needed more than one bottle for a tree. Maybe we used one bottle of oil for a big tree. You had to use aluminum wedges in the cut to make the tree fall against its natural lean. Sometimes you needed two or three wedges. As the tree lifted, you'd have to change wedges.

You'd try to fall the trees pointing near the spar tree—towards the spar tree. Sometimes you'd get a tree caught up in the branches of another tree, so you'd fall another tree onto that one. Sometimes the wind would keep them upright. And you'd have to watch where the buckers were. Sometimes we'd work as buckers too—we'd help them buck.

It was all mountainsides at Comox Lake! The logs would roll a long way and smash up, so you had to watch what you were doing. You couldn't just fall them anywhere. You had to fell the tree onto the ground, not on another tree or it would smash up. They broke a lot of timber ... If you broke too much timber, you'd be out!

I just worked at the lake, never at Headquarters or Ladysmith. We lived at Cumberland when I was at the lake. We lived in a little house here in town. They had a bus run from Cumberland to the lake, and then a boat up the lake. In the morning you got in the crummy and away you went!

Chuna Tobacco died in 1997. His son Terry Tobacco was well known as a Canadian track star, winning silver and bronze medals at the British Empire and Commonwealth Games in 1954. ■

Getting the Hang of It

Jack Berkeley told his son Don another story in their 1995 interview:

Jeez, another time we had a tree—this was way up the Alberni valley, that's the valley to the left—and this tree was high, maybe 160 or 170 feet I guess, something like that, pretty limbery you know. It was a standing tree; it wasn't a raised tree. So we got the guylines all skinned out and got slack pulled on the end, to pull enough slack on the guylines so there's enough to reach up the tree, you know.

So we were working early shift; it was barely daylight when Buster [Hart Brown] and I got up there. He was awful nervous on a tree, Buster was! It was a cold morning and he was shivering. You got to hang the guyline on the tree to correspond to the way you're going to yard. In other words, if you don't have it right on the tree, when your mainline comes around, it'll start sawing your guyline before you get to the guyline stump to get by it; so you got to put the bight on it, figure out which way you're going to do it.

So we—usually one each side—when you're on a tree with nothing on it, just the weight of pulling the guylines up, it's flopping around pretty good! So we got one guyline tied on; and we got two on; and we're taking the third up, and these guys are pretty dopey on really early shift. This goddamn guy got a guyline hung under a root or a chunk or some bloody thing. You don't notice it right away. All of a sudden I look down and the butt of the tree is over here! And Buster is hanging on! So I stopped and went, "Wake up, you so-and-sos!" So one of these guys ambled over and grabbed the guyline to give it a yank, and away we go the other way.

Jesus, oh God! Buster is hanging on for all he's worth. I'll never forget that. That was Annand—Gordon Annand. They couldn't care less! They were on the ground, you see. Yeah, Buster was panicky on a tree. ■

WILMER GOLD PHOTO

> **Skidder 3 at Comox Lake, *c.* 1935. Chris Holmes's caption was "Old Wattie Whyte and his dog Brownie." Suspended behind are two "chunk logs," which are counterbalances to the hayrack boom. "You buck a piece of butt log 10 feet long and 2 feet through," explained Sam Telosky, "and that's the chunk log, and you suspend it from the chunkline. You see, now it's got the weight of the boom, so all they have to do is slack the haulback [line] and they can move the boom anywhere." Later, for safety, they always hung a piece of wire rope, known as a tail, from the chunk to warn people below.**—JOHN HOLMES

< **A Clyde three-drum cold decker at Comox Lake, *c.* 1934. Top row, left to right: Bob Strain, Eddy Hughman (rigger), Ray Dawson (engineer). Front row, left to right: Dick Harding (whistlepunk), unknown, Vic Parkin (second rigger), unknown. The fully rigged spar tree is at right, and the landing pile is at left. "This Clyde is a cold decker, not a rig-up goat," observed Sam Telosky. "That's the skidder tree behind. They're either just moving the cold decker in or just moving it out. It's off the flatcar and into the bush!" Note the chain guard (marked "Clyde") designed to conceal the machine's connecting rods. "Connecting rods sure pack some punch! Break your leg as easy as winking. Ought to have a guard at the side of it and stop a guy getting hurt," editorialized "Old Man Safety" in June 1928 in the *Comox Argus*. Stan Hanham, who worked with Eddy Hughman at Headquarters, Bevan, and Comox Lake, recalled that "Eddy, as high rigger, was also in charge of the yarder, the engineer, the rigging slinger, chokermen, and whistlepunk—a six- or seven-man crew."**—WILMER GOLD COLLECTION, UNITED STEELWORKERS LOCAL 1-80, DUNCAN, BC

CHRIS HOLMES PHOTO

BROWNIE THE COUNTING DOG

Wattie (Walter) Whyte of Comox Logging was never without his dog Brownie, the famous counting dog, as visiting journalist Hugh Weatherby wrote in May 1948:

I've heard of dogs who could count, but the only one I've ever seen is 14-year-old Brownie, who belongs to Wattie Whyte, watchman for Comox Logging and Railway Company, at Comox Lake.

Brownie is showing his age now. He can hardly hear at all, but when Wattie Whyte holds up a certain number of fingers, Brownie will bark out the number. His limit is five, however, and he refuses to count beyond that. He probably figures that if you aren't convinced that he can count by the time he gets to five, you never will be.

Brownie learned to count when he was a pup. He used to go out into the woods and listen to the signals the men were passing back to the donkey. He'd bark one for every "yo." He even got good enough to follow the staccato signal for slacking off.

Brownie doesn't feel very envious of the younger, spryer dogs who occasionally come around, dogs who can hear every little sound, and whose legs don't creak when they trot, for after all, he can count to five, and how many of them can? ■

CHRIS HOLMES PHOTO

CHRIS HOLMES PHOTO

Above: A high rigger pauses for the camera on his way up a back spar, Comox Lake, 1937. The caption on the original print was "Tree-topper going up. Rigger going up to top tree."—JOHN HOLMES

Above, centre: A high rigger reaches the top of a back spar, 1937. Chris Holmes's caption was "Rigger climbing to top of tree 165 feet up. Comox Lake." Pat Williams (*née* Holmes) clearly remembered one such tree-topping: "Once they were rigging the spar pole, and the whole camp went out to watch it. And the rigger stood on his head! It was before the first Little Lake, within easy walking distance of camp." Emil "Sailor" Lehtonen was probably the high rigger who performed the stunt. As John Holmes recalled, Lehtonen "could climb a tree like a monkey and used to stand on his head on top of a tree! He'd put on exhibitions." —JOHN HOLMES

Above, right: Jack Berkeley rigging a spar tree, Comox Lake, 1937. "He's putting the guylines on. That's the first thing you do," recalled 87-year-old Sam Telosky. "I looked at those pictures and I'm just rarin' to go logging again! I was a hook and rig [combination of hooker and rigger] on a cold decker. We got a piece of three-eighths rope and a rope coil. First you top the tree, then send the rope down, then get the small pass block up. It weighs eight pounds. You hang that, then send the three-eighths strawline down and they send up the guylines."—JEAN BERKELEY

AN UNLIKELY SPAR TREE

In 1995, Jack Berkeley told his son Don the story of topping an oversized Douglas fir at the head of Comox Lake with his brother-in-law Hart "Buster" Brown:

So I had to top this thing! It was 32 inches in diameter where we topped it. It looked like a high stump, almost! It wasn't very high—it was about 120 feet, I guess.

Buster and I went up. The belt rope was 16 feet long, and I was holding on to the very end of my rope when I started up. Wasn't enough to tie on you! And Buster was the same. I was on this side, and he was on that side, and we helped each other's rope up till we got high enough to tie it on, you know. Then we got up there and sawed the limbs off—limbs like this! A foot across, sawed them off, got up to the top, and when we got up to where we were going to top her, we started hacking away. Chopped in the undercut and the side-notches, and we started to saw the back with a five-foot saw, I guess. Just got in partway, when the bloody wind got up! There's treacherous winds in that valley, it's a deep kind of place, and the cut would open up that much, two inches, and come right down like that, open up, then squeeze down, and then open up two

CHRIS HOLMES PHOTO

Skidder 3, one of the six venerable Lidgerwood workhorses introduced to the Comox Valley in 1911 by Comox Logging and used extensively for logging the mountain valleys above Comox Lake. Foreman Len Harding is leaning on one of the drums in this c. 1935 photo, taken at Comox Lake. Both levermen are also visible: the loading leverman is standing at left, and the yarding leverman is leaning on a drum behind. Recalled leverman Mack Grant, "The steam skidders had a loading engine on them and a skidding engine, each with an operator. One engine was to haul the logs in, the other to load the logs. You could load and skid at the same time."—John Holmes

inches again! And we had limbs up above us that were hanging down to maybe 30 feet below where we were standing on the tree.

And Buster looks around, and he was nervous as hell: "W-w-w-w-what do you think?" And I knew it was one of these offshore bloody things, I didn't think it was going to stop, so we tied the saw up on a limb and came down, and we had to go up the next day and top the bloody thing! Christ! Yeah. It was only 32 inches at the top where we topped it, but it was over six feet on the butt. Len Harding picked that—why he picked that goddamn thing! There were other trees around there, you know. He picked that for a spar tree! ■

> Making a cold-deck pile beneath a back spar, which is out of sight at upper right. The airborne log has been yarded in from the left and is about to be dropped onto the cold-deck pile. "This is when they're starting a cold-deck pile. It would get a lot bigger!" observed Sam Telosky. Sven Ell wrote in a 2000 letter, "A cold deck happens when there is timber felled beyond the reach of the skidder, so a yarder is then put to yard these logs to a skidder to reach."
—Jean Berkeley

Large skidder and cold-decker crews were needed. The levermen included Eddy Boyd, Ray Dawson, Oswald Harmston, Jim Mathers Jr., Norman Parkin, Eino Patama, Bob Phye, Wallace Piercy, and Harry Sheppard, with firemen Martin McLoughlin and Wattie (Walter) Whyte. Skidder and cold-decker crews also included Ron Beaumont, Oscar Berkenstock, Lyle Coleman, George Colwell, Dudie Coombs, Bob Dawson, Tommy Gray, Harold Hames, Ed Lefley, Roy Morrison, Bill Parkin, Bill Stant, Bob Strain, Andy Telosky, and Herb Webber.

A crew was also required on the lake. Bert Radford, from Comox, skippered the *Little Eva* on Willemar Lake in 1936, and Sam Telosky and his friend Riley McLaren, from Haney (on the Lower Mainland), worked as boom men at the Willemar load-out.

A large complement back at Lake Camp, notably the sawfilers, blacksmiths, and cooks, kept the show going.

AT THE FAR END of Toma Creek, Comox Logging's Block 32 abutted the large Block 74 owned by the Alberni Pacific Lumber Company, based at Port Alberni. It was here in about 1935 that Jocko, a dog of mixed Labrador retriever and German shepherd descent, mistakenly boarded a logging locie belonging to Alberni Pacific and ended up on the far side of Vancouver Island. "He rode flatcars. Jocko was known affectionately as the second brakeman," recalled Sam Telosky. "Jocko belonged to head brakeman Dick Delin. Dick and the dog just lived together, you know." John Holmes told the story:

> That's the dog that went to Alberni! … They were logging up at the head end of Comox Lake. Alberni Pass was up behind, and Comox was logging adjacent to the claims of Alberni Pacific. The two claims came together at one place, and they could hear the Alberni Pacific whistles in the distance. Jocko, who was a big Lab cross, used to ride locie all the time, but one day he was nowhere to be found. They lost him for a while, and then they heard that a dog was riding the E&N or Canadian National out of Alberni! They made inquiries, heard that this dog enjoyed riding in the locie, decided it must be Jocko, and sent some fellows down to Parksville to bring him back. Jocko heard the whistles of Alberni Pacific! He went up on one locie and came back on another.

Holmes's sister, Pat Williams, continued the story. "Word got around among the train crews about this train-loving dog, and Alec Chalmers heard about him, checked him out, and sure enough, it was Jocko. When Jocko got back to Courtenay, one of the train crew took him home and looked after him. So he went back to the lake." Gord Blackburn added the following: "Our claim and MacMillan's [MacMillan Bloedel, which later bought Alberni

> Steel rails and railway ties were brought to the top of Comox Lake by Comox Logging's car float, or "transfer wharf." Here, the transfer wharf, carrying the Ten-Spot and two flatcars stacked with railway ties, prepares to leave the load-out at the bottom end of the lake, *c.* 1932. Out of sight and in front of the load are a tug and the company's crew boat *Mary*. "The raft that used to take flatcars, oil tankers, locomotives, etc., was huge. I really don't know how many logs were used for it," said Bill "Bronco" Moncrief. George Yates Jr. recalled, "Later I wrecked that raft. It was a monstrous rig with line attached to it. We moved a Madill tower [a kind of mobile steel spar] on it, but it was so heavy, and the transfer wharf was so waterlogged, that more of the tower was below water than above!"—Bob Stelfox

Trestle at Comox Lake, probably the new trestle over the Puntledge River, completed early in 1937. Sam Telosky identified the man as bridge crew boss Bill Eastman. "They had to be good men and were the highest paid men on the claim," Wallace Baikie wrote of the bridge crew in *Rolling with the Times*. "I remember when they were getting one dollar an hour back in the 20s. That was good pay. Bill was a hard boss, but the men respected him."
—John Holmes

Comox Logging's locie No. 17, a gas-powered Whitcomb, built at Headquarters in 1928 and used for shunting equipment around the town. Here it is pulling the crew boat known as the *Ark*, skippered at Willemar Lake in the 1930s by Bert Radford. Darrell McQuillan recalled, "Despite its sluggish appearance, the *Ark* was powered by two big engines placed side by side."—John Holmes

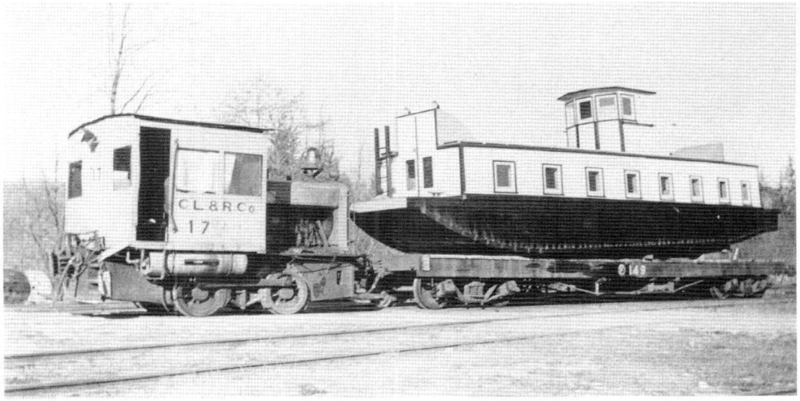

Chris Holmes photo

Ben Horbury photo

Pacific] bordered, which is why he got on the wrong train. I was firing locie there and knew Jocko well. He liked the sound of the logs being loaded, and he'd run up and down on the flatcars, and he'd go down to the log dump to see the logs being boomed." Sadly, Blackburn noted, "Later, Jocko got run over by a logging train at the Cruickshanks."

The larger point of Jocko's story is what it reveals about the rate of industrial logging on Vancouver Island. It had taken almost 70 years of continual cutting for loggers—and dogs—working from Courtenay to meet loggers working from Port Alberni. That is, it took seven decades for direct industrial contact to be made between the Strait of Georgia and the west coast of Vancouver Island.

THE TOP END of Comox Lake was also the site of Comox Logging's successful experiment with "Cat and arch" yarding, that is, yarding using Caterpillar tractors and logging arches. Chris Holmes wrote in his Comox Lake scrapbook: "In 1933 Caterpillar tractor logging was successfully experimented with at Comox Lake which I believe was the first operation of this nature on Vancouver Island—possibly in British Columbia." This technology promised to revolutionize logging because, instead of relying on a skidder or yarder requiring at least 1,100 feet of highlead, miles of line, and transport right to the site by expensive railway spurs, it required, for yarding, only a mobile Caterpillar tractor with a few feet of highlead and not much more line than that. In Cat and arch logging, a line from the drum at the back of the Caterpillar tractor was run through a fairlead (an enclosed roller that guided the line), fed through three or four hoops on the arch, and finally attached to a log, which then dragged over the ground. "The idea," said Neil Martin, "was that you ran the line up through the wheels on the arch so you got a bit of lift. It gave you a little bit of highlead."

This innovative venture started in 1933 on Comox Lake itself, on the steep sides of the lake in the neighbourhood of Lake Camp. John Holmes recalled a comical incident that took place at this time: "Comox Logging had the Cat 75s coming in by barge to the Comox Lake camp at the head of the lake. These were Caterpillar Cats, the forerunner to the D7 and D8 Cats. I remember the day they arrived! Everyone was saying, 'The Cats are coming up by barge!' Nellie Piercy—Mrs. Ward Piercy—an English lady who answered an ad and came out here, went down to meet the barge with a bowl of milk!" When asked about this story in 2002, Nellie Piercy's daughter, Donna Edwards, admitted, "That is just the sort of thing I can imagine my mother doing."

The original Cat driver ("Cat skinner") was Oswald Harmston. "He was the first man to log with 'cats' in the Courtenay area.

A Cat logging show at Willemar Lake, *c.* 1937. Here, Comox Logging has combined Cat and arch technology with a traditional spar tree and donkey engine. The donkey has yarded logs into a cold-deck pile at the spar tree, and the Cat is busy yarding them down to the lake behind. "They cold decked there all one summer, and that winter they Cat logged up on a ridge," recalled Sam Telosky.—JEAN BERKELEY

> Cat logging at Willemar Lake, 1935. Left to right: Allan Woodrow, Arthur "Spit" Quinn (Cat driver), Oscar Berkenstock, Emil "Sailor" Lehtonen (rigger and hooker). "Spit Quinn was quite a guy; he ran the Cat at Comox Lake," said Bill Lyttback. The Cats proved their worth early in 1935 as snowploughs during the record snowfall. "Spit Quinn ploughed the highway right here in Merville in 1935 with one of them. He was brought down specially to do so," recalled Hi Churchill. Born in Nanaimo to German parents in 1897, Oscar Berkenstock started with Comox Logging in 1919. The July–August 1981 *Crown Zellerbach News* quoted him as saying that he'd done "every job there was to do in the woods from skidding to high-rigging." He spent most of his career as a head loader, but he also worked as a bucker, boom man, and loading engineer. He died in 1992.
—WILMER GOLD COLLECTION, UNITED STEELWORKERS LOCAL 1-80, DUNCAN, BC

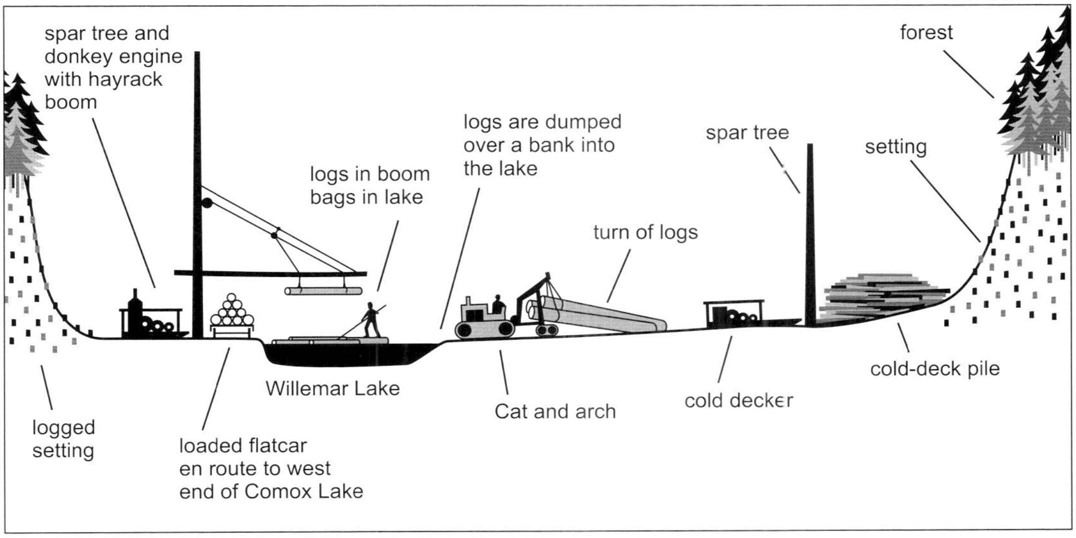

Cat logging around Willemar Lake, inland from Comox Lake, *c.* 1935. Between 1933 and 1937 Comox Logging pioneered logging by "Cat and arch," a system in which Caterpillar tractors yarded turns of logs behind them by means of a steel arch.

WILMER GOLD PHOTO

Gentry at the Cat Show

The *Comox Argus* printed a story entitled "Sailors See 'Cat' Logging" on August 13, 1936:

A party of about forty officers and friends from H.M.S. Apollo *enjoyed the hospitality of the Comox Logging & Railway Company on Tuesday at the Lake camp and watched logging operations. Among the party from the ship were Admiral Sir Matthew Best, Lady Best and Miss Best, and Captain M.J.C. deMeric.*

Local business men went down to the Comox wharf at half past seven and drove them to the wharf at Bevan, where they were taken on board the tug Eva *and were taken to the end of the lake under the care of Len Harding, who had charge of the arrangements.*

Harry Grieve took them out to the Little Lakes operation on the 8 Spot. They saw some trees felled and Eddie Hughman topped a 125 foot fir tree for them. Then the party went down the lake [Willemar Lake] *in "Little Eva" handled by Bert Radford to see the "cat" perform. Norman Parkin handled the 90 h.p. Caterpillar very expertly. The flexibility and power of the unit caused wonder and admiration in the party. The tractor can pick up and walk away with a load of between four and six thousand feet over the roughest kind of going and flip them into the lake ready for booming.*

The final exhibition was that of yarding the logs with Ray Dawson on the machine. By this time it was long past the luncheon hour and the guests greatly enjoyed a loggers' lunch at the floating camp at the lake. Thence home on the Eva *and back to the ship a little late for afternoon tea.* ■

CHRIS HOLMES PHOTO

Royal Navy sailors watch Skidder 3 bring a turn of logs to the skidder landing, Comox Lake, 1937. British and Canadian sailors docked at Comox were frequent visitors to the company's inland operations. "Once, 20 or 30 sailors came up to the lake pulling steel," recalled Eddy Berkeley. "They loved it! They thought it was a great thing. They had to undo the bolts and take up the spikes. Men were scarce; they worked in the summer. They thought it was great to get off the ship in the nice weather."—John Holmes

In this *c.* 1933 photo, the second loader unhooks the loading tongs at Comox Lake while visitors look on. The head loader's job was to arrange loads between the cheese blocks on the flatcars. "Head loading was a big job!" recalled Sam Telosky. "The outside log has to be good because there's no high stakes, only cheese blocks to hold the logs on. The head loader says, 'Give me that one over there.' Lefty Larson was head loader at the lake. He'd say, 'Give me that one behind you!' He tells this to the second loaders, and they run out with tongs. With cheese blocks you had to have a *lay*—the logs had to *fit*. Now with stakes it's not a worry." Here, the head loader has placed a large log on each side of the flatcar, snug against the cheese blocks. Between them he has placed two smaller logs. The next row of logs will consist of three large logs, followed by two large logs, followed by a final top or "peaker" log. This formula was followed for thousands of flatcar loads of old-growth logs. The head loader was assisted by two second loaders, who hooked and unhooked logs at the landing pile or trackside. One of them, standing on the landing pile, is visible at far left. "There was only one head loader," recalled Gunnar Jonsson. "He was always on top of the car placing logs." The loaders worked closely with the loading leverman on the skidder, who swung the hayrack boom constantly between the landing pile and the flatcars. Meanwhile the yarding leverman worked at the back of the skidder; he is seen here bringing in a turn of logs from the cold-deck pile. The turn can just be seen in the distance at far left. —Dorothy Tukham

> A turn of six logs arrives at the landing pile, Comox Lake, *c.* 1935. Originally, Comox Logging's skidders yarded logs with tongs, but they switched to chokers in about 1921. "It took several years before the fellows smartened up," recalled Joe Cliffe in a 1980 interview. "I would say that Jack McQuinn was the man that threw away the tongs and put on chokers ... As the timber got smaller we had to bring in five or six logs, depending on the size of the logs. It was nothing to see five or six or seven logs coming in. So you had to do away with the old tong deal." Many visitors remarked on the impressive sight of semi-airborne logs in transit. "Seeing three or four 30-40 foot logs come in over the skyline is like looking at giant snakes, for they seem to writhe as they slither, half dragged, half hoisted, over the rough ground," wrote journalist Renny Englebert in 1947.—John Holmes

CHRIS HOLMES PHOTO

CHRIS HOLMES PHOTO

Chris Holmes's caption: "Cherry picker Comox Lake 1930s." This cherry picker has come in to take away the last remaining marketable timber at a skidder setting. The spar tree is still standing, but the hayrack boom has been taken down and moved to the next setting. The cherry picker is "cleaning up the landing," picking up stray and fallen logs and the spar trees themselves, which have been cut into 40-foot lengths. The accessible timber has been logged to a distance of about 1,100 feet from the head spar; only stumps and slash are left. Next to the cherry picker are the brow logs, still to be retrieved; they were used at skidder landings to help the loading engineer guide logs onto the flatcars. "Generally, you always have brow logs to help the engineer bounce logs. They used brow logs a lot when loading with tongs," said Sam Telosky.—John Holmes

CHRIS HOLMES PHOTO

Comox Logging was the first large logging company in B.C. to log with Caterpillar tractors ("Cats") and arches. Here, the car float, or transfer wharf, moves Cats on Comox Lake, 1936. According to John Holmes, who later drove a Cat at Ladysmith, the model was a Cat 75, which was the forerunner to the D7 and D8. The car float was intended to transport flatcars and locomotives up and down Comox Lake, and here Cats and arches have been placed on flatcars.—John Holmes Collection, Courtenay & District Museum & Archives, 990.67.43

Willemar Boom Man

Sam Telosky (born 1915) recalled his early years at Comox Lake:

I went to Comox Logging in late 1935 or early 1936. I got started at the Royston boom, with Ab Stewart as foreman, and then Len Harding grabbed me! At first, I went right up from Royston to Little Lakes pushing logs. One of them was Willemar Lake and there were two other small lakes up there too [together called Forbush Lake]. *We called them all Little Lakes. Joe Cliffe had been a rigger at Little Lakes. Len Harding got me the job of pushing logs down the Puntledge River from the Little Lakes into Willemar, where we had a loading works similar to those at Comox Lake. We only worked at Willemar when the freshets ran.*

Comox Lake was electric but the Willemar load-out was steam—a powerful 10 × 12 steam donkey. Baikie Brothers had a contract from Comox Logging to cold deck logs and dump them into Willemar Lake.

I also worked as a boom man at the load-out at the bottom end of Comox Lake. My brothers Andy and Pete were there too, but after about a year I went to Cruickshanks, where Bob McQuillan broke me in as rigger. Altogether I spent five or six years around Comox Lake. I'd go out on weekends and stay at the Riverside Hotel and play baseball.

Sam Telosky ended his career as superintendent of Elk River Timber and now lives in retirement at Campbell River. ■

Cat logging at Comox Lake, *c.* 1933. The Cat, a tracked, V8-powered vehicle made by the Caterpillar Tractor Company, pulled a steel arch behind it, and a line from the Cat ran to the logs via the arch, which lifted the front end of the logs for easier yarding.

NORMAN PARKIN PHOTO

Cat logging with an arch, Comox Lake, 1933. Here, a Cat 75 pulls a load of logs down to Comox Lake. The Cat operator is Oswald Harmston. The grade shown here was far too steep for railway logging, and superintendent Bob Filberg was excited by the possibilities of this mobile new technology that allowed logs to be yarded down steep slopes without the need for expensive logging railways reaching right to each setting. This early experiment took place on the Comox Lake shore within a couple of miles of Lake Camp.—FRED PARKIN

That was when Comox Logging bought them in 1933," the *Crown Zellerbach News* noted in January 1960. Harmston was also Cat foreman at Comox Lake. He was soon followed as Cat driver by Arthur "Spit" Quinn, formerly a Marion shovel operator. In 1933 Sailor Lehtonen was hooker, Ed Lefley was chaser, and the chokermen were Lyle Coleman and Vic Dodds. By 1935–36 Norman Parkin, George Robertson, and Lyle Coleman were also driving the Cats, and Oscar Berkenstock and Allan Woodrow had joined the crews. Hi Churchill recalled that in the 1930s Comox Logging had four Cats working at Comox Lake, of which three were used solely for yarding. The fourth, driven by Norman Parkin, had a blade and was used for road making and snow clearing.

In 1935 the Cats were put to work on a large experimental logging show at Willemar Lake. This was cold-decker and Cat logging. Fallers went in and felled trees, a rigging-up goat rigged the spar trees, a cold decker yarded the logs into cold-deck piles, and finally the Cats yarded the logs down to the lake. At Willemar Lake the logs were boomed, then towed by the *Little Eva* or allowed to drift down to a load-out at the east end of the lake.

In September 1935 a reporter from *The Timberman* magazine described the Willemar (Trout Lakes) show of Comox Logging:

> Operations have been pushed on to the next chain of inland waters, known as Trout Lakes, three in all and each smaller than Comox Lake. A railroad has been completed from the end of Comox Lake to the first of the Trout Lakes. The shores and slopes along the lake are being logged with diesel tractors and arches, and where the grades are too severe for the tractors, gasoline yarders will reach up 1,000 to 1,500 feet for the timber. When some 50,000,000 feet of the low-lying timber is removed from the first lake, a dam will be thrown across the stream forming the outlet into Comox Lake and the three smaller lakes will be flooded to form one big waterway.
>
> On Trout Lake Mr. Filberg does not bother with rollways and dumps for his tractor fleet. He has picked out settings where the shoreline recedes gently into the lake, and the tractor and arch wade right in with the load until the entire turn will float. The unhooker wades out, almost to his waist, and frees logs on the surface of the lake, where the wind carries them down to the hoist, where they are lifted onto the cars. Should you visit Trout Lake, the motor launch "Little Eva" will land you right on the diesel tractor, which will take you ashore.
>
> Thus it will be seen that the logs from Trout Lake have quite a journey to tidewater. There is first a tractor or gas skidder [cold decker] haul to the water. Then a mile or two drive or drift to the end of the lake where they go aboard the cars. Next comes five miles of rail haul to Comox Lake where the logs are

WILMER GOLD PHOTO

Logging photos taken in the narrow valleys of Toma Creek, Puntledge River, and Cruickshank River tend to show more of the operation than those taken on the flats. For example, this 1935 photo of Skidder 3 bringing in a turn of four logs shows the skidder, the roads leading to it, the back end of the setting, and five members of the crew. The railway is completely obscured by the logging debris. "See the pond at the back?" pointed out Bill Woods. "That was common at Comox Lake. And there's no back spars on the hill behind: they used stumps, which they notched and fastened with railway spikes." Sam Telosky agreed: "On steeper ground we always used stumps. We called it stump rigging. You had to have a good stump, and you'd need to have a stump or two behind your good stump just to make it secure." —Wilmer Gold Collection, United Steelworkers Local 1-80, Duncan, BC

CHRIS HOLMES PHOTO

< Aerial hazards, Comox Lake, *c.* 1935. "Near disaster. Headspar dropped about 8 feet into ground," wrote Chris Holmes on the back of this photograph. The top of the spar tree has snapped off and dangles above. Only the buckle guys keep the tree from toppling, and Skidder 3 toils to keep it all in the air. "It broke! I'll be darned! It broke in between the top guys and the buckle guys, by the look of it," said Sam Telosky. This mishap was caused by intense pressure, constant vibration, and a soft forest floor. "Raised spar trees were sitting on chunks," explained Bill Wilson. "They'd put a mat underneath it. Once, at Oyster River, they raised a tree and it sank right up to the block! Everything was hanging loose. It sank into the blue clay almost all the way up the hayrack, 15 or 20 feet. This showed the weight of the tree."—John Holmes Collection, Courtenay & District Museum & Archives, 990.67.16

Chris Holmes photo

This is the same setting as the image on page 210 but the collapsed spar tree has now been replaced and loading is well under way. Note that, meanwhile, the falling gangs have removed many of the trees in the distance. These two photos illustrate the vertical limits of skidder logging at the west end of Comox Lake, which involved the gradual removal of valley-bottom timber. The higher forest of the mountainsides would be safe until the truck logging era of the late 1940s.—John Holmes

A Mountain Conveyor Belt

Comox Logging's log production system required an extended railway conveyor belt from stump to dump, hill to spill. In other Canadian resource industries (staple trades), the conveyor belt consisted of flatcars or boxcars containing ore, coal, potash, grain, or even crates of fruit. The transport technology was the same, but the commodity was different.

Surviving Comox Logging photos show empty flatcars waiting at skidder settings in remote mountain valleys, or skidders yarding logs into mountain settings and loading them onto flatcars. This was truly the "edge of the world economy." The uncut forest is sometimes visible in the background of the photos, and the flatcars are always covered in splinters of wood and pieces of bark from previous trips to the log dump at Royston.

The wide flatcar beds were horizontal utility structures. They provided a level, movable, and somewhat unlikely industrial conveyor belt in the wilderness that connected the timber at one end with the log dump and market at the other. The skidder landing was the nexus of log production. Some 400 men helped produce logs and nudge them along the way. They felled, bucked, measured, yarded, loaded, hauled, dumped, sorted, and boomed logs, or toiled on the grade, steel, bridge, or section gangs. Every task promoted the movement of logs from the very edge of the world economy to its more accessible margins. ■

dumped back in the water. Now ten miles [eight miles] of tow to the Comox loading station, where they are again lifted onto flatcars for another fifteen-mile trip to the main dump and booming grounds at Royston. After this comes a fifteen-hour tow to the sawmill at Fraser Mills, near Vancouver.

Among the men at Willemar Lake were two Finnish fallers, Otto Niemi and his brother, Eno, according to an April 2003 letter from Otto's son Arfie (Adolf, born 1928). Like those of many Comox Logging boys, Arfie's stories involve fishing: "My Dad & my Uncle Uno [*sic*] bucked logs on the second lake of the Willemar Lakes. One day my Dad took me & my brother to work with him so we could fish off the logs. It was good fishing. I was 10 or 11, my brother was 11 or 12." Others at Little Lakes were riggers Joe Cliffe and Harold Hames; Hames worked at Forbush Lake in 1938.

For a time there was a bit of a community on Willemar Lake. In about 1936 the *Ark* replaced the *Little Eva* as a crew boat, taking men from the load-out up to the Cat show. A newlywed couple, Eddy Lefley (born 1912) and his wife, Bessie (*née* Watt), moved into a floathouse on Willemar Lake so that Eddy, a Cat mechanic, could be nearer his work. Bessie told the story to Betty Annand, who recorded it in *Voices from Bevan*:

> At that time the train ran from the top end of Comox Lake to the Little Lakes. Since Eddy serviced the machinery for the logging company and had to work on the machines when the camp wasn't running, it was convenient to live as close to his work as possible. Bessie considered living on the water an adventure and took it in her stride. It wasn't unusual for one of the crewmen to knock on her door and call out, "Take your clothes line in, Bessie. We're going to move you further up the lake."

Cat logging was such a success that Cats descended from the mountain shows to other places around Comox Lake and the Comox Valley. In about 1935 a Cat show was held on Comox Lake at Half-Way Rock, in an area later known as Little Italy. "Little Italy was partway down the lake on the shore where the land levelled out," said Fred Blackburn. "The Italians from Cumberland lived there." Spit Quinn drove a Cat, Ed Lefley maintained it and worked as watchman, and Ray Dawson ran a cold decker with a crew that included Davey Janes and Albert Trotter. The Cats were then sent to Headquarters and Camp 3, where they operated for several months. "It was 1936 or 1937 when the Cats came to Camp 3," recalled Hi Churchill. "George Super's wife was stuck in Camp 3 in heavy snow. Nothing went up or down the highway here for a week. The only Cat with a blade on it was at Comox Lake, so they sent Spit Quinn down with it. When Super's wife [a Croatian] saw

CHRIS HOLMES PHOTO

Skidder 3 at Comox Lake, *c.* 1935, with a row of loaded flatcars alongside. "The number of flatcars at a setting depended on the grade," noted Walt Edwards, who started at Comox Lake in 1937. "If the grade was tough, as at Comox Lake, they'd bring seven flatcars. That was the Comox Lake number." Note the two tongs dangling from the hayrack boom, and the uniformity of the log loads. Every car follows the formula of 4, 3, 2, 1: four logs on the bottom tier and a peaker log on the top. When the logs were really large, a formula of 3, 2, 1 was followed. Until the late 1940s everything else was left in the bush. "When the pulp mills came in, they took the small stuff, but Fraser Mills was geared up only for big logs," recalled Eddy Berkeley. Once the logs were securely on the flatcar, the chaser stamped them with the company's "Circle F" log stamp. Once stamped, these 40-foot logs became both private property and a marketable commodity. "They say that the impression of the log stamp went in for 10 or 12 feet," remarked Art Beech. This deep impression deterred beachcombers.—JOHN HOLMES

Lost at the Load-out

Third loader Davey Janes (1912–2002) lost a leg at the Comox Lake load-out on November 10, 1936. He recalled the incident:

I was at Comox Lake in 1936 when I got hurt. I was working at Ladysmith, but they were going to shut down so I came up to Courtenay to help dig the Courtenay slough with a Marion shovel. On November 10 I was sent up to the bottom end of Comox Lake as third loader for a few days till they shut down.

Hallie Dixon was loading leverman at Comox Lake when the loading boom got away from him and mangled my leg. All day long all you'd hear was a chunk-chunk-chunk-chunk as Hallie loaded logs onto the flatcars, he was so quick. He wanted to give the poor second loaders a break—they worked all the time, and he wanted them to have the chance to have a smoke, so he picked a load up and left it sitting at the landing, and the vibrations from the engine shook the load loose. The weight log got away from him, and it swung the boom right against me. I was running along beside when it came loose.

After getting hurt, I went back to Ladysmith, where I ran shovel, and then I was back at Cruickshanks running a loading shovel. I'd see Hallie at dances in Merville, and he'd cry about what he'd done.

Known inevitably as Peg Leg, Davey Janes had a long career as a machine operator at Ladysmith and Comox Lake, and later as a gyppo logger and beachcomber on northern Vancouver Island. "What he did with a peg leg!" exclaimed his friend Neil Martin. In his 40s Janes married Sally Sealy, the cook on a boat at Gold River, and they had four children. ■

CHRIS HOLMES PHOTO

The load-out at the bottom end of Comox Lake, *c.* 1942. Here logs were lifted and swung by hayrack boom onto railcars for transport to Royston. "At Comox Lake, the log lifting rig is operated by electric power, and is comprised essentially of a 150 h.p. electric motor on an old steam loader drum and frame," noted *The Timberman* in September 1935. High rigger Sam Telosky remembered greasing the blocks on the spar tree. "I used to go up and grease those blocks once a week. I'd go up at noon—you couldn't hold up everybody!"—Filberg Heritage Lodge & Park Association

WILMER GOLD PHOTO

A Cat dumping a turn at Willemar Lake, 1935. Left to right: Lyle Coleman (chokerman), Norman Parkin, Oswald Harmston (Cat driver), Bob Dawson.
—Wilmer Gold Collection, United Steelworkers Local 1-80, Duncan, BC

the machine coming, she thought it was the Germans for sure and ran under the bed!" The same Cats proved valuable in salvaging logs north of Camp 3 after the Great Fire of 1938.

The Cats proved their worth in a number of spheres and showed, in theory, that skidders were redundant. A combination of donkeys and Cats could do all the yarding and loading required. Skidders were no longer the only means of yarding logs from the back ends of settings.

The success of the Willemar Lake experiment led Comox Logging to open its large new logging show at Ladysmith in 1936 with Cats, donkeys, and trucks or, as the *Argus* put it early in 1937, "the unique combination of tractors and motor trucks." Indeed, two Cat experts from Comox Lake were transferred to the Ladysmith shop: Vic Dodds, who became master mechanic at Ladysmith, and Cat mechanic ("Cat doctor") John Luscombe.

WHETHER THEY CAME from Toma Creek or the Puntledge watershed, all of the company's logs were dumped into Comox Lake near Lake Camp, directly into large circles of boomsticks known as "bag booms." Swifter logs (boomsticks placed across the top of a log boom for stability), essential in turbulent ocean towing, were not required on Comox Lake. Tyke Surgenor, who worked on the boom in the 1940s, said they'd "just put a bunch of sticks around them, a big circle around the logs," and take these bag booms to the bottom end of the lake. At first the booms were towed down the lake by the tug *Eva*, but by the late 1930s they were left to float down the lake on their own. At the end of the lake, the *Argus* noted in July 1933, "Halley Dixon lifts them out of the water with his electric hoist and dumps them on Canadian Collieries flat cars, which amble their way through Bevan ... to tidewater at Royston."

By the late 1930s much of the low and accessible timber of Block 32 at the top of the lake had been logged, either with conventional methods (railway and skidders) or with innovative Cats and arches. In 1937 Bob Filberg ordered the removal of Lake Camp and the entire logging corps to Block 31 at the mouth of the Cruickshank River, which led to the last blocks of timber that Comox Logging would log around Comox Lake before the war. ■

Corporate Contact

Alec Chalmers had the distinction of working for all three of the "Big Cs" in early-20th-century British Columbia: Canadian Collieries, Canadian Pacific, and Canadian Western (Comox Logging). Born on Denman Island in 1901, Chalmers was the son of an Orkney Islander and grandson of Thomas Piercy of Denman Island. He worked as a boy for Canadian Collieries before joining the CPR's British Columbia Coast Steamship Service, where he worked on the *Princess Mary* and ended up as quartermaster on the SS *Charmer*.

Mechanically inclined, Chalmers then became a locie engineer with Bloedel, Stewart & Welch at Union Bay, and then for Thomson & Clark at Deep Bay, and finally moved to Comox Logging in about 1924. He worked at both Bevan and Comox Lake as engineer of the Two-Spot and the Ten-Spot respectively until June 1936, when he resigned to become the local agent of the Standard Oil Company of British Columbia. One of his jobs was providing oil to Comox Logging's tank cars. Standard Oil rented part of Brackman & Kerr's feed store on the Courtenay River, and Chalmers had to separate the bales of hay and sacks of oats destined for the valley's farmers from oil barrels intended for Comox Logging.

Bill Chalmers, Alec's younger brother, became Comox Logging's superintendent at Ladysmith and Nanaimo Lakes. ■

WILMER GOLD PHOTO

Stump to dump: Comox Logging's Ten-Spot comes down to Comox Lake from the Willemar Lake area, 1935. Left to right: Alvin Parkin, Alec Chalmers (engineer), Dick Delin. The Ten-Spot, a 65-ton geared Heisler locie built in 1912 for the Columbia River Lumber Company, was brought to Vancouver Island by Comox Logging in 1929. In the 1940s this locie was transferred to Comox Logging's new railway at Nanaimo Lakes. It was scrapped in 1952. Alvin Parkin (1906–78) left the company in the 1950s and owned the Campbell River Lodge. —Wilmer Gold Collection, United Steelworkers Local 1-80, Duncan, BC

CHRIS HOLMES PHOTO

All quiet on the western dump: the humdirgen waits on the trestle at the top end of Comox Lake, near Lake Camp, *c.* 1935. Logs from Toma Creek, Puntledge River, and Willemar Lake were dumped here, bag boomed, and towed or left to float the eight miles down to the load-out. "The Alberni line is at left, Mt. Hennessy is at right, and the Willemar line is to the right of Mt. Hennessy," said Sam Telosky.—John Holmes

A great many boomsticks were needed at Lake Camp to encircle the logs dumped there and shepherd them down the lake to the load-out. Here, an unidentified man operates the boomstick winch, *c.* 1937. The winch was used for positioning boomsticks so that holes for boom chains could be drilled in each end with an auger. Note the Douglas fir bark piled in the foreground. Such bark was salvaged as fuel for the many wood stoves at the camp. Even a drafty kitchen wood stove, if piled with bark last thing at night, would contain embers in the morning.
—Jean Berkeley

CHARLIE STROULGER PHOTO

Boring boomsticks at Lake Camp, *c.* 1937. The boomstick winch is at right, and the men in the centre of the photo are drilling boomsticks. "He's got boomsticks and a gas auger, a four-inch auger. The chain goes through that. They're making boomsticks right there," said Sam Telosky. The boom man at left has a pike pole over his shoulder.
—Jean Berkeley

WILMER GOLD PHOTO

The log dump at the head of Comox Lake with the large cookhouse of Lake Camp in the distance, 1935. The trestle was canted to allow the logs to roll easily into the lake. The cab of the locie is just visible beyond the log loads. The cheese blocks (steel supporting wedges) have been removed from the sides of the flatcars nearest the lake, and the loader at far right (probably Harry Grieve) is using an L-hook, which was half a tong, to trip the loads into the lake. Boomsticks forming a bag boom can be seen at left.
—Wilmer Gold Collection, United Steelworkers Local 1-80, Duncan, BC

< Boom man Sam Telosky at the Comox Lake load-out, *c.* 1936. From his start at the boom, Telosky progressed rapidly through the company's ranks at Comox Lake and was a foreman in the Queen Charlotte Islands by the 1940s. Among the other boom men at this time were Bob Aitken and Sam's brother Pete Telosky.—Charlie Stroulger

Boomsticks on flatcars on the Comox Lake load-out trestle, *c.* 1938. Most sawlogs were 40 feet long, which was the length of a flatcar, but longer loads had to be spread over two or more flatcars. Boomsticks, which were between 66 and 72 feet long, took up two flatcars. These boomsticks are supported by platforms, known as false bunks, set at each end of the flatcars. Note that only one such bunk has been installed beneath the front of each load. The false bunk allowed the load of boomsticks to swivel as the train rounded corners on its way to the log dump. A Douglas fir stake has also been inserted into a pocket on the side of the flatcar near the false bunk to stabilize the boomsticks for the trip to Royston. Sam Telosky was impressed by this photo. "There's three loads of boomsticks there. Everyone was always hollering for boomsticks! You needed nice logs for boomsticks. When you load boomsticks, you always put on a false bunk. The flatcar bunks are 40 feet, so you can't put a boomstick on a 40-foot bunk."—John Holmes

Bag booms at the bottom (east) end of Comox Lake, *c.* 1932. "The wind did all the towing down Comox Lake because it blew that way," recalled Sam Telosky, who started working there in the mid-1930s. "When they got a boom, they'd round it up and let it go—there was always a prevailing wind and the logs would end up down there. These are Douglas fir booms."
—Bob Stelfox

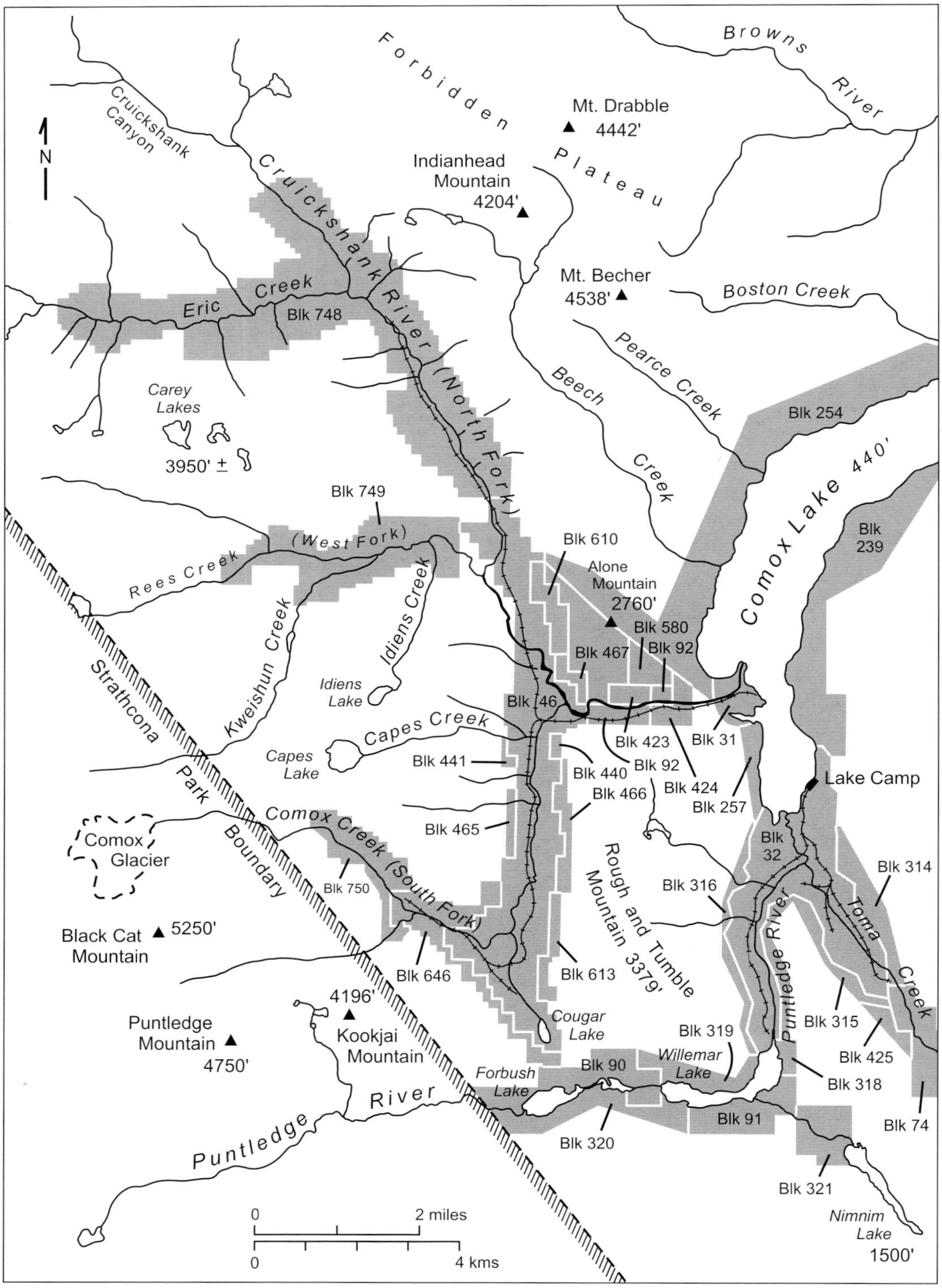

The Cruickshank River system, site of Comox Logging's last steam and skidder show, 1937–46.

Earlier operations, for the most part, lay on relatively easy ground in the vicinity of the central camp of Headquarters. Logging is now being conducted under entirely different ground conditions.

—*THE TIMBERMAN*, "LOGGING AT COMOX," SEPTEMBER 1935

7

UP THE CRUICKSHANK

IN THE 1930S JOHN D. ROCKEFELLER JR. OF NEW YORK SOLD SEVERAL large timber holdings on Vancouver Island, all of them originally in the Dunsmuir land grant, to Canadian Western, Alberni Pacific, H.R. MacMillan, and others. In January 1937 Comox Logging's parent company, Canadian Western, bought several valuable new properties, large valley-bottom timber blocks on the Cruickshank River, from Rockefeller: Blocks 46, 748, and 749, which together extended from Block 31 and Alone Mountain right up to the headwaters of Rees and Eric creeks and almost to the edge of Forbidden Plateau. "It will extend the life of the Comox Logging operation on Comox Lake," noted the *Comox Argus* on February 4, 1937. "The new property is up the Cruickshank Canyon between Comox Lake and the Forbidden Plateau, which was cruised last year by the interests concerned. At the present rate of operation this timber would take many years to log."

In the spring and early summer of 1937, most of Lake Camp was towed a mile and a half from the head of Comox Lake to the mouth of the Cruickshank River, where everything was recreated except the school and married men's quarters. It was the same camp in a different place. The new camp was strategically situated in Block 31, the company's last original and unlogged timber property in the Comox Valley. The falling, grade, steel, yarding, loading, and shop crews all moved, in a large redirection of human energy, to a new setting and to a forest never before touched by the sharp teeth of the "Swedish fiddle" (crosscut saw).

Same camp, different place: Cruickshanks Camp (view to the southeast), *c.* 1939. The undulating slope of Rough and Tumble Mountain towers over the camp. "After the logging was finished at the head of the lake, the floating camp was moved down to the Cruickshanks, and that area was logged," recalled former Lake Camp schoolteacher Pat Williams (*née* Holmes). The float consists, from left to right, of bunkhouses, the long cookhouse, and office. The row of six bunkhouses would be put ashore in the 1940s. Already on the shore at right are the foremen's houses and a small A-frame that, according to Bryson Whyte, was for "cutting wood with a steam winch, which picked up the wood from the water and put it on the float, where it was cut by another steam-run saw. Heated water was stored in a raised tank." The donkey engine in Mike Kisisky's steam plant and woodyard is belching smoke at left.—Dorothy Tukham

SAM TELOSKY PHOTO

In a 1984 interview, Reg Pidcock told Martin Hagarty that Comox Logging used to take 40 men up Comox Lake in the crew boat in the 1940s. This *c.* 1940 photo shows the camp boat *Mary* (at top) towing Comox Logging's scow up Comox Lake from the load-out. "They needed the scow for transport because Cruickshanks Camp just got too big," recalled Sam Telosky. "The 1938 fire burned a lot of uncut timber, so they had to move people out of Camp 3 to Comox Lake and Ladysmith. They were putting on a double shift and closing down Camp 3."—Sam Telosky

Best foot forward. A lake-level view of the trip down Comox Lake on the *Joan*, 1943, with the south shore at right. The road from the load-out was not extended to Cruickshank River until the late 1950s. The commute was a good time for chatting, and Bob Muir recalled that he sat next to veteran logger Eddy Berkeley and engineer Toynbee Harmston (1893–1972). Harmston was a Comox Valley Homeguard who joined Comox Logging in 1911 and later ran skidders at Camp 1, Camp 2, and Comox Lake.
—George Yates Jr.

Woods Work

Bill Woods (born 1927) recalled the seasonal and daily work routine at Cruickshanks in the early 1940s:

A lot of guys my age from Cumberland went to work for Comox Logging. They would work spring, summer, and fall in the woods and go back to the mines in the winter. In April they went to the woods; in September they quit. Between September and November it would start to snow, up to a foot or two feet at a time, and it [logging] *would shut down for the winter. By the end of September there was half an inch of snow in Cruickshanks; two or three weeks later it would start staying on the ground, and the lake would shut down in late October. There was rain and snow and cold; it was freezing. Caulk boots are the hottest things in summer and the coldest things in winter! A lot of guys went to rubber boots.*

By December or January, there'd be 14 to 16 feet of snow in the Cruickshanks and they'd open up the bottom of the lake for log loading, and I would third load there. Cruickshanks opened in March, and there was still snow in March.

The crummy would leave Cumberland at 6:10 a.m., when it was still dark. Skinny Colling drove the crummy—it was an old school bus—from Cumberland to the head of the lake. There was also a bus from Courtenay to the lake, and they used to beat us, so once we jacked up their crummy and went right past them! It was all gravel and the dust came into the bus. If we hit a deer, Dick Sgarbossa would gut it and we'd pick it up on the way back. We went out the Bevan Road to the bottom of the lake, where we met Austie Blackburn and the Joan. *That thing was packed full!*

Then we went up the lake. After a 20-minute boat ride we reached Cruickshanks. It was still dark. We went right to the commissary to get a pack of tobacco and a chocolate bar. Then we went to the South Fork [Comox Creek], *where it was either raining like a bastard or there was two feet of snow. At lunch we needed a fire going for the first and second loader to toast sandwiches. Then we went into the woods for eight hours, and we got back in our houses in Cumberland at 8:15 p.m. We were away for 13½ hours every day, soaked to the ass. Punking whistles for $2.35 a day.* ■

BOB FILBERG PHOTO

Captains of industry: directors of the Canadian Western Lumber Company on the *Joan*, 1940. At left is Henry Mackin, president of Canadian Western. Born in New York State, Mackin worked in sawmills and box factories at Portland and Astoria, Oregon, before being hired as sales manager at Fraser River Sawmills in 1908. He later advanced to president and general manager of Canadian Western (parent company of Comox Logging).—Filberg Heritage Lodge & Park Association

The tug *Quinsam* being lowered into Comox Lake, c. 1940. From its arrival in the mid-1930s, this diesel tug was employed on the booms at the top of the lake and at Cruickshanks Camp.—Sam Telosky

Cruickshanks Camp, as it was known, was really an extension of Camp 2. Camp 2 was a good deal more transient than its former rival, Camp 3, whose proximity (within a mile or two of Camp 2) between 1918 and 1926 fostered competition for log production records, safety standards, and achievement in loggers' sports. Camp 3 remained in the same location, 12 miles north of Courtenay, on the flats near Northy Lake, from 1922 until its closure in 1942. By contrast, Cruickshanks was the sixth location for Camp 2 since 1926. The camp had been sent on a timber odyssey through the unsettled fringes of the Comox Valley. In just over a decade, Camp 2 and many of its personnel had moved from Oyster River to Bevan, Quadra Island, Bevan again, the A-frame, Lake Camp, and now Cruickshanks Camp. Comox Logging's steam and skidder "Cruickshanks show" continued from its inception in 1937 until the abandonment of skidder logging in 1946.

Sam Telosky's rigging crew at the North Fork, c. 1940. Left to right: Gordon Blackburn, Sam Telosky, Chan (Chandler) Tame, Mel Bowen, Alec Nelson. "We had just rigged the big skidder boom behind," said Telosky. "They'd always send these young guys up with me, to teach them. I did pretty well everything on skidders, but mainly I was high rigger."—Sam Telosky

COMOX LOGGING NOW had accessible timber along about ten miles of the North Fork (properly known as the Cruickshank River, though the loggers called it Cruickshanks) and about eight miles of the South Fork (properly Comox Creek). It would take until 1950 to cut these long sides in confined valley bottoms.

The company wasted little time getting started after the Rockefeller purchase. On July 1, 1937, the *Argus* announced that logging was about to move into the Cruickshank River valley: "Active construction is going ahead. A grade is being established up the Cruickshank River. A new log dump trestle is being built, and there is some heavy rock work on the line and many bridges. The road will tap a large quantity of timber in the Cruickshank watershed. Logging will start on this operation early in July." It took a year, however, to ready the ground for this major railway show. At the end of July 1938, the *Argus* reported that fallers, buckers, and the steel gang were still at work: "Up the Cruickshank there are fallers and buckers ahead of the steel gang clearing the right of way. There are actually more men on construction than when the camps are going full swing." Grading was still under way in September 1938.

Marion shovels under grade foreman Tom McQuillan made the railway embankment up the Cruickshank. Sam Telosky recalled that Arthur "Spit" Quinn and his brother, Mose (Moses), were on the Cruickshank grades in the late 1930s. They were assisted by young helpers Frank Davison and Malcolm Carwithen, as well as a great many members of the grade crews and section gangs. "Quite a few Yugoslavs worked in the crew," Bill Moncrief recalled. They included a Croatian named Joe Stemac, who became the Cruickshanks section gang foreman in the 1940s, and grade crew

CHRIS HOLMES PHOTO

Felled and cleared right-of-way through uncut timber, Cruickshank River, 1938. A falling gang and grade crew have been busy, and a Marion shovel is just visible in the distance. The shovel has pushed the marketable timber to the sides, powdermen have blown up the stumps, and the grade awaits the steel gang to dump gravel and lay tracks. "All the timber at Cruickshanks was huge," recalled Bill Woods, who started working there in 1941. "You'd swear you were in a park! The moss was everywhere. The logs were 40 inches across at the top—all of them." —John Holmes Collection, Courtenay & District Museum & Archives, 990.67.41

A canyon on the Cruickshank River, *c.* 1940. This river led directly into glacier country. "That Cruickshank is always cold—it comes off the glaciers, eh? Once we had to walk across that river on a log, and [the river level] came up by a foot every hour," recalled Bill Woods.—Laurie Mathers

Building the main line up the Cruickshank River: a Marion shovel poised on the cribbing for a railway bridge, spring 1937. The shovel is sitting on a mat on top of the bridge timbers. "It's all cribbed for a bridge," said Sam Telosky. This was part of the work that the *Comox Argus*, on July 1, 1937, described as "heavy rock work on the line and many bridges."—Jean Berkeley

member Martin Domovich, also a Croatian, who lived in a bunkhouse at the mouth of the Cruickshank River until he quit woods work in 1939.

COMOX LOGGING CHOSE not to move the school or married men's houses from Lake Camp to Cruickshanks, a decision that ended the two-year teaching career of Patricia Holmes (later Williams). "They closed the school," she recalled. "That's why I left. They opened up Cruickshanks Camp and took all the single men's stuff [the cookhouse and bunkhouses] over there, and the married people just left. The school finished in June 1937." Holmes, who turned 20 that year, moved to Vancouver.

The absence of women and families seemed to be the first sign of difficult times for the camp, which had been a family camp in its various incarnations since 1918. "Cruickshank had no married quarters," noted Bill Woods and Frank Davison, with some disappointment; they worked at the camp as young men. The demise of the dance hall further marked a camp in decline. Built in 1934 at Lake Camp by men's voluntary labour, the dance hall was towed to Cruickshanks and converted into a blacksmith shop. Eventually it was lost in a flood. "It's a shame it met such a miserable end," Pat Williams lamented.

Stirring outside events and new technology alike affected the camp. Comox Logging's new Cat and truck show at Ladysmith, where 275 men were employed in 1938, drained resources away from the lake. At least two dozen men who had worked at Comox Lake moved to Ladysmith, enticed by the irresistible prospect of higher wages and a promise of many years' work. Among the Lake Camp men to go to Ladysmith in 1936–37 were Harold Cliffe, Joe Cliffe, Vic Dodds, Eddy Hughman, Davey Janes, John Luscombe, George Robertson, Bill Walker, Bill Westwood, Bill Wilson, and many Finnish and Swedish fallers, including Viking Johnson (Osterblad), John Jungaro, Bill Lyttback, Vic Olson, Gus Rundquist, Eric Sundberg, and Ivan Wicklund. Others went later, and many Camp 3 and Headquarters men and families also moved to Ladysmith.

Side-push Oswald Harmston replaced Len Harding as foreman when Harding went to the Queen Charlotte Islands with the Aero Timber Products Company (a separate company from Comox Logging, but with Bob Filberg in charge). Harding and Filberg recruited a good many Cruickshanks men to join Aero Timber in 1941 and 1942, including Andy Barr, Gordon Blackburn, Mel Bowen, Len Lester, Dan McKenzie, Hughie McKenzie, Chan (Chandler) Tame, and Sam Telosky.

In the early years of the Second World War, the company had trouble finding men to work at the isolated Cruickshanks show,

SAM TELOSKY PHOTO

The steel gang, *c.* 1939, puts the finishing touches on a stretch of logging railway that, according to Sam Telosky, was "not very far" from the camp on the Cruickshank River. Foreman Joe Asselin is at left, and a wheeled cart containing gravel is behind the men. The men are shovelling gravel between the ties, which, like the steel, have been reused. "Joe Asselin would always say, 'Come on, you guys: dig, dig, dig!'" recalled Telosky. "They laid it out, they laid steel, and they doubled as the section gang. They dumped gravel: they had a little jack and they'd tamp gravel under it so it was all solid." A member of the section gang from 1938 to 1940 was future writer and fisherman Loggie Grieve (born 1916), son of Kenny Grieve, who farmed near Merville School. "It was half prairie, half stump ranch. We lived on fish and game," recalled Loggie. He elaborated in a 2004 letter, "I got the job only because the track boss Clay Walsh knew my dad and how badly we needed the money. I would have kissed asses to keep the job. Actually I needed only to do the work to their satisfaction—as do loggers then and now."
—SAM TELOSKY

SAM TELOSKY PHOTO

Cruickshanks was a railway show, which meant that everything and everyone had to come in and out by this industrial corridor. In this *c.* 1942 photo, Comox Logging's Ten-Spot has just pushed some empty flatcars (right) into the setting. Left to right: Tommy Tame Jr. (fireman), locie engineer (possibly Harry Grieve), Tommy Gray (head loader). "Grayhound! Tommy Tames [*sic*]!" exclaimed Sam Telosky. "Grayhound was after the water bag—they always had a cold water bag on the side of the locie."—Sam Telosky

A broken flatcar at Cougar Lake, South Fork, 1943. In the background is the lake, which was right at the end of the railway line. "I thought that a log dropped on it, but no," explained George Yates Jr. "I showed this to Bill Moncrief, and he said that a cold decker was being taken off or put on this flatcar, and it slipped and smashed the flatcar. I thought a big log came out of a tong, but I was wrong." The flatcar has been pushed to the end of the track and left for the time being. "We used to swim across Cougar Lake because if the timber was down from the spring [had been felled in spring], the hornets would get into it, and they would come out and sting you behind the ears," recalled Bill Woods. "We'd have to jump right in Cougar Lake with the hornets chasing us. When the hornets came, I ran right past Elmer Orr. This was so the bees would go to Elmer!" —George Yates Jr.

GEORGE YATES JR. PHOTO

Rigging crews were quite willing to take donkey engines across rivers to reach the back of skidder settings. Here, Sam Telosky's crew turns a donkey into a watercraft on the South Fork, *c.* 1940. "Crossing the Cruickshanks River with the rig-up donkey," said Telosky. "That's the back rigging crew on the rig-up goat. We were getting ready to go up the hill." Left to right: Sam Telosky, Gerry Empey (engineer), Oliver Doucette, George Stemac.—Sam Telosky

Learning the Ropes

When Sam Telosky became a high rigger, he moved to the front line of the whole operation. He topped and rigged spar trees and moved cold deckers.

Before any logging was done in the Cruickshanks, I worked as a chokerman out of Headquarters for Norm Hammond, who was a hook and rigger [combination of hooker and rigger] *on a cold decker. Norm was at the marshalling point one morning, and he needed a chokerman, and I thought, "Jesus, I'd sure like to be a high rigger like him!" He was a big guy. I also worked as a chokerman for Ed Hughman on a cold decker, and later I worked with Ed on the rig-up crew. Then the same year* [1936–37] *I worked with Vic Parkin, who was a head rigger on a skidder out of Headquarters.*

In the spring of 1937 I went with the skidder from Headquarters to the new Cruickshanks operation of Comox Logging, and soon I became a second rigger. I can always remember that! What happened is that Bill Braes, who was second rigger for Bob McQuillan, got hurt on the first setting on Cruickshank. He fell and it was rough ground, and he hurt his back.

Bob McQuillan came across the river and he said, "Hey, Sam! Wanna try it? Go ahead and try 'er!" I'd only been halfway up a spar tree before, and it was 80 or 90 feet up. It was the second rigger's job to take down the rigging. I got to 40 feet, and then I looked down and felt sick, so I decided not to look down again. I said, "To hell! Never again!" and I went right up to the top. From then on I was okay.

You had to get rid of the guylines by hand, and there was a trick to it. So I became second rigger under Bob McQuillan. He broke me in there as rigger—he was a good teacher, a good guy. I learned more from Bob McQuillan than anyone else. ■

> "Loggers could do all kinds of things, find all sorts of solutions to get the logs out," Sven Ell recalled. Rigger Sam Telosky (left) was in charge of moving this cold decker to the back of a skidder setting on the South Fork, *c.* 1940. He recalled that this photo was taken soon after the photo on page 226. His caption was "Moving rig-up goat up mountain to rig a back spar." Engineer Gerry Empey is in the shadows inside the machine. "I always had to be right in front, right on the fairleads, to steer it," recalled Telosky. "You have to move the mainline [cable] from runner to runner [in order] to steer. It's the only way to steer them. You don't want to get *over* a stump, or you could spend a whole day getting off it." A logged setting is visible at right, and a cherry picker with vertical boiler sits on the main line at bottom right.—Sam Telosky

Splicing line on the North Fork, *c.* 1940. Left to right: Alec Nelson, Sam Telosky, Gordon Blackburn, Chan Tame. "It's an eye splice," recalled Telosky. "We'd spike the line to the end of the log for an eye splice. For a long splice we'd stretch the line out." The tools of the trade shown here include marlinspikes, sledgehammer, and double-bitted, short-handled axe.—Sam Telosky

Three Back Spars a Day

Sam Telosky (born 1915) recalled his work as a rigger at Cruickshanks in the late 1930s:

At Comox Lake we used to double shift the cold deckers. Dave Aitken was on one shift and I was on the other. Aitken was hard! He was older than I was and he used to get on me quite a bit. I was maybe 21 and he was 55 or 60, and he didn't like that very much. He'd say, "You young bugger! You didn't do that right!" He'd give me heck every day. But he showed me quite a bit.

I got ahead pretty good. I topped and rigged back spars for quite a while, and I had my own second rigger, Chan Tame. I was rigging back spars so that a donkey could yard the logs into a cold-deck pile. The skyline between the back spar and head spar was always around 1,200 feet, and you always went clockwise to the next road.

High riggers used a double-bitted axe for limbing and topping the tree. The axe had to be short to get in front. It was called a topping axe, and it had a curved, rounded blade so it would not get stuck. A regular falling axe has a flat blade. The first thing you do is put the guylines on. You top the tree, then you send down a piece of three-eighths rope, then get the small pass block up. It weighs eight pounds. You hang that, then send down the three-eighths strawline, and they send up the guylines.

There are about 20 or 30 trees [back spars] around a skidder setting: lots of trees! We topped three back spars a day: two before lunch, then we had lunch, and did one in the afternoon. Jack Downey and I fell a lot of back spars, but Reg Taylor did nothing but top back spars for Comox Logging. He was based at Headquarters, and he was about the only guy doing it. Once I topped steady for a week with Reg, but he didn't say much. I figure I topped about 200 back spars a year for three or four years, but Reg Taylor must have topped 4,000 or 5,000 back spars with Comox Logging. ■

partly because of the manpower demands of the war and partly because of the defections to Ladysmith and the Queen Charlotte Islands. For those who didn't live at the camp, travel to the site was long and complicated. Two bus crummies came to the end of the lake each morning, one from Cumberland and one from Courtenay. Darrell McQuillan, who started working in 1942 as a chokerman on the North Fork, recalled that "the crummy left Courtenay at 5:20 a.m., and then we drove to Comox Lake, took the boat up the lake, and took the caboose to the operations. Every day involved four hours' travelling on top of the usual hours." After a few months Darrell left to join the Royal Canadian Air Force (RCAF).

The magic had gone from the camp, which reverted to a basic work camp: remote, harsh, utilitarian, seasonal, and more in keeping with the "hiring hall" approach of coastal logging than the Homeguard tradition. "Cruickshank work was hard—the camp was hard, so the men were not married and we could never get a cold-deck crew up there at Comox Lake," said Sam Telosky, who worked there from 1937 to 1942. "Nobody wanted to work at Cruickshanks, and Bob McQuillan came to me and said, 'We gotta start breaking men in here. We gotta have a cold-deck crew.'"

Tragically, the Cruickshanks Camp was also plagued by unusually frequent accidents, injuries, and deaths. Serious injuries were suffered by riggers Jack Berkeley and Bill Parkin and second loader Pete Telosky. Faller Osborne Bell, the *Comox Argus* reported in July 1937, "was struck by a side-winder [a tree knocked to the side when cut] on his shoulder and side while falling a tree at Comox Lake camp. He was taken by boat and ambulance to hospital." Four Cruickshanks men died in logging accidents in three years: hooker George Surgenor, in May 1940, while working at Port Neville during the shutdown; chaser Charlie Nybrann at Alone Mountain, in November 1942; and head loader Bobby Colling and bunkmaker Sid Cresswell, in May 1943, on the South Fork. By contrast, there were no deaths at Bevan or on the A-frame, and the only death at Lake Camp was that of Fred Tuddenham, a chaser killed in 1935.

Moving a rigging-up donkey down the railway tracks at the North Fork, c. 1940. Mobile cold deckers went in ahead to yard logs into cold-deck piles that were within a skidder's 1,100-foot reach. Left to right: Eddy Lefley (engineer, in shadow), Sam Telosky (sitting on fairleads), Gordon Blackburn. Sam Telosky explained the photo as follows: "I'm moving the donkey across the track. See the big skidder boom at left? I'd just rigged that tree—that's what I'd done. I raised that tree. We used pretty near all raised trees. We always put a mat under them, which was made of logs lashed together, and we set the butt of the tree right on the mat ... The speeder at left was our crummy."—Sam Telosky

FORTUNATELY FOR COMOX LOGGING, the camp continued to appeal to young unmarried men from nearby Bevan and Cumberland and to commuters from Courtenay and elsewhere in the Comox Valley.

Cruickshanks increasingly attracted Cumberland boys. "Mostly Cumberland guys worked at the Comox Lake division of Comox Logging," said Bill Woods, who worked there for most of the 1940s. Comox Logging had started hiring Cumberland and Bevan men in the late 1920s, and it continued to do so at Comox Lake. Bevan

Cold comfort camp: Sam Telosky's rigging crew at the North Fork, c. 1940. Left to right: Mel Bowen, Ed Lefley, Chan Tame, Jack Harris, George Grieve, Gordon Blackburn, Alec Nelson.—Sam Telosky

Give the Kid a Job

Bill Woods recalled his mining background and his first work in the woods:

I was born in Cumberland in 1927. My father was from Lancashire. He worked for Canadian Collieries; he started [mine work] *in England and worked underground for 39 years. He went down and got up the dead Chinese after the big explosion* [in Cumberland, February 1923; 17 of 33 miners killed were Chinese]. *My mother was a Dickinson. They were miners too. My great-grandfather came out before the turn of the century from Chorley in Lancashire, where the mines are. When they wanted miners from England, they'd show them pictures of fruit trees in Canada! Any time you wanted, you'd go out and pick apples and peaches!*

If you were big enough, you went to work—you didn't go to school. I went to work aged 15 in the pit for Canadian Collieries. I was paid 25 cents an hour, six days a week. This was at No. 5 mine. But I never had intentions of working in the mines. Why not? Did you ever walk in and look down that hole? I was down there, and you see the cage coming up and down the hole. Get on that bugger and go down? Besides, mining kept coming and going. It was unreliable work.

A guy called Mcrpole fired me when he heard I was only 15, because the Compensation Board would complain. I got a job in Inkster's Mill in Cumberland for a while, and then I started for Comox Logging aged 15½ ... The side-rod was Bob McQuillan, and Oswald Harmston was superintendent. I went six or seven times up the lake to get a job.

I would say, "Where's Mr. Harmston?" "He's already gone." I'd wait and wait for him. Finally he'd return and I'd say, "Mr. Harmston ..." And he'd say, "Yeah, what do you want?" "I want a job." And he'd say, "Do you wear a belt, kid?" "Yes." "Do you roll your own smokes?" "Yes." "Well, you can't work here because you'll either be pulling up your pants or rolling smokes, and you won't get any work done!"

Finally, Reg Churchill said to Harmston, "For Christ's sake, give the kid a job!" "All right," Harmston said. "Go punk whistles on Skidder 3." ■

boys included the three Aitken brothers (Bob, Connie, and Dave), and the Webber boys (Bill, George, and Herb). From Cumberland came John Buchanan, Eddy Lefley, Rod McAllister (previously of the A-frame and famous for playing his alto sax in camp), Bill "Bronco" Moncrief, Dick Sgarbossa, Chuna Tobacco, Eddy Williams, and Bill Woods.

A friendly rivalry developed between Comox Lake men and men from the company's Tsolum (Headquarters) division. They saw a good deal of each other in the 1940s, when Tsolum got out of skidder logging and sent several crews to Cruickshanks. Bill Moncrief recalled, "We were known as the Cumberland coal-heavers, but we took our revenge by calling the boys from the flats the Tsolum flat-landers." Later, Comox Lake employees of any origin were known as the "Comox Lake sidehill gougers," and they continued to disparage the Tsolum flat-landers. "They called us the 'sidehill gougers' because of the steep country," continued Moncrief. "We were trudging along the hillside! Gouging meant digging into the hill, for example, digging a trail out. It was easier walking on a trail."

There remained a community of sorts at Cruickshanks, despite the absence of women and children, the desecration of the dance hall, and the loss of many long-time Camp 2 workers. With Len Harding gone, Oswald Harmston moved his house to Cruickshanks from Lake Camp; Homeguard Bob McQuillan, a rigger, became his side-push. Timekeepers Chris Holmes and Pete McLoughlin gave the camp a competent and friendly administrative presence. The Chinese cookhouse staff also remained from Lake Camp. "The Chinamen got up at five and got the fire going," recalled Bill Woods, who started at Cruickshanks in 1941. Amelia Yates (*née* Williams), whose family cottage was nearby, recalled that she used to go over to the camp for big sugar cookies: "The Chinese cook would wave us over to the dock and we'd go over in our rowboat. If you befriended them, you were a friend!"

According to Norman Leung, of Leung's Grocery in Courtenay, one of the Comox Lake cooks opened up the Royal Café in Courtenay in the off-season, and Bill Moncrief recalled that in winter some of the Chinese cooks moved down to Cumberland and cooked at the old Union Hotel. The Chinese cooks never lost touch with China, Bill Woods recalled: "When the Chinese Revolution started [in 1945], the head cook at the Cruickshanks said, 'I go fight on side of Chiang Kai-shek!' But he never even got there—they say he was robbed and killed."

Another character at the camp was the perennial Wattie (Walter) Whyte, fireman and watchman. His grandson Bryson Whyte and two of Bryson's friends came to stay with him and unintentionally

Gordon Blackburn (left) and Sam Telosky splicing line at Cruickshanks, *c.* 1940. They are long splicing, that is, joining two lines by weaving together the separated strands at the end of each line. This process involved stretching both lines out for a square fit. "When you break a mainline, you have to put a long splice in it," explained Bill Woods. "I was on one end of a marlinspike. You *push* in the marlinspike with your shoulder: it's over one and under two, and then roll them all in."—Sam Telosky

In 1944, Bob Filberg's 19-year-old son, Rob, went up on a pass line to a back spar at Cruickshanks and took this photo of the cold decker sitting on felled and bucked timber. Most of the timber has been bucked to the standard 40-foot length. A mass of straps, blocks, and line has been piled onto the front end of the cold decker. A figure is standing at the bottom end of the cold decker and looking up at Filberg. "This is at the foot of a tree they're going to rig for a back spar," said Bill Wilson. "The head spar was way out to the top of the photo. See the felled trees everywhere! Nothing's been yarded in yet." The cold decker's job was to yard the felled and bucked timber into a cold-deck pile for the skidder to yard in to the landing.—Filberg Heritage Lodge & Park Association

Robert McCormack Filberg photo

Cold-deck pile, Cruickshank River, *c.* 1940. Bill Woods commented, "Comox liked those fir logs. There was beautiful timber at the lake. See the way they're all tangled? It's almost a crow's nest, we called it: every time we pulled one out, we weren't sure if they'd all come down."—Sam Telosky

played a cruel trick. "One of the boys grabbed a handful of blackberries and rubbed it in another's face," Bryson recalled. "And Wattie came running out and said, 'God, oh God! What have you done to the boy!' Old Wattie thought it was blood. He scrubbed and scrubbed, looking for the cut."

Also at camp was a "car-whacker" or car-knocker, that is, a man who repaired and maintained flatcars. This was Jimmy Mathers Sr., a mustachioed Merville veteran and electrician who had emigrated from Northern Ireland just before the First World War.

In their day and a half off every week, Comox Logging men took advantage of the opportunity to hunt, fish, and climb. Sam Telosky hunted deer with head brakeman Dick Delin in the swamp behind Alone Mountain. "We'd make noises in the bush to run the deer out for Dick, and once he got a nice big three-spike." Some ten miles up the valley from camp was a huge alpine area, including the edges of Forbidden Plateau and Strathcona Provincial Park. The camp was the base of a rescue operation in 1939, when Telosky and others helped a prospector who was stranded on a ledge high in the mountains.

THE CAMP FEATURED what was then a nearly moribund technology. Cruickshanks was the last hurrah of skidder logging, the last major skidder show ever opened by Comox Logging. In 1942 trucks were introduced at the company's Tsolum (Headquarters) division, and the remaining skidders and skidder crews were gradually transferred to Cruickshanks, where they were used until 1946—ten years after the company introduced Cat and arch, prebuilt loads, and trucks at Ladysmith. The company retained highlead skidder logging at Cruickshanks because of the large investment in the old technology and because of wartime restrictions on trucks and truck parts. In short, the Cruickshanks crews were using a technology more suited to 1912 than to the 1940s, but they had no choice, and the skidders were proven workhorses.

The familiar technology meant that the traditional stages in the stump-to-dump process were followed at Cruickshanks until 1946, though none of the main logging photographers—Walter Montgomery, Wilmer Gold, and Charles Sillence—ventured up the Cruickshank River to witness this last fling of skidder logging at its high-mountain show. The trucks at the Tsolum and Ladysmith divisions had a good deal more novelty value.

A key member of the Cruickshanks show in its early years was Sam Telosky (born 1915), a farm boy from Haney (in the Fraser Valley) who had come to Royston as a boom man in late 1935 or early 1936. After a brief career on the boom at Royston, at the Comox

> Lunch break on the Cruickshank River, c. 1939. Sam Telosky, who took the photo, was rigger on a cold decker, and George Surgenor was hooker (hooktender). Left to right: Ed Linder, Jock (John) Small, Dave Balmer, Bert Saunders, George Surgenor, Oliver Doucette, George Stemac, John Loden, Gerry Gareau, Alec Nelson. Telosky explained this photo as follows: "Surgenor was hooktender. The hooktender has five guys working for him. My crew is there too. This is both crews having lunch! The whistlepunk always did the lunch fire. Bert Saunders is the whistlepunk—he did the fire. Linder, Doucette, Stemac, Nelson: they were all my back riggers." Bill Wilson told a skidder lunch story from the 1920s or 1930s: "They were sitting around eating lunch when the chain broke and the bull block [large steel pulley at top of spar tree through which mainline passes] landed right in the fire in the middle of the skidder landing. Eight or ten of them were sitting there. Hard hats weren't even known."
—FRED PARKIN

SAM TELOSKY PHOTO

Surgie's Last Trip

George Surgenor (born 1916) worked for Comox Logging as a hooker (hooktender) on the Cruickshank River. Like other young, unmarried loggers he preferred to keep busy during layoffs, as his friend Walt Edwards (1918–2003) recalled: "Comox shut down for a couple of weeks and George left ... Jeez, he was a nice boy!" Another friend, Sam Telosky, told the tragic story:

When Comox Logging closed down for a month or two, we'd go to Vancouver to hire out ... I worked at Pitt Lake, Baxter Poles, Abernethy & Lougheed, Sproat Lake, McNab Creek—I was at 14 camps altogether!

In about 1940 Comox Lake shut down, so George Surgenor and I went to Vancouver to Black's hiring hall. We took our packsacks with us. They had a board on the sidewalk that said, "Wanted: Hooktender at Port Neville." This was a rigging job, and Surgie took it. They only needed one guy. They'd give you a ticket right there to get a boat. At the same time I went to Sproat Lake to a rigging job. We were best friends so we agreed to meet in a couple of months.

Two weeks later George was killed there, by God! He was shipped out of Port Neville in a pine box. I was at Sproat Lake when I heard that. A log hit him. They brought him out in a pine box, and they put him off the boat at Union Bay. That was pretty bad, you know. I knew George really well. I used to stay at the Surgenor house on Saturday nights. The Surgenors lived at Mission Hill near Courtenay. I stayed there for a month once. Their other boy, John, died later at Ripple Rock [between Vancouver Island and Quadra Island]. *Bill Beech was running the boat, and he hung on when the whirlpool turned the boat over. Bill and another survived, but nine guys drowned at Ripple Rock.*

George Surgenor died on May 11, 1940, aged 23. ■

ROBERT McCORMACK FILBERG PHOTO

Vertical challenges. Nineteen-year-old Rob Filberg climbed up the side of a mountain at Cruickshanks in 1944 to record one of Comox Logging's last skidder settings. In the absence of level ground, the skidder, centre, has been placed on a level bed of cribbing. The skidder has just started to yard in the timber, which is still strewn on the forest floor in 40-foot lengths. Six flatcars wait at the setting, and the skidder is loading them by hayrack boom from a small landing pile. When Bill Woods was shown this photo he said, "That's the big wooden trestle where I started working in 1941! That's Rough [and Tumble] Mountain behind. They've left the creek banks unlogged—they sure left a lot of saplings along the riverbanks." Sam Telosky explained, "Around the creeks you'd leave the timber standing. Around the creeks the timber would be defaced: it would get a lot of high water when the freshets came, and boulders. It looks like the creek came straight down from the mountains, so they bypassed it and put the next road on the other side of the creek."—FILBERG HERITAGE LODGE & PARK ASSOCIATION

SAM TELOSKY PHOTO

Sam Telosky had the same idea as Rob Filberg (photo opposite page) and took this distant view of a skidder setting deep in the Vancouver Island Mountains (South Fork), c. 1942. The difference between the two photos is that this setting has been logged: the felled and bucked timber has already been removed. A locomotive (at right) prepares to pull five loaded flatcars out of the setting while the skidder (at left) finishes loading the last flatcar. As in the scene shown in Filberg's photo, Comox Logging is leaving a certain amount of the forest uncut. "It was a bit swampy, probably, in the middle or the trees were small, so they'd bypass it," explained Telosky. "And so you'd go to the next patch of good trees and pull it down. It was probably swampy ground and poor punky timber, or small scrub timber like scrub cedar, so you'd just leave her." The blurred parts of the photo resulted from condensation on the lens caused by Sam Telosky's exertions on the way up the mountain. "They left a lot of timber behind, the corners," said former Ladysmith faller Sven Ell. "They'd leave it. They figured it wouldn't pay. Now they go back and get the corners."—SAM TELOSKY

Getting Hurt Too Much

In 1943 high rigger Jack Berkeley was badly injured on a skidder setting on the Cruickshank River. He was working on the rigging when an engineer's mistake threw him 25 feet. In 1995 Jack and his wife, Jean, reminisced about the ordeal in an interview with their son Don:

Jean: *His uncle Oswald Harmston brought him down Comox Lake in an open boat!*

Jack: *An eight-cylinder boat, it could do something like 45 knots. Blackburn's old tub, the crew boat, would have taken all day. It went about eight miles an hour.*

Jean: *And then Oswald brought him here, and he wanted to come home and change his clothes before he went to the hospital! He hadn't been to the doctor's office, but he wanted—*

Jack: *Oh, it was dirty out in the woods! Christ!*

Jean: *Oh, they were used to that. So anyway, the next day—*

Jack: *I just wanted a smoke—shows how stupid you could be in the shape I was in!*

Jean: *Ed* [Berkeley] *took me out to the hospital the next day, and I passed out when I saw his face!*

Jack: *She fell right on the floor!*

Jean: *I never slept all night. And his dad was dying at the time, and then he was worried, and Jack said, "You'd better go down and see him, or he'll think it's worse than it really is." And then of course after he got home from the hospital, he said one day, "Son, you'd better not go back to the woods." But he did.*

Jack: *He said, "You're getting hurt too much, you'd better get out of there!" I got up in the morning down yonder here, in the washroom, and I looked in the mirror. I didn't know myself at all! I was just black all over with a big white turban on my head.*

Jean: *Your head was cut open here, a big cut over your eye, nose broken, ribs cracked. And he survived all that!*

Jack: *I had my head down, and it just felt like somebody had poured a bucket of cold water over my head.*

Jean: *And the guys in the bush, like young Filberg, they wanted to pour iodine in it! And they were going to put cotton batten in it!*

Jack: *My* [skull] *was showing, you know, across here, right between the eyes, and when the nurse was cleaning it out I could hear the bone grind here, just right here, just below my eyes. And I was all scabbed from the waist ... this one arm fell over the line, then the line tightened when I fell onto it, I couldn't get off it, and it just beat my ribs until I was black and blue.* ■

Lake load-out, and at Willemar Lake, he started at Cruickshanks as a second rigger in 1937 and stayed, off and on, until 1942. Then he joined the Canadian Army, but Filberg pulled him out and sent him to the Queen Charlotte Islands with Aero Timber Products, where he became a foreman. He was at Cruickshanks in the first years of logging. "I rigged the first tree on the Cruickshanks in 1937. That was up through the cut. We started on the South Fork, and then we went to the North Fork," he recalled.

Cruickshanks in its skidder phase (1937–46) was purely valley-bottom logging with cold-decker extensions up the lower mountainsides. Tony Turner's survey crews and Bill Eastman's bridge crew were kept busy laying out skidder settings and making them accessible. The railway main line was pushed up the river valley, as foreman Mike Poje put it, to "catch the lower timber" (*The Timberman*, March 1961). As with the Toma Creek and Puntledge River valleys, the Cruickshank River (North Fork) and Comox Creek (South Fork) valleys were remarkably gentle grades, progressing gradually from 500 to 1,000 feet above sea level. These mountain valleys were also narrow, as forester Erik Piikkila observed: "Some nice ground was to be found in the valley bottoms with nice Douglas fir, but the valleys weren't very wide, and soon the steeper ground came into play." The maximum reach from trackside was about 2,000 feet. Sam Telosky recalled:

> A skidder reached out 1,200 feet, or maybe 1,250 was the length of their skyline. Anything beyond that, we'd cold deck it: bring it in to 1,200 feet and swing it in on the skyline from there. I always rigged the back spar; I topped back spars for quite a while. The back spar was also known as the cold-deck tree. A cold decker could also get out about 1,000 feet, but we tried to keep it under 1,000. This meant we could get more than 2,000 feet from the track if we had to. After that? Well, Cruickshanks was really not much wider than 2,000 feet: you were up into the rock bluffs by then.

Rigger Chan Tame standing on a cold-deck pile at Cruickshanks with a back spar behind and Clyde donkey below, *c.* 1940. The cold-deck pile would shortly be yarded to trackside for loading. In 1942 Tame joined Aero Timber Products in the Queen Charlotte Islands before joining the Canadian Army, going overseas, and ending the war in a POW camp in Italy. A Comox Homeguard through his grandmother Louisa Beech, Tame was closely related to the Lever, Winger, and Woodruff families.—Sam Telosky

In 1941, when Bill Woods started as a whistlepunk, Skidder 3 was working at South Fork and Skidder 5 was at North Fork. Every skidder required at least one cold decker to provide it with logs from the back end, so yarder and skidder crews multiplied, especially after the third skidder arrived when Tsolum division switched to truck logging. Machines and personnel were conveniently recorded in August 2002 in a letter from Bill Moncrief, who started as a whistlepunk on a yarder at Cruickshanks in 1942:

> In August 1942 I started blowing whistles on back of the Alone Mountain on Clyde No. 3, a cold decker. Bill Parkin was rigger, Ed Lefley was engineer, and rigging slinger was Ole [Olaf] Storr. Our chaser, a Finn named Charlie Nybrann, was

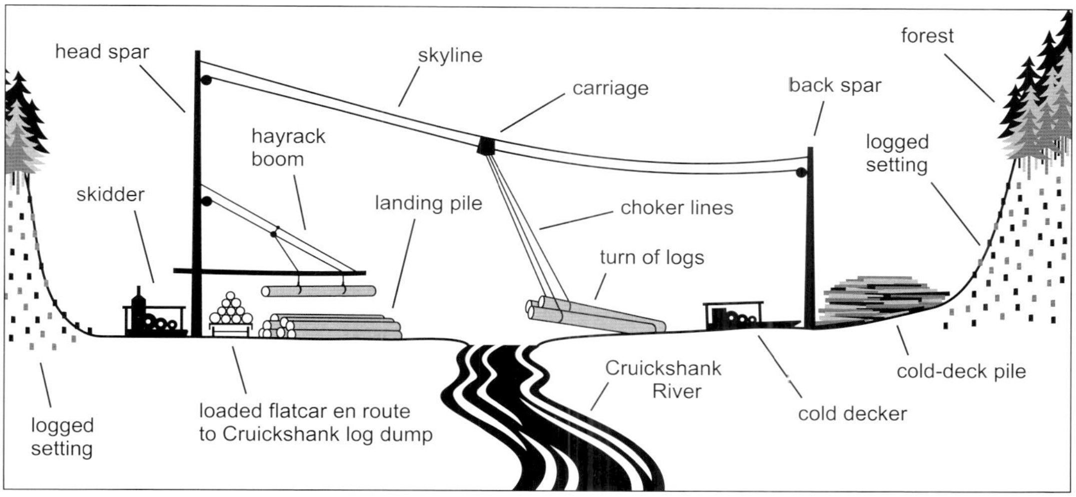

Logging a high valley: Comox Logging's rail and skidder logging show in the Cruickshank River valley, Comox Lake area, 1937–46. This steep and isolated mountain valley witnessed the last appearance of Comox Logging's venerable Lidgerwood skidders, used in the Comox Valley and nearby islands and mountains since 1911.

Hooker George Surgenor and his five-man yarding crew on a cold-deck pile, Cruickshank River, *c.* 1939. "Look at those sticks behind those guys!" exclaimed Bill Woods. Sam Telosky recalled, "They're waiting for the rigging to come back." Left to right: George Sowden (chokerman), Gerry Gareau (second hooker), Dave Balmer (chokerman), George Surgenor (hooker), Bert Saunders (whistlepunk), Jock (John) Small (chokerman).—Sam Telosky killed on November 4th 1942 on the back of Alone Mountain. When moving to the next setting Bill Parkin was hurt—he had his arm pulled into the haulback fair-lead while moving the Clyde. He ruined his arm. Harold Hames replaced Bill. Jack Harris replaced Ole Storr [who had broken his leg]. Jock Horbury replaced Charlie Nybrann. We did three cold decks on the back of Alone Mountain, and then we moved down to trackside, and then we moved on to West Coast Mountain.

When I started at Comox Lake, the operation consisted of three cold deckers (two Clyde yarders and a Twin V8 yarder) and two steam skidders. The riggers of the Clyde yarders were Jack Downey, Bill Parkin, and Harold Hames. Jack Downey's machine, another Clyde, was across from us on the back of the Cruickshank Mountain, then below us on the West Coast Mountain. Bob Aitken was rigger on the Twin V8 yarder.

Later, just before the railroad operation was finished, No. 5 skidder and a Clyde cold decker were brought in from the Headquarters operation. They were what the Comox Lake sidehill gougers called the Tsolum flat-landers! On the skidder, the yarding engineer was Mack Grant, loading engineer was George Cliffe, rigger was Franklin Smith, and hooker was Ron Muir. On the Clyde, Dave Aitken was rigger, Jack Berkeley was rigging slinger, and Eddy Boyd was engineer. Skidder No. 5 swung all four of the cold-deck piles (two Hames, two Downey). When skidder trees were to be rigged, an oil-fired steam machine was also brought in from Tsolum. Head rigger was Tony Sabin, Tom Tames [Tame] Sr. was engineer, and fireman was Neil McDonald.

By 1942, despite the demands of Ladysmith and the war, a good deal of new blood had been recruited. Local men whose careers would span the post-war decades cut their teeth on the skidders and yarders of this mountain valley. In addition to those mentioned by Moncrief, men who worked at Cruickshanks between 1937 and 1946 included fallers Gordon Annand, Waino Jylha, Toivo Koski, and Otto Niemi; levermen Ray Dawson, Elmer Orr, and Harry Sheppard; fireman Frank Davison; hooker George Surgenor and second hookers Gerry Gareau, Gerry Merrill, and Bill Westwood; chokermen Dave Balmer, Alec Graham, John Loden, Red Robinson, Dick Sgarbossa, Jock (John) Small, George Sowden, and Bill Woods; whistlepunks Bert Harris, John Holmes, Carl Lidberg, Fred Parkin, and Bert Saunders; head loaders Oscar Berkenstock, Bobby Colling, Wilfred "Skinny" Colling, I.J. "Lefty" Larson, and Albert Trotter; second loaders Eric "Red" Doucette and Pete Telosky; third loader George Yates Jr.; chaser Tommy Gray; and bunkmaker Sid Cresswell. The cold-deck engineers included Gerry Empey and Dick Woodrow; Ward Piercy, known

Comox Logging's No. 2 on the trestle at the Cruickshanks log dump, *c.* 1939, with the view southeast towards the flank of Rough and Tumble Mountain; the swampy delta of the Cruickshank River is behind. "I have lots of memories of Two-Spot!" recalled Sam Telosky. "It was the main machine in the Cruickshanks. One day the boxcar crummies were taking 25 to 30 guys to work. We were going along at about 25 miles per hour, and the train goddamn jumped the track on us! We went right into the bushes among the stumps. We were all piled into the end. The other guys were a pretty good cushion!"—John Holmes

Cold-deck crews changing shift at a skidder landing on the Cruickshank River, 1942. "This is my cold-deck crew," commented Sam Telosky, "which was double shifting with Dave Aitken's cold-deck crew, going through the skidder landing at noon." The engineers are Norman Parkin and Ray Dawson, and chaser Tommy Gray is also shown. These men have just arrived by rail at the skidder setting and will make their way out to the cold decker.—Sam Telosky

The Danger Guy

Sam Telosky told the story of the injury that nearly cost the life of his brother Pete in about 1939:

I always remember those six whistles. Once we were logging in Cruickshanks, about ten miles out, and my brother Pete was second loading on a Lidgerwood skidder when the haulback line broke and caught him across the back of his neck and threw him in the air and broke his neck. When you're second loading, you're the danger guy: the skidder lines are always up above, and the receding line broke. Ray Dawson was the engineer. I heard it break! It was like a rifle shot. I knew what had happened right away, and when I heard the six whistles I knew it had to be one of the three loaders, and sure enough, when I got there it was Pete lying there.

Len Harding generally had the speeder, but he wasn't there that day. By the time the locie got there from the lake, it took eight or nine hours to get Pete to hospital. It took two hours to take 10 to 15 flatcar loads to Cruickshanks dump and back, and my brother lay there on the track with a broken neck waiting for three hours! Finally we started to carry him down the track. We got him to the Comox hospital [St. Joseph's Hospital] *at 5 p.m. He was hurt at 9 a.m.*

Dr. Mooney set his neck and he was in traction for a month, and when they took the traction off, his neck was crooked. Bob Filberg came out and said, "Sam, Pete isn't doing that good and we're gonna send him to Vancouver tomorrow." They had to set it again in Shaughnessy Hospital. Filberg was pretty good. From that day on he did the worrying for me—he was that kind of guy—and two months later my brother came back, and in a year he was back to work. Filberg was a good man, a good-natured guy.

After that I didn't like working with my brothers. ■

GEORGE YATES JR. PHOTO

Third loader George Yates Jr. (born 1922) took this photo of Skidder 5 on the South Fork in 1942, shortly before he joined the navy. The flatcars await the turn of logs being yarded into the landing at left. Yates's caption: "No. 5 skidder. South Fork area. Operator (Shep) Sheppard. Loading levers: Elmer Orr. Head loader Bob Colling." Bobby Colling and Sid Cresswell were killed near here the year after this photo was taken.—George Yates Jr.

The Agony Whistle

On May 4, 1943, head loader Bobby Colling and bunkmaker Sid Cresswell, on the Skidder 3 crew, were killed at Cougar Lake on the South Fork. Colling was 31 and Cresswell was 51. "They were sitting on a log," recalled the third loader, who preferred to remain anonymous. "A turn was coming in and the line hit a sapling, and then the line hit them both in the back of the head as they sat there."

Bobby Colling's brother-in-law Bill Woods also recalled the incident:

[My sister and Bobby had] *been married for less than three years. I was not at work when Bobby was killed. He was hit and so was the bunkmaker—you needed a* [false] *bunk if a stick was over 40 feet. They were killed near Cougar Lake. They were all sitting on a log with their back to the rigging. There were injuries too. They killed two and put one in hospital. It was dangerous work. Lots of guys got banged up in skidder logging. It was terrible hard work!*

It was called the Agony Whistle. You could hear it all over the steel because it used to ricochet down the tracks. The Agony Whistle was seven great big longs.

They said everything goes in threes, and then Shorty got killed on Skidder 5 at North Fork. He was killed by a siwash in a line. Three guys were killed in two weeks. Two out of three Colling boys got killed in the woods: Bobby and also his twin brother, Ernie. It was terrible bad luck in one family to have two sons killed. Skinny Colling survived—and just died recently. ■

Twice Dead on South Fork

Bert Harris recalled the accident that claimed the lives of Bobby Colling and Sid Cresswell in 1943:

I was born at Dove Creek and lived there while I worked at Comox Lake. I walked down and rode to Comox Lake every morning with Sid Cresswell and another fellow. Sid was killed, and so was one of the Collings, and the fellow who drove me was injured, all on the same day, so I had no ride to Comox Lake. They dropped a turn at the landing and got three of them. I wasn't with them when they were killed (I was on another machine), but I was transferred to Tsolum because of that accident.

Norman Parkin provided the black details:

Two guys were killed near Comox Lake when a log came dislodged. They were eating their lunch when the engineer yelled, "Look out!" They stood up and the line hit them like a scythe. They were both killed instantly. They put the bodies on a big table in the railway crummy, but then the crummy derailed. They found their bodies in a heap on the floor at the end of the crummy. Someone said, "Those guys sure are lucky. If they weren't already dead, they would have been killed for sure." ■

Rigger Sam Telosky heads up a back spar, South Fork, *c.* 1939. In the distance, some 1,000 feet away, are the head spar, skidder, and logging railway. The cold-deck pile is behind the photographer. Telosky explained why he was going up the tree: "At Comox Lake half our trees were rigged without topping them. We were logging with tops on! Bob McQuillan didn't bother with topping back spars. The hooker, George Surgenor, said, 'We can't work anymore because them limbs are coming down! No goddamn way! You'd better do something, man, we pretty near killed a guy.' So I was waiting with my belt and climbers at dinnertime and I made good time. I only held up the crew for five minutes. I had to get over the skyline and the big receding block." Note that the road between back spar and head spar is clear apart from small debris. Everything else has been bowled over by the semi-aerial transit of logs.—Sam Telosky

SAM TELOSKY PHOTO

Rigger Sam Telosky took this photo at a skidder setting on the South Fork, *c.* 1939, with the view back towards the back spar shown in the photo to the left. A turn of three 40-foot logs is being yarded across the valley bottom from the back spar, which is some 1,000 feet away. The cold-deck pile is visible at the base of the back spar in the distance. The ties and steel of the logging railway can be seen at the bottom of the photo, where the chaser and loaders are waiting. The skidder is out of the photo at top right. Overhead, guylines extend from the head spar to stumps. Snags, saplings, and scrub timber remain standing, and a few logs have fallen along the "road" during yarding and will be picked up later. "At right is the oil tank for the skidder," explained Telosky. "At left the water tank. You always had to have water."—Sam Telosky

Punking Whistles on Skidder 3

Bill Woods recalled his first day as a whistlepunk at Cruickshanks in 1942:

At 15½, I was the youngest one in the whole of Comox Logging. My wage was $2.35 a day. I was sent to the commissary to get my caulk boots, and it was run by Pete McLoughlin, who came on the same boat as my dad from the old country. He was also from Lancashire. He was bookkeeper and commissary operator, and we called him Whitey because of his really white hair.

I was sent to the South Fork of the Cruickshanks to punk whistles on Skidder 3. I started working beneath Rough [and Tumble] *Mountain at a setting next to a big wooden trestle. When I started, Skinny* [Wilfred] *Colling said, "Hide behind a stump if a log gets away!" I had 1,500 feet of whistle wire around my neck when I saw a log coming down the hill at me. I jumped behind a stump and I watched that log go past the stump, and the stump was smoking where it touched the* [log]*! I saw Skidder 3 bust up those logs and toss them around like matchsticks.*

Side-push was Bob McQuillan, who was known as Rocky Mountain, or just Mountain McQuillan, because he was always coming down from the mountain. The high rigger was Bud Carwithen. If he was up the tree and made a mistake, he could swear for 20 minutes without using the same word twice, but then he turned into a Jehovah's Witness. Elmer Orr ran the loading levers. Head loader was Bobby Colling, who later married my sister. Sid Cresswell, the bunkmaker, got killed with Bobby Colling. Skinny Colling was second loader, Red Doucette was second loader, yarding engineer was Old Shep—Harry Sheppard—he ran the skidder. He always used to say, "Hey, kid, no mistakes today!" He was a tall, thin guy who used to live in Comox. Second hooker was Gerry Merrill, who was a full-fledged chemist. Clifford was first hooker or head hooker. I can't recall who the chaser was. There were four chokermen including Eddy Williams, Alec Graham, Red Robinson, and eventually me.

Soon I was promoted to chokerman, which meant a pay raise to $3.65 a day. They used to fly four by four, which meant four chokers coming and four going. The engineer would drop four chokers in the bush. If you had a tight log, you had to get a hole under it; other guys would help you. In the '40s you had to work hard, by shit! Every time you turned around, the chokers would be there, or the tongs. They'd say, "Go out and break some lines!" This meant break some chokers. It meant "Get cracking." ■

Three Whistlepunks

Bert Harris, John Holmes, and Fred Parkin—all sons of Comox Logging employees—started as whistlepunks on the Cruickshank River in the early 1940s before moving on to skidder work. Their stories are similar:

Bert Harris: *I started in 1941 at the corner of the North Fork and West Fork* [Rees Creek] *of Cruickshank River blowing whistle on a little gas machine run by Bob Aitken. I started as a whistlepunk. Then I fired skidder, also at Comox Lake. This was on Skidder 3 or 4, I can't recall which.*

John Holmes: *Yes, I did work at Cruickshanks one year, which was the second year I worked for Comox, so let's see. I was born in 1924, so it would be around 1940 or 1941, and the camp was at Cruickshanks by then. I worked there when I was 16 or 17. I was blowing whistles on a yarder. I blew whistles and worked on the rigging. I spent only one day on a skidder, and it was just like a zoo with lines and chokers flashing up and down! I wasn't meant to be on the skidder, because it was too dangerous.*

Fred Parkin: *I started with Comox Logging up Cruickshanks, blowing whistles on a cold decker for Jack Downey. I was 15 in February 1943 and started in September when I was 15—but you had to be 16! Then I set chokers on skidders on the North Fork of the Cruickshanks. That was the end of the railway track. After that they put trucks in.* ■

> Looking east down Cruickshank Canyon, summer 1948, from the end of the trail from Lake Beautiful on Forbidden Plateau. The logging scars in the distance are within Comox Logging's Block 748 in Cruickshank Canyon (towards the end of the Cruickshank River), some six miles northwest of Comox Lake. Eric Creek is between the low ridges at right, and Moat Lake is just out of sight to the right. This photo illustrates the vertical limit of skidder logging, which depended on railways and cold deckers and was therefore best suited to valley bottoms and to the lower 1,000 feet of hillsides. Higher elevations would be logged by truck in the 1950s and 1960s.—Courtenay & District Museum & Archives, 989.191.4

Cribbing was required to keep logging equipment level on the lower slopes of mountains. Tommy Gray (chaser), at left, and Norman Parkin (engineer) on a Clyde rigging-up goat, Cruickshank River, 1942. To the left is a back spar with a cold-deck pile at its base. Dave Aitken and Sam Telosky were in charge of this machine. "I was double shifting with Dave Aitken," said Telosky. "That's our barrels of gas at left. We'd take up all the gas we could. Sometimes we'd be out there for two months, so we'd need a lot of it. Tommy Gray was my chaser there. That's not booze, that's his water bottle! We're on a sidehill—see how we had to build up to get the machine level. We'd pull in a couple of logs and move her up."—Sam Telosky

as Wardy, was rigging-up goat engineer. The riggers included Bud Carwithen, Bob McQuillan, Reg Taylor, and Sam Telosky; second riggers included Bill Braes and Chan Tame; and back riggers included Gordon Blackburn, Mel Bowen, Oliver Doucette, George Grieve, Ed Linder, Alec Nelson, and George Stemac. Shop crews, including gas mechanic Ross Tapp and machinist Walter Anderson, came from Headquarters to Cruickshanks periodically. "We were working on the yarder, a gas Clyde—we were overhauling the Clyde," Anderson recalled.

This railway show in the mountains featured at least four locies: the Two-Spot Baldwin, the Eight-Spot Heisler, the Ten-Spot Heisler, and the Fourteen-Spot Shay, which the company bought second-hand in 1942. Among the engineers were Bill Bourne, Martin "Pidgy" Davison, Bill Defty, Fred Frew, Harry Grieve, and Bill Rourke. Grieve ran a Heisler, Davison ran the Shay, and Rourke, Bourne, and Defty ran the Two-Spot. Train crews included Reg Churchill and Jack Mitchell (brakemen); Gord Blackburn, Bob Cessford, Frank Davison, Ralph Harmston, and Tommy Tame Jr. (firemen); and Dick Delin (conductor or head brakeman). Defty, according to his grandson Shawn Agar, was the last engineer to take out a load of logs from the South Fork. Many of these men would be transferred to the Nanaimo Lakes division, also a railway show, in the mid-1940s, when railway logging ended at Cruickshanks.

As elsewhere, the geared Heisler and Shay locies were used for the steep grades on spur lines into skidder settings. "They'd leave the empties [flatcars] every two to three miles on a siding, which was a piece of double track. You'd run your loads in there and take the empties back. Every setting had that," Sam Telosky recalled. The Two-Spot Baldwin served as the main-line engine, as Bill Moncrief wrote in his August 2002 letter: "The saddle tank locomotive now on display in Courtenay called the Deuce (Two-Spot) was used on the flat mainline runs, hauling from the main sidings in the north & south Fork to the lake dump." The Two-Spot brought the loaded flatcars down to a trestle along the shoreline. There the logs were dumped into the lake with the help of a humdirgen (steam-powered unloader), operated in 1946 by Toynbee Harmston. As at Lake Camp, logs were dumped directly into bag booms and then left to float down the lake on their own.

THE BIG LIDGERWOOD SKIDDERS went out of use at Tsolum in 1945, and the last remaining skidder (No. 5) was brought to Cruickshanks. In 1946 the company abandoned skidder logging at Comox Lake, tore up the track, and converted the railway grades into logging roads at the same time. "By 1947 it was all trucks," Frank Davison observed.

Hook, Line, and Sinker

Comox Logging employees were known for their pranks and initiation rituals. Bill "Bronco" Moncrief, who started at Cruickshanks in 1942, remembered being told, "Go and get a sack of choker holes!" and "Go and get a Mutt and Jeff!" (A "Mutt and Jeff" was a short chain with a hook on each end, but greenhorns would probably be confused, thinking of the popular cartoon characters.) A green whistle-punk might be told to "grease the main-line cable" (some 80 feet up in the air). Gordon Naylor, who worked periodically at Comox Lake, remembered that "a new, inexperienced man would quickly be put through the paces by the old hands, who'd order him to go fetch a bucket of steam. Or they'd say, 'Your job will be to oil the skyhook'" (*Crown Zellerbach News*, June 1964). There was no such thing as a skyhook.

Bob Muir, son of hooker Ronnie Muir, recalled a fishing adventure on the Cruickshank when he was 15 years old, in 1946:

When I started at Comox Logging, blowing whistles, Dad said, "You go down to the river"—the Cruickshanks, it was a hell of a hot day—"and fill the water bottles." On my way I ran into a bucker and he said, "Jeez, there's some nice trout in the river! Have you got time?" And I got some line and got three or four trout. When I got back, Dad laid into me! He said, "Fishing!?" So I put the trout in my lunch bucket and took them home to Mother. When I got home I said, "There's three trout in there." Well, Elmer Orr, the loading leverman, had opened up my lunch box and filled it with sticks! ■

Dead in the water. Locies in the November 1939 flood at Cruickshanks. At left is a Heisler, at right the Two-Spot. This flood played havoc with the Cruickshanks show, as Sam Telosky recalled: "We logged for one or two years at Cruickshanks, then they had the flood, which took out everything."—Sam Telosky

The Comox Lake flood, November 1939. "That's the year everything flooded," said Sam Telosky. "Towards the Cruickshanks Canyon it's pretty flat and everything floated up, including railway ties. The Cruickshanks flood lifted the track right up! They shut her down for six months, and I went to Headquarters on the skidders." —Sam Telosky

SAM TELOSKY PHOTO

Mountain timber. Loaded flatcars are following the Cruickshank River to Cruickshanks log dump, *c.* 1939. —John Holmes

In many ways the skidder show at Cruickshanks fizzled in an ignominious ending. The camp was cannibalized in turn by the demands of Cat and truck logging at Ladysmith, the Second World War, and Aero Timber Products based at Juskatla in the Queen Charlotte Islands. Indeed, the camp at Juskatla seemed to carry on the raw energy of Camp 2 and its successors. Comox Logging's other Comox District division, Tsolum, started converting to truck logging in 1942. Now off the beaten track, and marred by tragedy, Cruickshanks Camp became a technological backwater, a dumping ground for superannuated equipment. It was an inglorious end to the once-flourishing Camp 2. ■

Downhill all the way. The Two-Spot pushes six loaded flatcars onto the log-dump trestle at the mouth of the Cruickshank River, *c.* 1940, and the humdirgen at right tips the logs into Comox Lake. Here they will be boomed, sent down to the load-out, loaded again onto flatcars for the journey to Royston, and finally towed across the Strait of Georgia to Fraser Mills. This unloading facility was destroyed in the 1946 Comox Valley earthquake.—Sam Telosky

A much larger proportion of loggers of the Comox Logging & Railway Company make their home in the district than in the average of logging camps.

—*COMOX ARGUS*, JANUARY 7, 1937

8

HOMEGUARDS AND HIRING HALLS

< **Uno Forsman was almost as distinguished as his wife, Esther, world champion bucker, and won several coastal bucking contests. Here, he puts theory into practice around Camp 3, *c.* 1935. At left is Tom McQuillan (grade foreman); the man standing on the log is unidentified. The side-notch is at right and the undercut is at bottom; Forsman is standing on the slivers of wood left between the undercut and the cut from the other side (backcut). The *Crown Zellerbach News* noted Forsman's retirement in January 1964: "Uno Forsman, a veteran faller at the Courtenay division, retired this month. During his thirty-seven years in logging he felled an estimated 138 million board feet of timber. He started logging in the spring of 1926 as a back rigger but spent most of his logging days as a faller, first with cross-cut saw and then with power saw. Mr. Forsman plans to take a trip to his native Finland and on his return to farm his holdings north of Courtenay."**
—VICTOR OWEN

AFTER ACQUIRING NEW MOUNTAIN TIMBER IN THE 1930S AND 1940s, Comox Logging continued as the motor of the Comox Valley economy, but its position was threatened by larger events over which the company had little control. The company relied on external demand for its timber, but the Great Depression drastically reduced this demand and exposed the company's vulnerability. A decline in house construction and the generally poor market for lumber resulted in unsold inventory, both at Fraser Mills and at Canadian Western's lumberyards on the Prairies. Stagnant sales, in turn, translated into a lack of demand for logs from the woods camps in the Comox Valley and hard times for the 400 men employed there. "All timber logged during those years was sold at a loss," Bob Filberg wrote in 1968. "These were seven long, lean years [1929–36] ... Highly skilled logging labour, with a great wealth of know-how and experience, worked for low wages or could get no work at all."

The company closed down for six consecutive months in the winter of 1929/30. Bevan chokerman Stan Hanham recalled that he went back to his stepfather's farm at Merville in 1930 and never returned to the woods. Comox Lake leverman Norman Parkin remembered one year—probably 1930—when he worked for only five months. In May 1930 the *Comox Argus* declared that logging was not profitable in the current market and that Comox Logging would therefore "curtail operations" until better prices were obtained. "It is better to close in the summer when there is alternative employment," asserted the *Argus*, referring to local farm work, the stopgap of the Comox Homeguards. Comox Logging also reduced the men's pay. "Some companies will close, others will cut wages," the article continued. "The Comox Logging Company, with which the Courtenay District is chiefly concerned, will adopt the latter course and their camp will remain open." Norman Parkin and Gunnar Jonsson both recalled this substantial wage reduction. "The company cut wages to keep the men working," Parkin said. Jonsson, who had arrived from Sweden in the spring of 1929, faced plummeting wages on the rigging crew at Camp 3:

> In the spring of 1930 I got $4.00 a day, which was good wages, but shortly afterwards they cut the wage to $3.50 and then to $3.20 a day. The big boss—the superintendent [Filberg]—comes out and says, "If you fellows want to work, you'll have to take another cut, the price of logs is so low." Wages got as low as $2.50 a day. Rigging men and chokermen both got the same wages. I cleared $50.00 to $60.00 a month after paying for compensation and buying a set of gloves a month at $2.00 to $3.00 a pair. They'd pull your hands to pieces if you didn't have them! If you stayed in the bunkhouse, they charged a dollar a day with grub and all.

The year 1931 was not much better. "The Comox Logging Company closed all its camps and workings on Saturday and regretfully informed its crew that there will be no more work until the end of the year at least," the *Argus* noted on October 15, 1931. Again the cause was the low price of lumber: "Market conditions have never been this bad." Altogether, 375 men were out of work in the cool and productive autumn, with no prospect of logging work until the following spring. Some of them returned to their farms and stump ranches in the valley, some went to Vancouver, and some registered with Captain George Bates for work at the government relief camp at Oyster Bay. In 1932 conditions worsened when an American protectionist duty on Canadian lumber, the Hawley-Smoot tariff, went from $1 to $4 per 1,000 board feet, which cut coastal mill production by 20 per cent.

Logging families still talk about the personal hardships of the Depression, hardships that started with the feet and spread from there. Comox Lake skidder man Bill Chalmers "always had two lumps on the back of his feet because he wore hand-me-down boots," recalled his daughter, Nola Eastman. Blacksmith Fred (Fritz) Dalby, a Norwegian, unemployed in the Depression because there was not enough work in the shops at Headquarters, dusted off an old skill and started making shoes. His son, Ralph, recalled that all he needed was an old pair of soles; he could make the rest. Fred Dalby, "too stubborn to go on relief," went to work as a blacksmith for a logging company at Horne Lake (near Qualicum Beach).

Skidder 5 hooker Ronnie Muir with one of his homemade rowboats on the Dyke Road, Comox, 1943, with Royston in the distance. Muir made the trailer from a Model T Ford at Camp 3.—Bob Muir

ROGER BULLOCK PHOTO

A good hair day. Don Goodsell of Merville and Mildred Hawkins of Comox, Fifth Street, Courtenay, *c.* 1940. Roger Bullock's original caption for this photo was "Hawkins girl and subsequent husband." Like most people in the Comox Valley, Don Goodsell and Mildred Hawkins were tied in many ways to Comox Logging: Don's sister, Jenny, married faller Gunnar Jonsson, and Mildred's brother was Jasper Hawkins, a head loader on a skidder. "You could walk down the main street and know everyone," recalled Bill Woods.—Roger Bullock Collection, Courtenay & District Museum & Archives, 997.532.23

Ronnie Muir at Shelter Point on the Campbell River road, *c.* 1937, with his 1936 Ford. Born in Surrey, England, in 1911, Ronnie Muir came to Merville with his parents when he was seven. His father, Jim, a veteran of the Great War, received a Soldier Settlement farm on Kitty Coleman Road. As a boy, Ronnie worked on the Janes farm but left to join Comox Logging in 1928, when he was 17. In 1930 he married Lizzie Manson, daughter of faller Fred Manson of Camp 3, and spent much of the next 17 years as a hooker on Skidder 5 and also on a cold decker at Comox Lake. Mechanically inclined, in 1947 he became one of Comox Logging's first power-saw fallers with his partners Gus Nystrom and Andy Barr. He was an expert boat builder and sawfiler, a keen fisherman, and a good soccer and baseball player. He retired in 1969 and died in 1983.—Bob Muir

Cedar Stump Stimulus

Al Knappett told the story of his father, Herbert George Knappett (1888–1971), whose resourceful use of remnants of old-growth forest helped his family survive the Depression:

Dad was a carpenter. He was born in England and came to Victoria in 1890, aged two. His father had come first, and then the rest of the family came. Dad knew nothing about England and never went back. He worked on the Island at first, and met my mother in Cumberland when she was teaching high school and he was putting up a building. When the Depression hit, he was living in Marpole [part of Vancouver] *and working as a contractor on the Marine Building in Vancouver. That ended in 1930 and he couldn't get another job. He had no money for gas, and he walked all over Vancouver looking for work.*

There was vacant property in Marpole covered in lots of old cedar stumps 20 feet high. He and his neighbours were cutting them up for firewood. These were remnants of the original forest that had covered Vancouver. Dad thought it was too bad to be burning such fine wood in his fire, so he got the plans for a 32-foot cabin cruiser and started to build it with his neighbour. They planked it with old cedar from those stumps! The hull was finished, and the cabin was on, when a guy came along and bought it. He hired them to finish the boat and install an engine, and then he gave them $500 for it. That was a lot of money in those days.

After that, Dad bought three registered Jersey cows and tethered them on the boulevards in Marpole—the boulevards weren't cut in those days—and then in the summer of 1931 he got a job on the Island. There'd been a fire in Cumberland and he helped rebuild it. He moved back to Marpole after that, but in 1932 we moved on the old Princess Mary *to Royston, where we started a dairy farm with the Jersey cows Dad bought for selling the boat. My brother and I used to peddle milk and butter around Royston. By 1932 the Depression had softened a bit. People on the Island didn't really know the Depression was on! They had small plots of land, they had game, they hunted. Around that time Dad got a carpentry job with Canadian Collieries that lasted 25 years.*

Two or three years later Dad decided to take a walk down to the Courtenay slough, and there was his boat! At that time it belonged to the provincial government. The guy who bought it from Dad had sold it to the government, and it was being used for hauling property assessors around the coast.

The Knappett family now owns a large construction company on Vancouver Island. ■

Headquarters (view to the northwest), *c.* 1939. By this time cars were much in evidence at Headquarters, but when the logging community was built in 1911 it was accessible only by rail. Second from right (with dark hair) is timekeeper and postmaster Roger Balmer, in front of Comox Logging's office, store, and post office. At left, Frank Biss is refuelling his father Lewis Biss's 1938 Ford. The high building in the distance is the roundhouse, where the locies were kept and serviced.
—Frank Biss

The house of George Sr. and Edith Yates at Millard's Beach between Courtenay and Royston, *c.* 1940. Their purchase of a new car must have been the ideal antidote to the Great Depression, and in the late 1930s cars start to figure prominently in Comox Valley photo albums.—George Yates Jr.

Eva Bailey (*née* Manson) with her niece Norma Muir at Camp 3, 1936. Eva Bailey's husband, Dick Bailey, was a chaser on Skidder 5 at this time, and Norma was daughter of hooker Ronnie Muir and Lizzie Manson. Eva Bailey died in Duncan, B.C., in 2009.—Bob Muir

Waldo "Adam" Adamson filing a crosscut saw, *c.* 1948. Adamson's skill would soon be immaterial; by 1955 crosscut saws were superseded by power saws (chainsaws). "This is my step-grandfather, Adam, in his filing shack at Little River," said Victor Owen. "In the background is a grinding wheel operated by a foot treadle, similar to one which my sister and I played with as kids. We would sharpen nails and play with the sparks. Waldo ended up as a barber, cutting hair at his house." Adamson, who was possibly Russian, worked as a fireman on Skidder 5 before 1920 and as a faller at Camp 3 in the 1920s.
—Victor Owen

CHARLES SILLENCE PHOTO

Five generations in the Comox Valley, *c.* 1937: the Beech-Lever-Winger-Tame family at Charles Sillence's photographic studio, Courtenay. Left to right: Louisa Winger (formerly Lever, *née* Beech, 1875–1943); Hannah (or Anna) Beech (*née* Cook, 1849–1944) holding her great-great-granddaughter Darlene Tame; Ruth Tame (*née* Lever, born 1894); Tommy Tame Jr. (born 1917). Hannah Beech's husband, William Boswick (1832–1931), was a Royal Navy sailor from Birmingham, England, who changed his name to Beach or Beech when he jumped ship. He came to the Comox Valley in 1862 and started a farm on the Upper Prairie Road at Grantham. Numerous members of the Beech, Lever, Winger, and Tame families worked for Comox Logging. Tommy Tame Jr.'s grandson Matt Ellison (born 1983) played in the National Hockey League for several seasons and now plays in the Kontinental Hockey League for Dinamo Riga, in Latvia.
—Ray Davis

The workers' experiences of the Depression varied greatly. Bevan's Bill Lyttback, who had arrived from Finland in 1926, was unemployed for only seven weeks, and Gunnar Jonsson was lucky enough to work right through the Depression, although with extended seasonal layoffs and wage cuts. On the other hand, Sven Ell, who came from Sweden in 1928, spent 16 months in a relief camp at Harrison Mills (on the mainland) in 1932–33, making $7.50 a month. The difference was that Lyttback and Jonsson already had secure logging jobs in the fall of 1929, whereas Ell was still working as a general labourer and had not yet entered Comox Logging's fraternal embrace.

The stump ranches proved their worth in the Great Depression. They were vital to survival during the extended fire and winter seasons. In the lean 1930s small farms combined productively with foraging and deer and grouse hunting. These activities meant, essentially, that needy families could cope better with the economic malaise by helping themselves to the natural provisions of company or Crown land. "I can remember depressions in which men would just go live upon the beach, fishing and digging clams, or exist upon their little farms, with deer meat. Nowadays they are all helpless when some payroll stops," Martin Grainger wrote, far too prematurely, in 1931 (Murray, *Riding the Skyline*). In fact, for subsistence many people in the Comox Valley turned to the logging slash in search of grouse, deer, and berries; to the sea for fish; and to the intertidal zone for bivalves. "We lived off the beach," recalled Doris Walker (*née* Marvin) of Royston. Her family ate clams and oysters and shot deer. Buck Pedersen, whose father, Axel, lost his job when the K & K Pole Company went bust in about 1932, said, "We had chickens and a garden. If one guy had a car licence he'd take lots of men to Point Holmes to dig clams." Anton Holm, a Swede-Finn who worked for Comox Logging after the Depression, camped on a beach on Salt Spring Island and, with his friend Carl, lived on crabs and fish.

Everyone had a story of the Depression. Over the winter of 1931/32 Sandwick farmer Eric Duncan provided room and board to a farmhand who had ridden his bicycle from Saskatchewan in search of work. Duncan wrote in 1932 that he had "never had [his] shed so crammed with wood before in [his] life, besides other needed improvements around the place."

Gardening was also a valuable Depression-era strategy. Charley Chown (1886–1983), who had started gardening in rural Oxfordshire when he was six years old, arrived in Canada in 1908, settled in the Comox Valley in 1916, and worked during the Depression for Comox Logging at $2 a day to support his wife, Edith, and their six daughters. "Sometimes the logging operation closed down for the

Moral Hazard

Pole maker Buck Pedersen (1921–2007) recalled a low point in international corporate relations, when an American pole company exploited its Canadian staff:

The K & K Pole Company went broke in the Depression, in about 1932. They went from 90 workers down to about 6. My dad was a foreman for K & K at that time. They were in trouble for a year or two, and they kept telling the men to take a pay cut. The company said, "When we sell the tow, you'll get your money," "We'll pay you next month," etc. People accepted this because everyone knew times were tough, but the company fell way behind. In the meantime the men put together the largest tow of cribbed poles that ever went out of Comox. It contained 30,000 poles packed in cribs.

I have been a bit of a communist ever since because of what happened next. As soon as those 30,000 poles crossed the line into the United States, the K & K declared bankruptcy. It declared itself broke! They went broke the minute the tow crossed the border, and the men got nothing. They owed Dad four or five months' wages. K & K sold the horses, bunkhouses, and dishes, and then the Workmen's Compensation Board got the first cut of the money. I've been left of centre ever since then. ■

BOB FILBERG PHOTO

The steam and internal combustion engines coexisted for a time at Headquarters. Warren Bannerman, Headquarters gas mechanic, is shown here with a 1939 or 1940 Plymouth. "Warren Bannerman was an excellent mechanic," recalled Art Beech, who worked with him in the machine shop at Headquarters in the 1930s. Bannerman, whose mother was a Grant, had lost a leg in a logging accident at Comox Lake. "His leg was pulled into the fairleads on a yarder," said Bob Muir. Walter Anderson said, "He got it twisted in some lines and lost it. He worked in the shop after that."—FILBERG HERITAGE LODGE & PARK ASSOCIATION

Sailors return to a Royal Navy ship at Comox, *c.* 1930. The RN and Royal Canadian Navy used the firing station and rifle range at Goose Spit. Parties of sailors toured Comox Logging's operations; at least one local girl (Adelaide Willemar) married an RN officer; and several naval officers retired to Comox, lending the valley an imperial and cosmopolitan air. Two such officers were Geoffrey Spicer-Simpson and Sidney "Dusty" d'Esterre; d'Esterre bought the Elk Hotel in 1922. Another patriotic expatriate was Arthur Dundas (born 1860), "a staunch believer in the British Empire," noted the *Comox Argus* on his death in June 1937. "Whenever a ship of the British fleet was in Comox Harbour, the red ensign could always be seen floating over his house in Mission Hill."—D'ESTERRE COLLECTION, COURTENAY & DISTRICT MUSEUM & ARCHIVES, 998.219.246

summers, and there was no unemployment insurance, baby bonus, or welfare in those days—but we managed to live through it," Chown told the *Comox District Free Press* (August 13, 1976). He relied heavily on his fine large vegetable garden on the Dyke Road, Comox.

The demand for lumber gradually increased, and Comox Logging remained the essential local economic motor. The company increased wages by 10 per cent in August 1933, and by another 10 per cent in February 1934. In 1936 it opened a large new truck logging show at Ladysmith, where Lyttback, Dalby, Ell, Holm, and many others were employed for good wages.

The one that got away. Lance Berkeley, blacksmith at Bevan, Lake Camp, Toba Inlet, and Cruickshanks Camp, shown c. 1930, probably at his farm in Courtenay. His two sons—Eddy and Jack—also worked at Cruickshanks. Son of Agnes Piercy and logger John Berkeley, who came to the Comox Valley from New Brunswick in the 1870s, Lance Berkeley (born 1888) had a 50-acre Homeguard family farm in Courtenay. He died suddenly of cancer at St. Joseph's Hospital, Comox, in 1943, when his son Eddy was 23. "I never did forget this," Eddy said. "We lived on the farm then, and my father was making what he knew would be his last trip to the hospital. I was driving the Model A Ford, and as we drove over the Dyke Road he said, 'Go slowly—I want to have a last look at the valley.'" He died two days later.
—Jean Berkeley

As the 1930s progressed, demand for unskilled teenage labour rose in all branches of the company. Boys were hired when they were 15 or 16. "Many young fellows as soon as they could get out of going to school would get hired on with the company, which substantiates Bob Filberg's idea that the company did not have to hire men out of Vancouver when they could hire and train local boys," Wallace Baikie wrote in an undated typescript.

Malcolm Carwithen recalled that in 1937, when he was 16, he rode his bike from the family farm out to Headquarters daily for three weeks, looking for work. Finally the foreman said, "Well, Carwithen, I'm sick of looking at you around here every morning. You go into the commissary and get yourself a pair of caulk boots and get to work!" Malcolm Morrison recalled being picked up on the road by Filberg when he was a boy in the 1940s. "'Young fellow,' he [Filberg] said, 'when you're old enough you've got a job.'" Amelia Yates recalled that her brother, Eddy Williams, started work at Comox Lake in about 1940, when he was 15. "He said he wasn't going back to school, so my mother said, 'If you're not going to school you'd better get a job,' so he got a job with Comox Logging."

The abundant opportunities for employment in the woods worked to the detriment of farmers in need of seasonal help. Woods work drove up wages and acted like a magnet to farm boys. Neil Martin, who was at Tsolum School in the 1930s, recalled the choice made by his school friends, the sons of soldier settlers at Merville: "Their farms were not very good farmland, they had to work hard, the boys had to get up early to milk cows, and Comox Logging boys always made more money." Another farm to feel the pinch was William Duncan's at Sandwick, as his daughter, former schoolteacher Barbara Marriott (born 1906), recalled. When asked whether farming went out of fashion, she replied:

> Yes, prices were low and farmers had a hard time making a living. They sent butter to the creamery, grew wheat and oats, but one year it rained awfully and no oats could be gathered. Farming was a real gamble. So a farm boy would go and get a

Safety First

In the 1920s, with the company's emphasis on local hiring, safety, and first aid, Comox Logging acquired the best safety record of any large logging concern in British Columbia. There were no fatal accidents in 1927, as the *Comox Argus* noted on January 5, 1928, owing partly to the Safety First campaign conducted by a former Comox Logging scaler and timekeeper, Major Harry Steere-Clark of the British Columbia Loggers' Association:

For the past two years most of the largest logging companies in the province have been conducting a campaign of safety first. It was planned under skillful direction and the companies themselves went into it wholeheartedly. At the headquarters of most of the logging camps bill boards were erected upon which were presented spirited sketches of how accidents happened. They were drawn by very clever artists and never failed to attract attention. Then committees were formed among the loggers themselves with a view to cutting down the number of accidents and space was taken in the local newspaper to present a series of cartoons depicting the same subject.

Has this campaign been successful? Has it cut down the average number of accidents in logging companies? We have no figures for the logging industry as a whole but we do know that it has had a marked—even a remarkable—effect on the record of the logging company with which we are most conversant—the Comox Logging Company. The Comox Logging Company employs 400 men and carries on one of the largest operations on the Pacific Coast, yet during the past year there has not been one fatal accident nor yet one where a workman was permanently injured. This is a remarkable and most gratifying record.

The newspaper spoke too soon. In 1928 five men were killed in Comox Logging accidents: Russell Goodrich in May, Percy Anderton and Arthur "Skinky" Boyd in August, James Boyd in September, and Willard Fielding in October. As well, a boy, Pat Fairbairn, was killed in August when a log rolled off a train at Courtenay Crossing and hit him.

Harper Baikie, a close friend of Percy Anderton and Skinky Boyd, decided to quit, as he recalled in *A Boy and His Axe:*

This was a great shock to all of us, especially me as I was such a good friend to both men. It got me to the point where I went into the office and quit before I would be the next one to be killed. They tried to talk me out of quitting but I had to get away from it all. Besides I was getting pretty stakey, as they say in the woods, and needed a change. ■

More Than Just Good Luck

After the agony of six deaths in a single year (1928), Comox Logging did not lose a single man until 1935. During these seven long Depression years, the company occasionally took out newspaper advertisements extolling its exceptional safety record:

NO FATAL ACCIDENTS IN FOUR YEARS. *October 1st 1928 was the date the last man lost his life owing to an accident in our workings. That's a record we are proud of! As we chase Old Man 1932 over the edge of the world into oblivion, we want to thank all our employees who have made the remarkable record, above, possible. We do our best but it would be impossible to achieve such a result without the whole-hearted support of everyone concerned.* (*Comox Argus*, December 29, 1932)

FIVE YEARS WITHOUT FATAL ACCIDENT. *Of course there is much more to such a record than just good luck. No doubt but that the chief reason for so good a record is the personnel of our employees. Each man not only works safely but looks to the safety of others, and these boys know their stuff. We try to keep the equipment in good safe condition. The foremen in charge of the various departments watch the safety of the men under them most carefully.*

The personnel of the crews is the chief reason for the small number of serious accidents. When you find the number of men we have who take sufficient, intelligent interest in safety work to study First Aid to the extent many of our employees have, it tells a lot. Also there are many accidents which would be serious were it not for the efficient manner in which they are first treated by their companions. (*Comox Argus*, December 28, 1933)

The unblemished record was finally broken on June 13, 1935, when Fred Tuddenham was killed at Comox Lake. ■

> job in the woods. Any boy in Grade 7 who knew he wouldn't pass Grade 8 could get a job in the woods as a whistlepunk and could say to his teacher, "I'm making more money than you are," and they'd be self-sufficient.

LOCAL MEN of all backgrounds found ready employment with Comox Logging. "Almost everybody who lived in the valley worked for the company," observed Jim Muckle. The company's workforce reflected the main currents of immigration to British Columbia at this time and the main strands of rural and settler society. In such a society, dominated by recent immigrants, it is possible to speak of Italians from Cumberland, Scots from Bevan, Russian Mennonites from Black Creek, and the largely British immigrants who constituted the soldier settlers of 1919 and the Empire settlers of the 1920s. All ended up on the company's payroll, as did grandchildren of Hudson's Bay Company labourers of the 1850s, of gold rush arrivals of the 1860s, of Maritime arrivals of the 1870s, and of the Ontario and British settlers who arrived on the CPR in the 1880s and 1890s. Even Eric Flinton, son of a local Anglican clergyman, got drawn into this employment engine. It is simply a myth that West Coast logging was everywhere typified by marginal characters, itinerant bush apes and birds of passage from the hiring halls of Vancouver, and cardboard figures like Red Malone, Eight-Day Wilson, and Boxcar Pete.

A simple corporate demand for labour did not automatically produce a rough and homogeneous working class of the kind evoked by popular logging literature. Two factors worked against the formation of such a class at Comox Logging. First was the existence of the old and established agricultural settlement of the Comox Valley, with its rooted, conservative, respectable farming population of Homeguard farmer-loggers. Second was the varied nature of immigration to British Columbia at this time. This was an immigrant society, after all, with massive labour demands in the resource industries and staple trades. All kinds of workers were required: white-collar workers, tradesmen and skilled workers, apprentices, and unskilled and menial labour.

Many immigrants found a niche in the company's ranks doing more or less what they had done in Europe. The sheer variety of jobs in this company meant that an immigrant was just as likely to keep his old trade or occupation as acquire a new one. For example, most Swedish and Finnish fallers had done some woods work in the northern parts of Sweden and Finland. Joe Bergsma, who had poled barges on Dutch canals, got a job poling at the Royston boom camp. Blacksmith Courtney Ingram, from Devon, found a job as a blacksmith. Lewis Biss started by pushing a pen as a boy in a solicitor's office in Gloucestershire and ended his working days

Dynamite and Rat-Tail Combs

Harry Trotter started working for Comox Logging in 1937. "Everything was big except the pay," he recalled. He also remembered his work for the rival companies to the north:

Comox Logging never got loggers from the agency in Vancouver. If you went to another outfit, they'd call you a farmer even if you'd never farmed. If you came from Courtenay, they'd call you a farmer! At Campbell River they got all their men from Black's, the big agency. The companies were Elk River, International Timber, Lamb's. Hastings Mill [i.e., the British Columbia Mills, Timber & Trading Company's main timber property] *was at Rock Bay.*

They'd say to Black's, "Please send up three chokermen. If they have rat-tail combs in their back pocket, send six." A rat-tail comb was one with a handle.

The logging companies, Union Steamships, and Black's—all three had an arrangement for travel. On boat day at the northern camps they'd fire a whole bunch and get some back by the next boat! Sometimes they even got good ones. This went for the whole coast. On the coast, they'd throw [dead] *men behind a stump till quitting time!*

In 1943 I went to Port McNeill [on northern Vancouver Island]*—they were short of loggers, so I went. My crew was from Vancouver. One kid had no hand—he blew it off with dynamite.* ■

On the dock at Union Bay, *c.* 1938. Left to right: Bill Walker, Bill Burgess, Bill Chalmers, Frank Stevens, Bill Auchinleck—close friends from Denman Island and Union Bay. Bill Walker was a boom man with Comox Logging at Royston, Quadra Island, and Ladysmith; Bill Burgess worked on a skidder at Camp 2 and Bevan in the 1920s and was later at Ladysmith; and Bill Chalmers had a distinguished career as a logging foreman with Comox Logging and its successors. Born in 1911, as a boy Chalmers worked for Henry Bay Logging Company, a small railway outfit that logged on Denman Island in the 1920s. By 1928 he was on a skidder crew at Camp 2, and in the 1930s he was at Comox Lake. He moved to Ladysmith in 1936, where he drove a Cat, and in the 1950s he became superintendent of Crown Zellerbach's large Ladysmith-Nanaimo Lakes-Nitinat division.—GWEN GREENWOOD

THE HOMEGUARDS' TRIUMPH

In the late 1940s and early 1950s, Bob Filberg and his pre-1914-vintage American foremen, Len Harding of Camp 2 and Jack McQuinn of Camp 3, finally retired after long and distinguished careers—at about the same time as the last easily accessible old-growth was logged off. Filberg and McQuinn retired to Comox, while Harding moved to Campbell River.

No American foreman was replaced with another American. Rather, local men took over leadership roles at this logging company. Known for their work ethic, versatile skills, and farming connections, Homeguards had long been promoted, as a reporter for *The Timberman* noted in September 1935:

Mr. Filberg is proud of many things around his operation. Not the least of these is the unswerving loyalty of his crew, many of whom have been with the company continuously since its inception in 1909. Superintendents, foremen, and side bosses are elevated from the ranks, and the energetic young man who joins the company is certain to get his reward for faithful service in after years.

Members of the Cliffe and Grant families were especially favoured with advancement. Hughie Cliffe and his nephews Joe and Harold all became foremen, as did three Grant brothers: Arthur "Butch," Bob "Scratchy," and Harold. Other capable local men promoted to foreman included Bill Chalmers, George "Pinky" Davis, Johnny Duncan, Oswald Harmston, Norman "Mary" Janes, and Jim "Slats" Robertson. Slightly younger men who would become foremen at Ladysmith included Bruce King, Mauno Pelto, and Bill Wilson. These foremen were all picked out of the crews. Bob Harding said that was "part of the strength" of the company. ■

pushing a pen as a timekeeper and accountant for Comox Logging. Such people emigrated to better their lives, and they often bettered themselves within the confines, more or less, of their chosen professions. This is not to deny the power of corporate capital but to affirm it. Comox Logging cast a wide net, remade itself in new settings, recreated its internal personnel requirements at new camps, and brought a good many along for the ride.

The company's employment of local men of any background was, in fact, a mutually beneficial arrangement that profited both company and employees—and worked to the detriment of outside workers and the trade union movement. Local farm boys were politically innocent compared with the hiring-hall radicals of the turbulent interwar years. Local men had an existing tradition of woods work, both on their own farms and in formal employment by logging companies. In a sense, valley residents put the company to work for them and tamed it, domesticated it, localized it. Woods work provided financial stability to individuals, families, and communities. Local work allowed men to stay in the valley, retain their farms or buy new land, and bring up their children locally. The logging company provided seasonal but relatively steady work, higher wages than farming, and, crucially, a reliable income in a pre-welfare era. It provided employment for teenaged sons, who typically started off as whistlepunks, sparkcatchers, or junior members of the grade crew or steel gang.

FRED BLACKBURN PHOTO

Bernice Blackburn (Austin Blackburn's daughter) holding one of her Churchill nephews at the Blackburn farm, Courtenay, 1941. The car is a 1937 Ford.—FRED BLACKBURN

The Great Depression strengthened the company's desire to hire locally. It produced a closed-shop atmosphere in which available work was reserved for brothers, sons, cousins, or friends. Exclusion was central to the Homeguard tradition. By the 1930s local men were hired and increasingly favoured in many key occupations. Comox Logging devised its own local hiring policy and reserved the hiring halls for specialized or emergency work. Men from Comox, Courtenay, and Cumberland formed an interrelated group of 200 to 300 working men who excluded workers from neighbouring communities and Vancouver hiring halls alike. "It was a family company with a Homeguard," as George Yates Jr. explained. "If your dad worked there, you worked there. That's the way it went. That's why it was a family affair."

A closed-shop sense also applied to promotions within the comany. The American foremen of the 1920s—Len Harding, Jack McQuinn, Pete Donnelly, and Charlie Renecker—were replaced in later decades by Homeguards, especially by members of the Cliffe and Grant families. "If you were in good with the family, you were okay," recalled Harry Trotter. When asked who "the family" was, Trotter replied, "The Grants, McQuinns, Downeys, Cliffes." The head of the extended Comox Logging family was, of course, Bob

Edith Yates (1894–1970) and dog Nellie at Millard's Beach between Courtenay and Royston, *c.* 1940. Her husband, George Yates Sr., had been night watchman and hostler at Camp 3. "Mother was born at Rocky Point, Metchosin [near Victoria]," recalled her son George. "She was Edith Emily Vaughan Parker, from an old pioneer family in Metchosin. They all came around the Horn in 1852 on the *Norman Morison* to work for the Hudson's Bay Company. My grandfather Thomas Parker was born in Kent in 1848 and lived until 1932. I have sets of grandparents in St. Mary's [church graveyard], Metchosin, and in Ross Bay Cemetery [Victoria]. Lawrie Wallace is my cousin—his mother was a Parker. He was in charge of the 1958 B.C. Centennial."—George Yates Jr.

Flossie Parkin, one of ten children of Ike (Isaac) and Margaret Parkin, at the family home at Headquarters, *c.* 1932.—Fred Parkin

Fred Parkin (born 1928) at the Courtenay Fall Fair, Lewis Park, 1938. Son of Norman Parkin and grandson of Ike (Isaac) and Margaret Parkin, Fred grew up at Comox Lake camp. He recalled that, as a boy, he "picked potatoes at Urquharts' for two days—for a hat! All bloody day! Old Man Urquhart was bandy-legged and crippled. Then, after, he'd give you a sack of culls—culled potatoes. The Urquharts delivered milk in milk wagons. Mrs. Urquhart was a lovely old lady." After the war Fred worked for Comox Logging. He was a second loader at Tsolum, ran a Marion shovel at Bare Rocks on Comox Lake, and then left the company to work for Mike Andrews on Cortes Island. "Fred Parkin got away from the big outfits and worked for gyppos like Tiny Seymour," recalled Art Beech.—Fred Parkin

Filberg, whose pervasive presence and hands-on style were nicely captured in a story told by Charlie Stroulger (1915–2005), who was working at Comox Lake in 1937:

> One morning I missed the crummy, and I was in the Riverside Hotel having coffee and toast when a man slips into the counter beside me, and he says, "Good morning, Charlie, are you sick or hurt or something?" It was Bob Filberg! I should have figured he'd come in for coffee! "No, I slept in," I said. "Well, you make sure Andy or Sammy [Telosky] wakes you up next time so you can get on that goddamn crummy!" And I could have gone right through the floor. "Make sure you don't miss it again!" Filberg said.

Men connected to the company still talk about the Homeguard phenomenon and its sense of exclusion. Gordon Blackburn affirmed that "Comox Logging was a Homeguard camp. If people from up the coast wanted a job here, they couldn't get one, because everyone here was related." Bob Muir recalled, "It was hard to get on a machine, skidder or cold decker, because no one wanted to give up their job in those days." In the 1940s Roy Maki, while logging at Northwest Bay (near Parksville) for H.R. MacMillan, asked about a bulldozed right-of-way near the MacMillan camp. "That's Comox Logging—that's a Homeguard outfit," he was told.

Occupations dominated by local men tended to have a low turnover in personnel; this was true of yarding, loading, and rail crews; engineers; machine shop workers; and tradesmen. And so it was at Comox Lake. "Sometimes you'd come back after a few months' layoff and find the same guys on the crew," recalled head loader Eddy Berkeley. "You'd go to work and find *the same crew*! This went on for years and years." Less prestigious occupations, such as grading, steel laying, falling, and bucking, which paid piecework or by the job, tended to have a higher turnover and were dominated by newcomers and outside men.

Muriel "Moo" Martin at Comox, *c.* 1935. Moo was one of three daughters of veteran Comox Logging trainman Bob Martin and his wife, Mildred (*née* Pritchard); Moo's brother was Neil Martin. She married a sailor named Johnny Paone. —Muriel Holmes

Connie Martin, of Comox, picking berries in a logged-over setting, *c.* 1935. A daughter of Bob and Mildred Martin, Connie would marry Herb Bratt, a Comox Logging and Elk River Timber chokerman and later a logging surveyor.—Muriel Holmes

One result of the local hiring policy was a fine safety record. In a pre-union era, safety measures were more likely to be a priority when brothers, cousins, and brothers-in-law made up the 21-man skidder crews than when crews had been haphazardly recruited from Vancouver hiring halls. In this family company, the guy at the far end of the choker line might be an old school friend or your "42nd cousin." By the late 1920s Comox Logging had the best safety record on the West Coast at a time of high mortality in the logging industry.

Its record had not always been so stellar. Deaths were far too frequent in the company's early years—seven in 1918 alone. Men came to dread the six or seven long whistle blasts that meant an accident in the woods. "We heard the skidder whistles, seven long

A coronation float for King George VI, Courtenay, May 12, 1937. The figure at left is Britannia, with her helmet, trident, and Union Jack shield. She personifies Great Britain and the ties of the British Empire. In the middle is John Bull, personifying England, with top hat, bulldog, and maple leaf shirt. The two figures at right, with their maple leaf dress, headband, and tie, personify Canada. The boy at right with the tie is Jimmy Dixon. The interlaced roses and cedar boughs represent the union of England and British Columbia.—Fred Parkin

Bob Muir (born 1931) in the front garden of his grandfather's house at Camp 3, *c.* 1938. He lived there for a time with his grandfather Fred Manson ("Old Fred") and his uncle Fred Manson ("Young Fred"). "The Manson house was the last one down the track," said Muir. "It burned down one Christmas. There was an electrical problem. They got up and grabbed me and threw me out in the snow in my underwear! There was about three feet of snow on the ground. The only other thing Old Fred had time to throw out the window was his alarm clock. Young Fred ran back in for his new shoes—but he grabbed one old one and one new one! They lost everything." Bob's parents were Ronnie Muir, long-time hooker on Skidder 5, and Lizzie Manson.—Bob Muir

whistles, and the women would get panicky. The ambulance would come into Camp 3—it was a 1934 or '35 Ford—and people would wonder," said Bob Muir. In the absence of roads, the main-line locie was sometimes used as an ambulance or even a hearse to take injured or dead loggers from the woods. Brakeman Bob Martin used to tell his wife, Mildred (*née* Pritchard), "horror stories of accidents," according to his son, Neil. "He used to talk about people getting injured in the woods and taking them out on the train."

But Bob Muir pointed out that "there were not too many deaths for the number of men. There were big crews." Comox Logging set an astonishing record of no fatalities in 1927, and in 1929 through 1934. "Safety work is also a hobby with Mr. Filberg," commented *The Timberman* in September 1935. Under Filberg, safety committees were struck at every camp, and safety instructors Harry Steere-Clark, Percy Le Mare, and Jack Taylor were hired. Comox Logging embraced the Safety First movement and broadcast a safety message that said, essentially, "Chances are it will be your own fault if you get injured or killed." An emphasis on Comox Logging as a tight-knit "family company" runs through the historical record. In July 1948, journalist Hugh Weatherby interviewed Jack Taylor on his nine years of safety work with Comox Logging: "Mr. Taylor is a firm believer in the fact that many accidents can be attributed to psychological causes, and that employees who are satisfied in their jobs and happy in their home lives do not get hurt as often as others." By that time Taylor had trained about 1,200 Comox Logging employees and their families.

In the 1930s the company started corporate, or "paternalistic," benefits programs. "The hospital fee at Comox is $3 per month for married men and $2 for single men," *The Timberman* noted in September 1935. These fees were deducted from employees' wages. "For this sum, they get complete medical attention for themselves and their families. Any accident or sickness contracted as long as the man is on the payroll, regardless of time or place, is provided for in this arrangement. The only exceptions to this complete coverage are for dentists' and oculists' fees."

Some Comox Logging men had worked for companies to the north, usually Elk River Timber at Campbell River and Merrill & Ring at Rock Bay, and told horror stories of their dismal safety records, reliance on hiring halls, and transient workers. One of these men, Bob Brown (who like his brother and father was called Buster), recalled, "My brother, Buster [Hart], said that at Rock Bay in the 1930s they'd run the ass off you for 30 days [straight]. The joke was they'd have a crew working, a crew going, and a crew coming. After a month they'd just fire you." Beatty Davis, who worked for Campbell River Timber at Gooseneck Lake in the 1930s, asserted

Unidentified Comox Logging employees somewhere on Forbidden Plateau, *c.* 1938. Company employees were involved in opening Forbidden Plateau as a recreational area. The old Mt. Becher trail started at the top of the Bevan sidehill, and a network of trails linked the Cruickshank River with Strathcona Provincial Park.—Jean Berkeley

Deer hunting near the Cariboo (Woodhus) Creek line, Oyster River area, *c.* 1939. Logging railways gave access to timber resources and to all sorts of wildlife. "Deer and a garden—that's what you needed! No one would have got by without them," said Eva Bailey (*née* Manson). Sam Telosky's caption for this snapshot was "We all got our deer. Bringing home the winter meat." Sam Telosky is at rear, and the other hunters (not distinguishable) are Bill Eastman, Bob McQuillan, and Dunc Thompson.—Sam Telosky

"Looking down into Buttle Lake" was the original caption for this *c.* 1938 photo. Jim Mathers Jr. and Hi Churchill walked all the way west from Cruickshanks Camp to Buttle Lake (in Strathcona Provincial Park, some ten miles west of Croteau Lake on Forbidden Plateau). "Buttle Lake drained half that country up there," noted Sam Telosky.—Laurie Mathers

JIM MATHERS JR. PHOTO

We have to stop meeting like this: Penti Saari, son of Uno (who ran the Royston cookhouse for Comox Logging), with a six-foot salmon shark at the Royston boom camp, *c.* 1943. "They'd catch them in a little putt-putt boat and bring them in," recalled Laurie Mathers. "They'd catch them in 200 feet of water off Hornby Island." Six-gill sharks came from the same area. Later, these marvellous and harmless animals were sold for prawn bait.—Laurie Mathers

Spectators and a six-foot salmon shark at the Royston boom camp, *c.* 1943. The girl is Frances "Beebee" Stewart, daughter of Comox Logging's boom foreman Ab Stewart. "I remember the same shark," recalled Al Knappett. "When I saw it, it was on the beach by the old government wharf at Royston."—Laurie Mathers

that "the toilet was a hole in the ground with a two-by-four nailed across it."

The most common story concerns the man who was killed in a northern camp and was left behind a stump until the end of the working day, as Frank Davison mentioned:

> Elk River and Bloedel only hired from the hiring halls, but they got rubby-dubs—they were alcoholics, you know—and the foreman would fire them. There was always one bus coming and one bus going. Comox [Logging employees] pretty well came from local farms, but even if they were from Courtenay they were *country* people. There were no city slickers in Comox Logging. The hiring-hall guys were here today and gone tomorrow. It was all hard work—pretty highball. They never preached safety, it was just "go." I've heard stories that if somebody got killed, they put 'em behind a stump.

Such stories, told both by Comox Logging men and by those men who worked for the culpable companies, are remarkably consistent and reflect basic differences between the hiring-hall and Homeguard traditions. For example, Bill Woods (born 1927), who went to work for Elk River Timber in the late 1940s, told this tale:

> Comox Logging used to shut down in the winter, and my friends and I used to go and work for Elk River Timber at Campbell River. The boss there, Johnny Harembourne, used to call down in the winter for us when Comox was shut down. I loaded at Little Echo Lake for Elk River with old Gilman, the loading leverman. Boy, you needed wings on your feet to work in that bloody lake! They hurt quite a few guys. Gilman had about 21 [injured or dead] guys to his name! Elk River were not getting the really good guys who'd been in the woods for years and years. If you'd been in jail in Vancouver, they'd buy you a ticket when you got out and send you to Elk River. But the men couldn't stand the cold and the snow, and a bus would come one night and leave again the next morning. You could buy a used pair of caulk boots for $5.

THE LOGGING UNION found it difficult to get a foothold in a group of employees who were so aware of their employer's strengths and benefits and of the surrounding companies' weaknesses. "I can remember in the mid-twenties when I started in the woods, if you mentioned union in a camp, the men—not the operators—would run you out of camp. This condition went on for about twenty years," wrote Harper Baikie in *A Boy and His Axe*. Similarly, Bill "Bronco" Moncrief (born 1927), who came from a Cumberland mining family, recalled that "the old Homeguards wouldn't bother much [with the union] because they already had seniority and

Three members of the Alec Walker rescue party on a glacier above Cruickshank Canyon, 1939.—SAM TELOSKY

Back at the cookhouse at Cruickshanks Camp after the successful rescue of Alec Walker, 1939. Left to right: Alec Walker, Sam Telosky, Dave Aitken. "We're drinking some of Filberg's beer. I've nothing but good to say about Filberg!" declared Sam Telosky. "He had a case of beer ready after a week on the trail. It tasted good! Walker was nothing but skin and bones when I got him out." By all accounts Dave Aitken was notoriously severe. One of his favourite expressions, according to his nephew Ken Aitken, was "Hard work never killed anybody." "And he meant it!" Ken added.—SAM TELOSKY

In 1939 a prospector named Alec Walker got stuck on a mountain ridge above Cruickshank Canyon. Walker's prospecting partner, Dave Aitken, a donkey engineer with Comox Logging, arranged a rescue party made up mainly of men from Comox Logging's Cruickshanks Camp. Sam Telosky is at left, and Dave Aitken is in the middle wearing the Cowichan sweater. The other rescuers (not clearly distinguishable) are George Aitkenhead, Wallace Baikie, Dick Harding, Const. Jim Osman, George Surgenor, and Dunc Thompson.—Sam Telosky

Aitken and Goodwin

Dave Aitken (1886–1973) of the Cruickshanks show counted Ginger Goodwin as one of his friends. Born in England in 1887, Goodwin came to Canada in 1906 and worked as a miner in Nova Scotia and British Columbia. During the First World War he was a union organizer, Socialist Party candidate, and draft evader, and in July 1918 he was shot to death at his remote cabin on the Cruickshank River by a member of the B.C. Provincial Police—an incident that prompted the first general strike in Canadian history and turned Goodwin into a labour martyr. In 1939 Aitken took Sam Telosky over some of Goodwin's old territory, as Telosky later reminisced:

I knew Dave Aitken. He was a tough old bugger—real tough! He would beat a guy up if he didn't set the chokers right! He was a cold-deck rigger at Comox Lake, and he also prospected in the mountains. In 1939 his partner, Alec Walker, got trapped in a canyon in the Cruickshank River, and I was on the rescue party with Dave. Dave was from Cumberland and he was a legend at Comox Lake. He knew that country really well, but they got trapped up there. Alec Walker wouldn't be alive without him.

I spent many nights with Dave around the fire talking about Ginger Goodwin. Goodwin was shot in the back by the Provincial Police at his cabin on the West Fork of the Cruickshanks. This happened during the First World War. Dave was Ginger Goodwin's right-hand man! He packed grub in to Goodwin, and he was at the inquests.

When we were going out, we stopped at the cabin that the police burned down for evidence. Dave had helped Ginger Goodwin build that cabin. He said the cops used to follow him up to Comox Lake when he was packing food in to Goodwin. We had a cop named Jimmy Osman with us on the rescue party, and I was friendly with him. Dave said to Jimmy, "Right here, Jim, is where one of your goddamn fellow officers shot Ginger Goodwin in the back! Right here was where that cabin was, and right here your buddies shot him in the back!"

It was kind of embarrassing. ■

benefits." And Ray Stockand (born 1921), also from a Cumberland mining family, said, "A lot of Comox Logging employees were old farmers; they were Homeguards. They were farmers—they weren't the tough old working people, so they didn't need unions! The Homeguards wouldn't get the unions in. And the management at Comox Logging didn't recognize union men, though some employees were union men, a handful were. It was a real struggle."

Early in 1934 the LWIU (Lumber Workers' Industrial Union) called a strike to protest unsafe living and working conditions, and to obtain a wage increase, a new minimum wage, a recognition of camp union committees, and an end to Sunday work. Although all the other large camps on the Island voted to strike—Merrill & Ring at Rock Bay, Campbell River Timber, and Victoria Lumber & Manufacturing at Chemainus—Comox Logging voted in favour of work. The strike votes of March 19, 1934, revealed 210 men from the company in favour of work and 132 in favour of striking. One of the latter was 20-year-old Jack Berkeley, who later recalled, "I remember in '34, that's when I signed my first union card, the IWA [LWIU], '34, and that was undercover. If you were found with a union button—they used to wear them on the cap—or if you were found reading *The Lumber Worker*, that was worth your job: you were gone! That's the way it went."

Filberg argued that many of the men who had voted "Yes" thought they had voted to continue working, and he alleged that more than 100 union representatives from Campbell River Timber had acted as "agitators" at Camp 3. He also contended that the large number of fallers at Camp 3 had shifted the vote in favour of the strike. Filberg may have had good reason to believe that the Finnish and Swedish fallers, obtained largely at the hiring halls, were more radical than the local men. "The fallers were the instigators of the unions," recalled Harry Erickson, whose father, Fred (Sigfrid), was involved. "I can recall my dad having secret meetings at their houses. If it was done in the open, you'd be fired—you'd be down the road. This was in the 1930s."

Another strike vote in May 1936 resulted in 208 Comox Logging men voting to work and 45 to strike. A year later, the provincial Comox riding returned a CCF (Co-operative Commonwealth Federation) member, Colin Cameron. Clearly the voters of the Comox Valley were more progressive than the employees of Comox Logging, who were landed and conservative in comparison. Comox Logging became the last large company on the British Columbia coast to be organized by the IWA (International Woodworkers of America), a union formed in 1937 from the remnants of the LWIU. The Courtenay local of the IWA was formed in 1942. ■

> The Three-Spot, the main-line locie, at Camp 3 preparing to pull a load to Royston, c. 1930. Comox Logging's main line went right through the camp, and trains were a familiar sight to residents. "It's our yard in front, and to the right is Wilma McKenzie's yard," explained George Yates Jr.
—George Yates Jr.

> Jack Carthew, engineer of the Eleven-Spot locie, poses with seagulls on the trestle at Royston, 1938. The photo was taken from the oil tank car. At left is the Royston bunkhouse. The son of James Carthew, a carpenter and builder from Nova Scotia, Jack Carthew was born in about 1890, grew up on the family farm in Comox, and joined Comox Logging in about 1911. At first he ran the humdirgen at Royston, but he spent most of his career as engineer of the Three-Spot; in 1953 he hauled Comox Logging's last trainload of logs to Royston. In the off-seasons he returned to his sheep farm near St. Joseph's Hospital, Comox. He ended his career repairing flatcars in the car shop at Headquarters, and died in 1972.
—Trudy Mooney

Where are the gods?
Must we hunch through the books
upon these dark stained shelves
to find their names?
Can we not learn the answer
from the land
that feeds us, breeds us, buries us?

—FROM "THE GUARDIANS," BY ROBIN SKELTON (*LANDMARKS*, 1979)

9 LOGGING LIVES

< **Waiting at the Comox wharf, *c.* 1938. The Esquimalt & Nanaimo Railway had extended its line to Courtenay in 1914, but for many years Comox retained its wharf and its connections to the steamship services of coastal B.C. In 1929 Comox remained a village of 200 surrounded by harbour, fields, and forest. By 1941 the population had risen to 400. Many local farming families—Carthew, Cliffe, Davis, Grant, Hawkins, McKenzie, Pritchard, and others—had made their mark in Comox Logging.**—BOB BERKELEY

THE COMOX VALLEY WAS HOME TO MOST OF THE MEN AND families supported by Comox Lake logging shows. These people had survived the Depression through a variety of strategies: by capitalizing on their community strengths, relying on their Homeguard traditions of farming and subsistence, and taking advantage of vacant company and Crown land to hunt and forage. For its part, Comox Logging strengthened its safety record, promoted local hirings, limited the number of hiring-hall employees, and excluded the itinerant, politicized, or radical loggers of coastal lore. By 1937 the worst effects of the Depression were over and many Comox Logging families moved to Ladysmith, where the company opened a show in 1936. Ladysmith was also the base for the company's Nanaimo Lakes division, where work started in 1941.

The social character of the Comox Valley resulted from a combination of the valley's rich natural inheritance and the company's interwar strategies. The area had many distinctive characteristics: a large, mountainous hinterland, physical isolation, deep rural patterns of farming and woods work, a reliance on working the land (farming, logging, and foraging) to yield an income, an ongoing engagement with the stump-to-dump production and transport of logs, relative economic security, a clannishness and interrelatedness, a certain rural independence and toughness, a keen sense of community, and a specialized and arcane woods vocabulary.

Many of these qualities were related to the economic engine that was Comox Logging, with its virtual monopoly on the old-growth riches of the valley. By the 1940s the company had removed the most accessible lowland Douglas fir forest from the flats north of Courtenay, from the Bevan sidehill, from the shores of Comox Lake, and from the Toma Creek, Puntledge River, and Cruickshank River valleys. All future logging would have to be at a higher elevation and at ever more remote edges of the world economy, and would resort to using roads, trucks, and their associated and more mobile technologies.

In short, Comox Logging had removed the most valuable asset of this "Land of Plenty" by the 1940s. Indeed, the company's six Lidgerwood skidders, purchased in 1911, between them yarded and loaded some 4 billion or 5 billion board feet of timber between then and 1946. The community's character was based, first, on the existence of an abundant first-growth forest and, second, on the systematic obliteration of that resource.

Although men and families undoubtedly benefited, the price paid was the wholesale removal of the valley's most significant natural resource by an effective combination of yarders, skidders, and logging railways. The skidders' demise coincided with the introduction of power saws, logging trucks, roads, and steel spars in the late 1940s and early 1950s. There would be enough high-elevation timber accessible with these technologies for another generation of logging, but by the 1970s the local old-growth timber ran out completely, even at the higher elevations, and Comox Logging's successors required progressively smaller crews working progressively smaller areas. The logging era might have lasted longer, but the company ran into the edges of the 259,000-acre Strathcona Provincial Park, which imposed an effective western boundary to local logging.

ALONG WITH THE REST of the world, the Comox Valley emerged from the Depression only to enter the crucible of the Second World War. The war interrupted the woods careers of many young men and ended the lives of others. "Logging and farming were essential work, but a lot of men joined up from those occupations. Logging was considered an essential business," said Barbara Marriott. Many former employees recalled that they were simply "frozen in the woods" during the war. "I couldn't join up," recalled Chuna Tobacco, faller at Comox Lake. "We were tied to that job with the Comox Logging Company then." Others related, somewhat uneasily, that Bob Filberg kept men out of the armed forces by getting them exemptions or deferments. Filberg wielded real power on the British Columbia coast: he was superintendent of Comox Logging's large

R.H. MARLOW PHOTO

Bob (Robert) Filberg in his prime, 1939, aged 51. "Of medium height, he stands very straight and walks as if he had fire bark under his feet," the *Comox Argus* quoted the *Vancouver Sun* in November 1935. "His hair is creeping back from his forehead a little, but, to apply his own statement concerning British Columbia's forest future, there is no danger of ultimate depletion. He smokes cigarettes, and a lot of them." Of Filberg, Joe Cliffe wrote in "The Filberg I Knew," "He was pleasant looking, clever and had a quick brain; all the parts a manager should have besides being able to pack a big load of worries." Filberg was originally a logging surveyor from Seattle; his career took off after he married Flossie (Florence) McCormack in 1916. Her father, James McCormack, was then vice-president of the Canadian Western Lumber Company.—FILBERG HERITAGE LODGE & PARK ASSOCIATION

R.H. MARLOW PHOTO

Mary Filberg (1922–67) and her father, Bob, at her wedding, July 1944. "Mary Filberg looked like Judy Garland. She adored her dad," said Lorna Cliffe. Mary married Judd Whittall, a Vancouver financier. The original inscription on the photo was "Mary and Hughie [Cliffe] —My love to you both—Mary W."—LORNA CLIFFE COLLECTION, COURTENAY & DISTRICT MUSEUM & ARCHIVES, 2000.158.22

Bob and Flossie Filberg's house on the Comox waterfront, *c.* 1940 (view to the west). Designed in the Arts and Crafts style by Filberg and carpenter Bill Haggarty, it was built between 1929 and 1935. "My father didn't have much formal education," said Haggarty's daughter, Sheila Kennedy, "and there were no specifications for Filberg House. Dad and Bob Filberg wrote the plans on cigarette boxes and serviettes in the café of the Riverside Hotel." Among the builders, cabinetmakers, and blacksmiths who contributed to the house were Malcolm Burnett, Reg Carwithen, Lu (Lucius) Cliffe, Haldor Eirickson, Colton Haggarty, Hughie McKenzie, Willie (Vilhelm) Meier, and Morgan Shepherd. To clear the lot, builders had to remove a large portion of the great midden that once stretched along the Comox waterfront. The house is now the headquarters of the Filberg Heritage Lodge and Park Association, known for its annual Filberg Festival.—FILBERG HERITAGE LODGE & PARK ASSOCIATION

BOB FILBERG PHOTO

Barbara Crockett at Filberg House in 1943. Born in 1911, Crockett was house-keeper at Filberg House from 1940 until her death in 1962. She worked as the Filbergs' maid and also as the cook, with a specialty in pies and pastries. She never married but dated Eddy Berkeley for some time. Crockett grew up a farm girl at Grantham (north of Courtenay), the daughter of dairy farmers Jack and Winnifred Crockett. One day, Mrs. Filberg dropped by the Crockett family farm for some cream, as Bob Cliffe recalled: "At the farm they would skim the cream off the fresh pails of milk and bottle it right there. Once, Floss Filberg happened to drop by to pick up the cream, and one of the boys took her out to skim the cream. He found that a mouse had fallen into a bucket and had drowned in the thick cream, so he simply picked the mouse up by the tail, squeezed the cream back into the bucket from the mouse with his fingers, and gave Mrs. Filberg her cream. Mrs. Filberg never used their cream again!"—FILBERG HERITAGE LODGE & PARK ASSOCIATION

operations in the Comox Valley, around Ladysmith, and at Nanaimo Lakes, and he was boss of Aero Timber Products in the Queen Charlotte Islands. (For his work with Aero Timber, Filberg would be awarded the Order of the British Empire at war's end, for supplying spruce to build vital lightweight aircraft such as the Mosquito bomber.)

His power allowed him to play God. Ruth Masters recalled that he said to her brother, Bill Masters, a chokerman, "'If you don't want to go to the war, Bill, I can fix it up for you.' But Bill went anyway!" Howard McQuinn, discharged at Victoria at the end of the war, told the officer in charge that he was going back to work for Comox Logging. The officer told him that "the Comox Logging Company drew the highest exemption rate of any company in Canada. Filberg's war work was so important that he'd just have to say so-and-so's job was essential and [that person] would be exempted from military service."

Many young men quit the company to escape Filberg's long reach and enlist. Some tried to join up, only to be told that they were needed at home. At least six men from Comox Lake joined the army in Vancouver in 1941 but were pulled out of the parade and told to report to the Queen Charlotte Islands to work for Aero Timber Products. One of them was Eddy Lefley, whose story was told by Betty Annand in *Voices from Bevan*: "He was in his underwear standing in the inspection lineup when his name was called over the loudspeaker. An officer informed him that Mr. Filberg, the head of Comox Logging, had sent word that he was needed at home more than he was in the forces." Former rigger Sam Telosky told a similar story:

> There were five of us who decided to join. The others were Dan McKenzie, Andy Barr, Chan Tame, and Len Lester. This was my whole crew from Comox Lake! The five of us went down the lake and went down to Vancouver and joined the army on Monday morning, and we got all our shots. We were at the Vancouver Hotel, getting ready to go to war. We were there for a week. One morning we were on parade at 6 a.m. and they said, "Telosky! Tame! Barr! McKenzie! Lester! Report to the captain!" And that night we were on the boat to the Charlottes. Bob Filberg was behind this. We were just about ready to go to Quebec for basic training.

Andy Barr and Chan Tame defied Filberg by going back into the army anyway, and Tame was subsequently captured, as Sam Telosky recalled: "Chan was a POW for three years. We figured he was dead, but he was in a POW camp in Italy."

> Cliffe Avenue, Courtenay, c. 1946. Surrounded by farms on three sides, Courtenay was the largest urban centre in the valley, with three or four busy downtown streets, a cinema, and the Bickle Theatre, shown here at right. The town's population rose from 1,077 in 1929 to 4,000 in 1948. Serviced by the Island Highway, the Esquimalt & Nanaimo Railway, and the Vancouver Barge Transportation Company, Courtenay was a service and transport centre for the many logging camps, farms, and stump ranches of the Comox Valley. It was known for its creamery, milk condensory, jam factory, and Brackman & Kerr feed store. The Berkeley, Blackburn, and Bullock families owned farms on the southern edge of town.—COURTENAY & DISTRICT MUSEUM & ARCHIVES, 2000.136.12

> Grades 9 to 12 at Tsolum High School, 1941/42, dominated by children of Homeguard farming families and of logging employees from Headquarters. Most of the boys are sons of Comox Logging foremen and tradesmen. There are ten more girls than boys in this photo, a reflection of the attraction of woods work; most of the older boys in this photo have already worked a summer or two with Comox Logging. Front row (this and other rows identified left to right): Fred Harndon, Saburo Takata, Tony Schweeres, Cliff Rennie, Don McIver, Norman Sprout, Dennis Ware. Second row: Agnes Mitchell, Isabelle Godwin, Margaret Beaton, Doris Overton, Leona Crowell, Eleanor Bridges, Betty Porritt, Evelyn Harris, Janice Smith, Ella Hugo, Doreen Grieve, Audean Shopland. Third row: Miss Piggott (teacher), Alice Crockett, June Flinton, Anne Carwithen, Leila Grieve, Kay Blakely, Katherine McKenzie, Shirley Patton, Mildred Larkin, Eileen Willis, Janet Isenor, Lorraine Rees, Isabel Flinton, Lucy Carwithen, Jean Marsden, Lois Janes, C.T. Rendle (principal). Top row: Mac Sinclair, Jackie Tukham, John Grant, Jimmy Gower, Archie Milligan, Hugh Grant, John Holmes, Elmer Helm, Noel Sinclair, Dick Machin. Identified by Lucy Grant (*née* Carwithen) and John Holmes.—DOROTHY TUKHAM

Cliffe Avenue, Courtenay B.C.
Silvertone 109.
AUTO SUPPLY
OLDS

TSOLUM HIGH SCHOOL
1941-42.

Baseball and Bartending

Comox Logging's baseball team, the Circle F, was promoted by company superintendent Bob Filberg and fellow American Richard "Pop" Dixon, proprietor of the Riverside Hotel in Courtenay. Sam Telosky recalled his years as a player on the team:

We were six brothers in Haney. I was born in March 1915. We lived on a farm, and when I was a boy I learned to call in wild ducks. We had been baseball players in the Fraser Valley and got to be pretty good—in fact there's a Telosky Stadium in Haney named after one of my brothers. My friend Riley McLaren worked on the boom for Abernethy & Lougheed at Haney with me, and then he got a contract with Baxter Pole Company at Royston and he said, "Come with me," so I came to Courtenay in about 1935.

They always had a good baseball team here [Courtenay]. *There was a league on the Island with teams in Courtenay, Cumberland, Port Alberni, Nanaimo, Chemainus, and Victoria. We played on a semi-pro level. On July 1 they always had a tournament. It was as good baseball as you'd see. Pop Dixon owned the Riverside Hotel and he had a pretty good ball team called the Circle F, but he wanted it better. Dixon was very sports minded. He ran locie for Comox Logging for years, and when he retired he bought the hotel. Filberg sponsored the team, which had a Circle F on front of its shirts: everything was stamped with a Circle F. They're still using that Circle F log hammer to this day!*

I worked on the boom at Royston at first, but Dixon heard of us, and my brother Pete and I went to work at Comox Logging. Dixon had two or three good ballplayers working in his bar. One guy, Wilkie, went to the big leagues, to the Pittsburgh Pirates. We played touring coloured teams from the U.S., and Japanese teams. I don't think you'd see a better local ball team anywhere than the one in Courtenay.

Pop Dixon tried to keep up drinking with the Swedes. He was a pretty good drinking man. He'd get a skin full [get drunk], *and he wasn't a big man but he'd twist wrists with the big Swede fallers! Dixon was quite a man when he'd get a few drinks in him. I worked for six months in the Riverside. Two or three friends of mine, ballplayers, also worked in the bar. Beer was ten cents a glass! I had to wear a white shirt. Three loads a day went past the hotel. One of them was at 6 or 7 p.m., after everyone had finished work, and everyone would line up and watch it go through: 50 or 60 flatcars mostly from Camp 3. All the loggers in the bar would go out and have a look at that train going past!*

Later Dixon got sick with a stroke, and baseball just went dead in Courtenay. ■

> The Riverside Hotel, Courtenay, *c.* 1945. By this time the venerable Riverside, built in 1889–90, had been modernized with a café, office, and ladies' entrance (below). Situated on Fifth Street just above the main-line railway tracks of the Comox Logging Company, the Riverside remained the favourite loggers' pub until it burned down in 1969. The site is now occupied by the plaza in front of the Sid Williams Theatre. Comox Logging employees, active and retired, liked to sit and reminisce on the low, curved cement wall at bottom left. "There was a half wall in front of the Riverside Hotel," recalled John Hatfield, "and the loggers would sit there and talk. There was no Friday night because Saturday was a work day. There would be a Jehovah's Witness preaching, a drunk on the street, and everybody knew everybody else. You couldn't imagine the valley then. The valley was *clean*."—Stan Hanham

Sixth Street, Courtenay, *c.* 1947, with the view southwest towards Cumberland. Comox Glacier is at left and Mt. Becher is at right. At near right is St. George's United Church. Snow like this kept the woods at Comox Lake closed for the winter months, but it was good for the patrons of the Forbidden Plateau ski lodge.—Fred Blackburn

Champions of the ladies' bucking contest, Courtenay, *c.* 1936. Left to right: Esther Forsman, Anna Dahlbach, Ingrid Hagg. These women, all Swede-Finns, were married to Comox Logging fallers who had settled in the Comox Valley in the 1920s, part of the substantial migration from Sweden and Finland to coastal B.C. in that decade.—Dorothy Tukham

MYRTLE HERON PHOTO

World champion bucker Esther Forsman (1906–80) at her farm near Courtenay, *c.* 1940, with her seven-foot bucking saw, the razor-sharp teeth of which are protected by a saw guard. She had just returned from a competition in Tacoma, Washington. Esther was married to Uno Forsman, a Finnish faller who came to Canada in 1926. "She was also a Swede-Finn," wrote her nephew Michael Forsman in a 2003 letter, "and twenty-two years old when she landed in Halifax in 1928. I think she had known Uno in Finland and came over because of him." She was also champion of ladies' nail driving at Comox Logging's annual loggers' sports.—Michael Forsman

< Madge Sheppard (later Appleby) at Forbidden Plateau Lodge, June 1945. Madge was the daughter of Comox Logging leverman Harry Sheppard and Rosalie Williams of Williams Beach. "Madge and I were born a day or two apart in June 1924, and this was our 21st-birthday skiing trip to Forbidden Plateau," recalled Myrtle Heron (*née* McKenzie). —Myrtle Heron

> **Lu (Lucius) and Alice Cliffe, 1938. Known as "the first white child" born in the Comox Valley, Lu Cliffe (1873–1947) was a hunter, farmer, carpenter, barn and house builder, furniture maker, violin maker, photographer, watch repairman, and Mason. His farm and his many practical skills allowed him to avoid permanent work with Comox Logging. "Lu never worked for *anybody*, period," noted his grandson Bob Cliffe. Alice Marshall (1873–1952) came from England to the Comox Valley as a school-teacher and married Lu Cliffe in 1896. For many years they lived near Courtenay at Maplehurst Farm, which Lu had inherited from his uncle Willie Harmston. Two of Lu and Alice's sons, Joe and Harold Cliffe, became well-known foremen with Comox Logging, and Lu's brother Hughie was also a foreman.**—FILBERG HERITAGE LODGE & PARK ASSOCIATION

BOB FILBERG PHOTO

CHARLES SILLENCE PHOTO

Jack and Goldie McQuinn, *c.* 1945. Originally from Oregon, the McQuinns settled in the Comox Valley in 1919 and brought up their 11 children on a farm on Little River Road, Comox. Jack McQuinn retired in 1956, having run both Camp 3 and Comox Logging's show at Ladysmith. He died in 1963. Goldie lived to be 100; she died in 1993. Their daughters trained as nurses, and their sons went into logging.
—DICK MCQUINN

Roy Cliffe (1903–92) at Maplehurst Farm, Mission Hill, Courtenay, 1943. Earlier a champion light heavyweight boxer, Cliffe worked as a loader with Comox Logging before taking over the family farm from his parents, Lu and Alice, in the 1930s. He also worked as a campaign manager for several generations of members of Parliament. He sold Maplehurst Farm in the 1960s, and it is now Mulligans Golf Centre.—FILBERG HERITAGE LODGE & PARK ASSOCIATION

CHARLES SILLENCE PHOTO

First aid team of the Comox Logging & Railway Company, 1942. Front row, left to right: John Vaton, Fred Tukham, Percy Le Mare (captain), Jack McQuinn (captain), Carl Sundin. Back row, left to right: Eric Gray, Alec Tilleard, Jack (A.J.) Taylor (instructor), Wilfred "Skinny" Colling, John Holmes, Don McIver. A typical competition test scenario went as follows: "A man is working at a bush fire when he is struck by a falling snag, which is found lying across his chest. On examination it is found that his face is pale with respiration somewhat impaired. Further examination reveals swelling and deformity at the lower third of the left femur and also discoloration at the left ankle. The palm of his hand has a transverse cut about two inches long bleeding profusely. Pack patient to Company Office which is situated half a mile away; you will find the doctor has just arrived. Act as you should" (*Comox Argus*, July 7, 1932). Tukham, Holmes, and McIver were sons of office workers at Headquarters. "We received free training and took part in competitions, the winners of which received excellent prizes," recalled Neil Martin, who was a member of the team in the 1940s. "I still have many of these." The team shown here won the Comox Logging Cup, the B.C. Department of Mines Cup, and the Grand Aggregate Cup at the Pacific Coast Safety First competitions held in Nanaimo in June 1942.—JOHN HOLMES

Jack McQuinn, in gumboots, presenting a book prize to an unidentified young woman, *c.* 1945. At far right is Comox Logging's safety instructor, Jack Taylor, who was "*always* called A.J.," according to Harry Erickson. This photo was taken in Courtenay in Comox Logging's little first aid office on First Street, near where the main-line trestle crossed the Puntledge River.—DICK MCQUINN

Fred Blackburn (born 1931), *c.* 1943. "A.R.P." stands for Air Raid Precautions, a civil defence force that originated in Britain in the late 1930s and was extended to B.C. following the Japanese attack on Pearl Harbor in December 1941. Eventually the ARP enlisted the support of over 200,000 volunteers in Canada, including Fred and his father, Austin Blackburn.—Fred Blackburn

CHARLES SILLENCE PHOTO

Bob Muir in his Royal Canadian Sea Cadets uniform, late 1944 or early 1945. Muir started working for Comox Logging in 1946 as a whistlepunk at Comox Lake, and by the 1950s he was running a yarder at Tsolum. He moved to Nanaimo in 1960 and now lives there in retirement with his wife, Cathy.—Bob Muir

Jack Ware's Early Start

Born in Sussex in 1918, Jack Ware Jr. came to the Merville Soldier Settlement at the age of 18 months with his English parents. As a boy he obtained a 20-mile paper route for Ben Hughes's weekly newspaper, the *Comox Argus*; the route took him, on bicycle, from Merville to subscribers in Grantham, Sandwick, and Dove Creek. "Once I delivered it on foot in the snow," Ware recalled. With his paper route savings, he bought, in 1933, 63 acres of logged-off Block 29 land at Dove Creek, and he kept this land for the rest of his life. While still a boy he farmed and built a barn on his stump ranch, climbed on Forbidden Plateau, took packhorses to Croteau Lake with his friend Clinton Webb, and worked for Edgar Cliffe's Tyee Freight Company in Courtenay:

I was shovelling coal out into the basements of hotels, which had coal-fired furnaces. We delivered coal to the Riverside and Courtenay hotels. You'd get a whole truckload of coal and you'd shovel it in! They usually had a coal bin somewhere. The coal was in small pieces, nuggets, the size of the end of your thumb. Cliffe handled the mail too—he picked it up from the E&N at Courtenay and took it to the post offices in Courtenay and Comox. I was pretty young then.

I quit Tsolum School twice, in 1935 and 1937, the last time for good. When I left school, I got a job with forestry on the Courtenay lookout. I saw the 1938 fire coming with the range finder! I worked on that fire and got to know all those Comox foremen. I stayed with forestry for a year and then worked with the Woods brothers, Bill and Bunt, logging burnt timber on Constitution Hill after the fire.

Then I quit and went to Comox Logging and got a job in 1940, setting chokers on Skidder 4 at the north end of Constitution Hill, on the flat. Jack Downey was rigger and Norman Hammond was hooker. I was working on Skidder 4 when a log fell on me and broke my pelvis. What happened was, when they made the railway grade they always threw the logs all sideways, and we were choking those logs when somebody walked up a log that was lying on another one, and the log fell on me. It crushed my hip and some vertebrae. They said I'd never walk again! I was three months in Comox hospital [St. Joseph's]*: September 21 to December 21, 1940. I was supposed to get compensation for 18 months, but I got tired of doing nothing so I got a job running grader for Buck & Turner on the construction of the railroad between Ladysmith and Nanaimo Lakes. But the war was progressing. I got called up for the army, but they gave me a full discharge. I tried to join the air force, but they wouldn't take me. Finally I joined the navy in 1944.* ■

Art Beech said that he, too, had tried to enlist, but a woods injury kept him out. "I got as far as Vancouver and they kicked me out. They said I couldn't walk, but I'm still walking! [When I was injured,] I was moving a yarder, a back rigging donkey, up a little hill out of Camp 3. I gave the engineer a signal to stop, but he went ahead. I landed on my head on the ground from ten feet up."

Almost everyone got involved in the war effort somehow. Because of the shortage of men, women were hired for the first time at the office at Headquarters. Three of them were Grace Martin, Rosie Muir, and Edith Porritt, who worked for Chris Holmes. Women and children knitted, made jam, and prepared food parcels. A fruit cannery on the Dyke Road, Comox, made jams for Britain, as noted by Isenor and coauthors in *Land of Plenty*:

> Two large-type blackberry varieties were soon growing wild throughout the valley. The Himalaya Giant and the bramble blackberry took over the hedgerows of the area. During the 1939–45 world war, students from many of the schools of the Comox District picked these blackberries for the Jams for Britain project. In the fall of 1945 the ladies of the district produced 3,400 pounds of jam from donated fruit for shipment to Great Britain where jam was a luxury in wartime.

Eleanor Keller (*née* Richter) picked wild blackberries for this project in about 1942. "I got a letter of thanks from one English kid, and so did Donald Grant, Art's son," she said.

Many worked for the RCAF at the Comox air base, which opened in 1942, and boys joined the army cadets. One such boy was Fred Parkin, who picked potatoes on the Urquhart farm for the cadets. Others joined the Air Raid Precautions (ARP); Austin Blackburn was in charge of making blackout screens for all the windows in Courtenay. Most boys and men not of military age joined the local units (Companies 23 and 24) of the Pacific Coast Militia Rangers (PCMR), run by Captain Geoffrey Capes and Chris Holmes, veterans of the First World War and Merville settlers of 1919. Also involved in the PCMR were Eddy Lefley, Bill McKenzie, Norman Parkin, and local writer and naturalist Mack Laing. Among the many local boys were Fred Blackburn, Harry Erickson, John Hatfield, and Fred Parkin. "We used to meet at the first aid shack of A.J. Taylor," recalled Fred Parkin. John Hatfield wrote in 2001:

> I was also a member of the Home Guard from 1942 to 1944. We were a small force of army reservists known as the Pacific Coast Militia Rangers. This battalion was mainly a coastal defence force; there were about 15,000 of us, mostly loggers, farmers, businessmen, semi-retired and retired World War One veterans. Many of the rangers were young school boys—minimum age was 16 years, but some were younger. We were

Eva and Dick Bailey in Vancouver, *c.* 1942. Bailey was then chaser on Skidder 5 out of Camp 3. Davey Janes recalled, "I can remember Dick Bailey from square one when he first came onto the skidder. He was somebody special—he was really good." Bill Wilson commented, "Dick Bailey was a blondy-haired guy with the curliest damned hair! He got to be a superintendent for Tom MacKenzie at Cowichan Lake." Eva Bailey (*née* Manson) was born at Quatsino Sound (northern Vancouver Island), the daughter of German-born fisherman and Camp 3 faller Fred Manson and his Swedish wife. The reverse of this photo is stamped "Foncie's Fotos 550 Granville Street Vancouver B.C."—Bob Muir

Striding into the post-war era are Buck Pedersen (left) and Sid Crisp, Vancouver, 1945. Royal Canadian Air Force veteran Pedersen (born 1921) wrote in a 2007 letter, "We both worked for J.H. Baxter & Co and joined up about the same time, Sid in the army and me in the air force. The picture was taken in Vancouver when we were there for discharge, strangely at the same time four years later ... Some time later [Sid] was killed working at Lake Cowichan." Buck Pedersen died in Sidney, B.C., in June 2007.—Buck Pedersen

A Soft Spot for Hitler

Buck Pedersen (1921–2007) recalled his war years:

I joined the Royal Canadian Air Force when I was 20, in 1941, after two years in the woods. At that time the Comox Argus *announced every week that someone had been killed in the air force. The paper covered a huge area, of course. Four of my pals from Courtenay joined up right away in the air force in the Second World War, and they were all gone in ten minutes! They were Douglas Fairbairn, Henry Ellis, Stuart Wood, and Howard Elsey, whose parents owned Fishermen's Lodge. I knew 27 guys from the Comox Valley who were killed in the Second World War.*

I was in the air force for four years. It was a great education for anybody who didn't get killed or hurt! The Commonwealth Air Training Program gave you more training than you got air time. You learned clouds, maps, winds. Tall guys went on bombers and small guys on fighters, owing to different leg room. About half the guys got commissions. When I got my wings, I flew off the east coast [of Canada] *for three months, and then I was in Coastal Command and flew over the Irish Sea and the North Sea. On the North Sea, if a plane piled up they only gave you 20 minutes to live, even with flying suits, but the guys on the U-boats had the worst chances of survival. They'd stay out in the middle of the Atlantic to be out of range of the planes.*

I've always had a soft spot for Hitler. If it hadn't been for Hitler, I wouldn't have met my wife! She was English, and her name was Brenda Newman. She came out in 1946 as a war bride. She died in 2000. ■

Bob Filberg photo

The wedding of Betty Cobb and Howard McQuinn at Filberg House, Comox, 1942. Howard's father was Comox Logging foreman Jack McQuinn; Betty's father was Roger Cobb, manager of Elk River Timber of Campbell River. Howard McQuinn had started as a chokerman for Comox Logging, ran a Cat at Ladysmith in the late 1930s, and completed two years at the University of British Columbia before joining the Royal Canadian Navy. After the war he rejoined Comox Logging and spent many years as a master mechanic, and like his brothers Johnny and Bob, he ended up as a logging foreman.—Filberg Heritage Lodge & Park Association

trained with the regular army and navy. We were to give assistance to army defence forces who were poorly trained and not familiar with the coastal country. We were to report all Japanese bomb locations and put out incendiary fires that threatened our forests. Fortunately many of the incendiary bombs landed in the prairies and inland coastal U.S.A. Some of our rangers left for war and never returned.

War stories are numerous, and local historian Ruth Masters has compiled biographies of all the Comox Valley men and women who served in the war. The wartime experiences of some Comox Logging employees would fill a large book, running the gamut from Neil Martin's years with Bomber Command, to Buck Pedersen's service with the RCAF, to Jack Ware Jr.'s work in the Royal Canadian Navy (RCN). Ware survived being torpedoed on the North Atlantic only because he was wearing his "winter woollies." Machinist Pat Cliffe (1920–94) made 57 crossings of the Atlantic with the RCN—he was an "engine room artificer on the North Atlantic run," according to his wife, Inez. Jack Downey's son (also Jack) was killed at the Normandy landing. Noel McPhee was shot down in his plane and killed in the last month of the war.

George Yates Jr. of the Royal Canadian Navy, *c.* 1944. Born in Comox in 1922, Yates grew up at Camp 3 and started working for Comox Logging as a teenager. By 1943 he was third loader on Skidder 5 at Comox Lake but joined the RCN that year. He rejoined Comox Logging in 1945 on his release from the navy. After the war he worked as a machine operator for Comox Logging and Crown Zellerbach at Comox Lake, Tsolum, and elsewhere on the coast. He retired in 1984.—George Yates Jr.

"When the war ended, I recall hearing the steam locies blowing their whistles!" said Harry Erickson, who was then a boy of 14 working as a sparkcatcher at Constitution Hill (northwest of Courtenay). "The war was *over*!"

The end of the war coincided with the end of skidder logging, which in turn indicated the end of the accessible lower-elevation old-growth and the end of an era of abundance. Comox Logging transferred its headquarters to Ladysmith in 1948 and moved its major logging operations to Ladysmith and Nanaimo Lakes, which together employed 700 men in 1950. The company also embarked on a pulp venture at Elk Falls, north of Campbell River, in the late 1940s, in a belated realization that the big timber was running out and that the logging slash, which traditionally was burned every fall, had commercial value as pulp and paper. The Elk Falls pulp mill opened in 1952.

All of these changes meant a dispersal of the Comox Valley Homeguards. The valley remained a recruitment area for Comox Logging and its successor, Crown Zellerbach. By the 1960s Homeguards could be found working throughout the British Columbia coast, from Crown Zellerbach's southern Nitinat division to the northern Sandspit division in the Queen Charlotte Islands. These post-war logging shows, and the coastal diaspora of the Homeguards, will be the subject of a future book, *Pacific Timber*.

Pat Cliffe, *c.* 1942. Son of Comox Logging's George Cliffe and his wife, Winnie, Pat (George Cliffe Jr.) was born at St. Joseph's Hospital, Comox, in 1920. He weighed only two pounds at birth and was not expected to survive, but Winnie Cliffe took him back to Camp 3, wrapped him in cotton, placed him in a shoebox on the open oven door of the kitchen wood stove, and fed him with an eyedropper. In 1935, when he was 15, he started as an apprentice to Dave Stafford in the Headquarters machine shop. He went on to work for 50 years as a machinist with Comox Logging and its successors, and died in 1994.—Inez Cliffe

Stoker Ware's Winter Woollies

Jack Ware Jr. (1918–2003) joined the Royal Canadian Navy early in 1944, after working for Comox Logging as a chokerman north of Courtenay and as a machine operator at Ladysmith. He was working as a stoker in the engine room of HMCS *Esquimalt* on April 16, 1945, when the ship, hours out of Halifax, was hit by a torpedo fired by the German U-boat U-190:

I got up at 5:55 a.m. to check the instruments, and I was in the engine room at 6:20 when everything blew up! I reached out and grabbed the ladder and made it up a hatchway and slid right off into the water. The ship was already on its side. It sank stern-first two minutes after the torpedo hit. We didn't have time to release the lifeboat, but we got onto some Carley rafts, which saved our lives. Most of the guys had been sleeping and had nothing on but their underwear. The ship's cat went down with the ship after scratching a guy in the face trying to get off.

We paddled six and a half hours trying to reach a lightship out of Halifax. Many guys died of exposure in that time. Forty-four men lost their lives, and there were twenty-seven survivors. I was the only guy from the engine room to survive, and I was the only one who could still walk when we were rescued by HMCS Sarnia. *It was all because of the woollen socks, which I learned to wear to keep warm in the woods, and also my woollen underwear! And I had on one of those warm logger's undershirts.*

It took me 40 years to find anyone from the ship. I never talked about it. But then I met an ex-German soldier who said, "You know Frank Smith in Victoria?" I said no. And he said, "He was on the same ship as you were. He said you saved his life."

In the 1980s Jack met Frank Smith, and in the 1990s he met Werner Hirschmann, an officer on U-190 who had moved to Toronto after the war and was welcomed as a member of the HMCS Esquimalt Memorial Association. In 1997 Hirschmann wrote to Jack:

One simply cannot understand, why the shore authorities did not react earlier, and also why the Sarnia did not radio to the shore authorities much earlier. So all in all a great number of failures caused the loss of so many survivors. The board of inquiry did not seem to take the matter very serious and their final comments can only be laughed at. My dear Jack, you and your shipmates and all other survivors on both sides are lucky to have survived and nobody can alter anything about things, which should not have happened. Let's try to make the best of the remaining time, and try to forget the bad things. ■

From Cherry Picker to Bomber Command

Born in Comox in 1921 and raised at Headquarters, Neil Martin left Tsolum School in 1939. He worked as a whistlepunk on a skidder show and as a fireman on a cherry picker out of Camp 3 before joining up and going overseas with the No. 6 Group of the Royal Canadian Air Force in 1940. He flew missions with Bomber Command that targeted, among other locations, the city of Hamburg:

It was a meat grinder, that Bomber Command! About 10,000 Canadian airmen were killed in Bomber Command. I calculated from the official documents that I had a 46 per cent chance of surviving the missions I flew in 1944 alone. This percentage doesn't include training accidents or landing accidents. In earlier years the casualty rate was higher. From Tsolum School, Cedric Treherne, Toughie [Ian] *Willson, Ian MacDonald, and Norman Swan were killed in the air force, and from Comox, Henry Ellis, Doug Fairbairn, and Jack Owen; from Courtenay, Semon Tatar, Frank Carter, and Noel McPhee; from Cumberland, Allan Hutton. You either died—or you survived the war! If you survived it, you had great memories.*

After the war I worked as a scaler at Comox Lake, and later as a bull bucker and safety coordinator for Comox Logging. I still dream about the dreadful things I saw in the woods. I saw the same kind of loss of life in the woods that I had seen in the war. Four boys from Tsolum School were killed in Bomber Command, but an equal number were killed in the woods: Cyril Hodgins, Charlie Churchill, John Robertson, and George Surgenor, but of these, only Cyril was working for Comox Logging [when he died].

On his return from the war Neil married Eleanor Burritt, of the Vancouver carpet family, and brought her back to Headquarters in 1946. Neil Martin died in Comox in June 2008. ■

OLD-GROWTH IS A THING of the past on much of Vancouver Island, so it is fitting and impressive that those 1912 houses in Victoria continue to provide old-growth Douglas fir timber to local builders, demolition companies, and the Duluth Timber Company. Great value is now placed on that close-grained domestic timber, remnants of a splendid resource that was consumed so hastily in the first half of the 20th century. Comox Logging cut its property, shipped its best bounty to Fraser Mills, and did not replant a single tree until 1943. In this, the company was typical of the large operators on the Island. The technologies employed in their flatcars, log booms, and sawmills were suitable for only the largest logs. It is significant that protected areas formed from remnants of old-growth in the coastal Douglas fir and coastal western hemlock ecological zones—Cathedral Grove in MacMillan Provincial Park, Bright Angel, Eves, Rathtrevor Beach, Miracle Beach—generally depended on private gifts of land for their preservation. Comox Logging did its part in depleting what Bob Filberg called the "Garden of Eden for loggers"—in emptying it of some 5 billion board feet of lowland and mountain timber by 1946—and leaving the descendants of the first Comox Valley settlers with sterile, quick-yield tree farms.

Leona Dixon (*née* Lockner), left, and her mother-in-law, Edna Dixon (*née* Marsh), in Vancouver, 1940s. Edna was married to Hallie Dixon, Comox Logging engineer at Comox Lake. The reverse reads, "Enlargements of this photo can be had at Kandid Kamera Snaps, West Hastings Street, Vancouver, B.C."—FRED PARKIN

BOB FILBERG PHOTO

"Kerr and Fairbairn" was Bob Filberg's caption for this *c.* 1940 photo. Douglas Fairbairn was one of more than 17,000 members of the Royal Canadian Air Force killed in the Second World War. At least 20 local members of Bomber Command were also killed.—FILBERG HERITAGE LODGE & PARK ASSOCIATION

The easily accessible timber, on the flats and lower mountainsides of the Comox Valley, sustained about 400 men, and their families and communities, from 1910 to 1945. Although the old-growth trees and the isolated work camps that exploited them disappeared, established families remain as the most important legacy of Comox Logging. A glance at the modern North Island phone book reveals the persistence, and often the remarkable growth, of 19th-century families such as Anderton, Baikie, Beech, Berkeley, Blackburn, Carwithen, Cessford, Cliffe, Crockett, Downey, Grant, Grieve, Harmston, Hawkins, Janes, McQuillan, Machin, Parkin, Pidcock, Piercy, Pritchard, and Radford. Many of them had come to farm the alluvial riches of the lower Tsolum River valley in the 1860s and 1870s. The continued presence of these families—and of their many matrilineal branches—in the Comox Valley is a measure not only of the fertility and generous scale of the 19th-century pre-emptions, but also of the Homeguards' stable and lucrative 20th-century work for Comox Logging. The Homeguards' access to land and to woods work provided them with a large measure of local rootedness, income, and security.

But the Homeguards were not the only ones to benefit from this natural largesse and the massive engine of employment that was Comox Logging. Not just the steel gang, but Comox Logging as a whole—and the entire Comox Valley—was a "League of Nations" in the interwar years. A complete list would be very long, but among

The Salal Telegraph

Built in 1911, Headquarters was dismantled and abandoned in the late 1950s, and no one has lived there since. It is, however, still very much in people's minds, and the first Headquarters reunion was held in 2007. "You could never regain the camaraderie, neighbourliness, sharing feeling of those times and places," Elmer Helm (born 1926) wrote in a December 2002 letter. "A book could have been written on Camp 3 and Headquarters." (A significant part of *Island Timber* attempts to fill this void.) Social life at these logging camps survives in curious, comical, or telling anecdotes, all of them specific to the people and place. Neil Martin (1921–2008) recalled the closeness of the industrial village where he spent his childhood and youth:

There is a Headquarters network that keeps track of almost everyone who ever lived at the settlement. I call it the Salal Telegraph, and it's always busy. Anyone with any connection with Headquarters is part of the network. You get on the phone with them and it's "Yap, yap, yap! Did you hear that so-and-so was in a bus accident?" There's a closeness, and it's rather nice.

In the 1960s an unknown visitor wrote of the defunct community:

Headquarters no longer exists as the busy community it once was. The bungalows were pulled down or moved to other parts of the district. But the empty shell of the mill still stands beside the river, almost hidden by trees. There are varying stories on the Headquarters sawmill. It has been said that the building of the mill was one of the requirements for Comox Logging & Railway obtaining the timber rights. Some say it ran for one day, others that the machinery was never used. The roof has gone and holes in the floor show where massive square-foot beams have rotted away. Alder saplings have come through the gaps in the concrete, and branches push in through the glassless windows. The floor is covered in moss and leaves and below the floor is the darkness of swamp drip and gurgle. Everything removable has been taken away years ago. What remains are the monolithic concrete walls, looking like they will stand forever; a monument to the Comox Valley loggers of an earlier day. (Anon., "Headquarters, The Old Townsite")

The imposing cement walls of the century-old Headquarters sawmill, covered with obscene Anglo-Saxon graffiti and littered with the detritus of teenagers' parties, can still be found in the bush above the Tsolum River. A mill that never ran is all that remains of a town that never slept. ■

CHARLES SILLENCE PHOTO

A Headquarters family. Left to right: Jack Tukham, Signe Tukham, Fred Tukham, *c.* 1944. The eldest of eight children, Signe Anderson (1900–1950) came from a small farm in Sweden and moved to Vancouver Island in 1920 to be with her father, Fred Anderson, a welder at Headquarters. At first she worked as a nanny to the Filberg family, but in 1922 she married the Estonian-born timekeeper at Headquarters, John Tukham (Juhkam), who died in 1942. Her sons, Fred and Jack, were born in 1923 and 1924 respectively. Jack Tukham joined Comox Logging's forestry division after the war, while Fred (Godfried) Tukham became an executive in the oil fields of Saudi Arabia. He married an Englishwoman and retired to the English countryside.—Dorothy Tukham

the families to get a foothold on the Island at this time were these: from the *British Columbia mainland and other parts of Canada*, Churchill, Hodgins, and Telosky; from the *United States*, Filberg, Harding, and McQuinn; from *England*, Biss, Chown, Steere-Clark, Gray, Hilton, Holmes, Muir, Sheppard, Smith, Spencer, and Ware; from *Holland*, Bergsma; from *Germany*, Adamschek, Passmann, and Richter; from *Italy*, Padula; from *Croatia*, Domovich and Super; from *Denmark*, Pedersen; from *Sweden*, Anderson (Walter), Ell, Jonsson, Sundin, Wicklund, and Widen; from *Finland*, Aaltonen, Anderson (Herman), Erickson, Forsman, Hagg, Jylha, Koski, Lehtonen, Lyttback, Niemi, Pelto, Saari, and Storr; from *Estonia*, Tukham; from *China*, Leung. Comox Logging also helped keep on Vancouver Island a number of Cumberland-area families—including Moncrief, Sgarbossa, Stockand, Tobacco, Whyte, and Woods—left adrift by the moribund Canadian Collieries mines. Men from these diverse backgrounds found work with Comox Logging at every stage in the stump-to-dump production of saw-logs for Fraser Mills. Often these men married local women, and their descendants are now numerous on Vancouver Island and beyond.

Whether fallers deep in the forest or boom men on the water, they all worked on this western "edge of the world economy" as Comox Logging moved from the flats to the Vancouver Island Mountains between 1925 and 1945. What drew them to the Island, and saw them through these difficult decades, was mountain timber. ■

Left to right: Myrtle McKenzie and her friends Maj and Pat in Vancouver, 1947. Soon after, Myrtle would marry Peter Heron (1920–98), a former Royal Canadian Air Force pilot from Fort Fraser, B.C.
—Myrtle Heron

> Myrtle McKenzie (later Heron), left, and her friend Gloria in Vancouver, 1947. "This was taken before my wedding when I was working in an office in Vancouver," recalled Myrtle. "It was dead-end work. The photographer was on the street, and I think you'd pay a dollar for three photos, and then you'd pick them up later." Myrtle's father, Jack McKenzie Jr., was a Comox Logging high rigger who became a wartime foreman with Aero Timber Products in the Queen Charlotte Islands. After the war he ran a logging camp on the west coast of Vancouver Island for Harry McQuillan.
—Myrtle Heron

The Orr family at their cabin at Kye Bay, near Comox, *c.* 1940. Left to right: Elmer Orr, Janet Orr, Freddie Orr (boy), unidentified man. Elmer Orr was loading leverman on Skidder 3 at Comox Lake until skidders were scrapped in 1946, and later he ran a yarder at Comox Lake. A keen fisherman, he patronized Upper Campbell Lake as well as Kye Bay.—Fred Parkin

The Camp 3 Garden

The mile-long site of Camp 3 is now bisected by the Hamm Road connector of the new inland section of the Island Highway. Delton Henrich runs a bison farm on the northern part of Camp 3, while his half-brother Oscar Kreutziger runs cattle to the south.

Oscar went to school here as a boy in the 1940s, but now he can't find his way around this tangle; he can't make sense of it. An old-growth forest until about 1920, this place was then logged, the slash was burned, and a camp was built in 1922, burning down in the same year. It was rebuilt and partly burned in 1923, was razed by fire in 1925, and was threatened in the 1938 fire. In between the fires it was cleared and domesticated, and planted with gardens and fruit trees; about 35 families called it home. In the 1930s, hundreds of single or visiting men lived in the bunkhouses, but then the timber ran out in Comox Logging's huge Block 29 surrounding the camp. In 1942 the company closed the logging camp, which was then converted to an army camp where thousands of men trained. Later, that too was abandoned and the site became, temporarily, a dense forest of Scotch broom. After the war, the Kreutzigers' cows occasionally drowned in the old wells until Comox Logging sent a bulldozer out for two weeks to fill the wells. Between the 1950s and the 1980s the forest gradually reasserted itself, and now Oscar Kreutziger cuts the second-growth for firewood.

Domestically, Camp 3 was an extension of 19th-century Comox. The Comox genealogy goes back to families from Worcestershire (Hawkins), Staffordshire (Cliffe), and Lincolnshire (Harmston), who were among the first colonists of the Comox Valley. Originally farming families, by the 1920s they had, as young married couples, entered the logging camps of the level forested plain north of Courtenay and Comox. A thread of rural England still survives here: russet apples, pear trees, domestic vines, and foxgloves, all cowed and tangled within the exuberant second-growth canopy. Foxgloves were planted by Homeguard Jack McKenzie Jr. from seeds provided by his wife, Ella (*née* Hawkins). "She was great for planting all kinds of things," said her daughter, Myrtle Heron.

The first Camp 3 reunion was held in 2005. ■

glossary

Note: These terms are defined as they were used between 1920 and 1950. Some terms have evolved over time with modern logging methods.

A-frame
An A-shaped structure mounted on a large log float and used to yard timber down the steep sides of lakes and inlets.

Agony whistle
Six or seven long blasts of the skidder whistle in the event of an accident in the woods.

Arch
An arch-shaped steel frame resting on a set of tractor treads, and later on wheels, behind a Cat. The arch gives the lift necessary to elevate one end of a log off the ground for easier yarding. A line from a drum on the Cat leads through a fairlead, through three or four hoops on the arch, and to the end of the log or logs.

Back end
The part of a skidder setting farthest from the track; an area associated with back spars, cold deckers, cold-deck piles, and long corners.

Back rigger
A member of the skidder's yarding crew who helps rig the back spars. *See also* Skidder crew.

Back spar
A tree (generally standing, rather than raised) rigged up at the back end of a setting and connected to the head spar to provide lift (highlead) for yarding logs; also called cold-deck tree.

Bag boom
A circular or bag-shaped float of logs enclosed by boomsticks. Bag booms are used for log storage and transport on sheltered inland lakes. Unlike flat booms, they do not require swifters to stabilize the logs, because bag booms are not intended for ocean transport.

Bicycle *See* Carriage.

Bight
The dangerous area in the curve of a yarding line, which is under tension and may suddenly straighten out; loggers are frequently admonished, "Keep out of the bight!" *See also* Siwash.

Blaze
An axe mark made on a tree to indicate a surveyor's boundary or a spar tree.

Block
(1) A metal shell encasing pulley sheaves, used to guide lines in highlead logging. Blocks come in various sizes and shapes to suit the particular duty and load requirements (*see also* Carriage).
(2) A section of a bucked log suitable to be split into firewood (also called a round).

Blowdown
A tree or stand of trees felled by the wind.

Board foot
The standard unit of volume for timber, measuring one foot by one foot by one inch (144 cubic inches); also called foot board measure (ft.b.m.).

Boilermaker
A machine shop worker in charge of building and maintaining locomotive, skidder, and donkey boilers.

Boom *See* Bag boom; Flat boom.

Boom chain
A stout chain, generally six feet long, used to link boomsticks together.

Booming
The process of making log booms by sorting, stowing, and moving logs in the water.

Booming grounds
A protected bay, often a tidal estuary, where logs are boomed and stored.

Boomstick
One of the logs that surround a boom to hold it together. Usually 72 feet long, boomsticks have four-inch holes at the ends so they may be joined together with boom chains. *See also* Swifter.

Brakeman (Brakie)
A locomotive crew member who releases brakes, throws switches, and inspects loads and flatcars. He also diagnoses flatcar problems and leaves cars at sidings for repair by the car-knocker. Main-line locies have a first brakeman and a second brakeman.

Branch line
A secondary line on a railway, leading off a main line and into an area being logged.

Brow logs
Large logs yarded to each side of the track to help the loading leverman guide logs into place on the flatcar.

Buck
To cut or saw a tree into log lengths for yarding or for firewood.

Bucker
A woods worker whose job is to follow the fallers and saw the felled trees into standard log lengths.

Bucking saw
A sturdy, inflexible crosscut saw, generally seven feet long, used by a single bucker for cutting felled trees into lengths. *See also* Falling saw.

Bull bucker
The supervisor of fallers, buckers, and scalers.

Bullcook
A man (usually an old or maimed logger) who performs various odd jobs in a logging camp, such as sweeping out bunkhouses, making beds, and cutting firewood.

Bunk
The heavy steel cross-piece on the bed of a flatcar upon which logs are placed for hauling. *See also* False bunk.

Bunkmaker
A skidder crew member who makes false bunks for flatcars, sweeps debris from flatcars, and performs other odd jobs at the landing.

Cable *See* Line.

Car-knocker
A man who maintains and repairs flatcars.

Carriage (Bicycle)
In skidder logging, a block running along the skyline from which the choker lines descend. The carriage guides logs into the landing and takes chokers back out to the chokermen.

Car shop
A building used for repairing and maintaining flatcars.

Cat (Caterpillar)
A tracked, V8-powered vehicle made by the Caterpillar Tractor Company and used for yarding in Cat and arch logging, *c.* 1930–60. A line from the Cat runs to the log via an arch, which lifts the front end of the log for easier yarding. *See also* Arch.

Caulk boots
Boots with spikes on the soles for better grip when walking on logs and slash (pronounced "cork").

Chainman
A survey crew member who holds the measuring chain, carries the transit or compass, and performs minor jobs.

Chaser
A member of a skidder crew who unfastens the choker after a log has been dropped on a pile at the landing; or a member of a cold-decker crew who unfastens the choker after a log has been dropped on the cold-deck pile.

Cheese blocks
Steel wedges placed on the flatcar bunks to hold large logs in place (probably named for their resemblance to V-shaped wedges of cheese).

Cherry picker
A small, usually wood-burning donkey engine fitted with a loading boom, secured to a flatcar, and taken along the track to retrieve timber from rights-of-way after grades are put in. The cherry picker also logs trackside nuisance areas and picks up logs that have fallen from flatcars. It is manned by a crew of four or five: engineer, fireman, head loader, and one or two second loaders.

Choke
To pass a line (choker) around a log and pull it tight, like a noose.

Choker
A short loop of line with a hook at one end, attached to the skyline carriage and used to yard logs.

Chokerman
A member of a skidder or cold-decker crew who "sets chokers," i.e., straps chokers around logs for yarding to the landing.

Cold decker
A mobile donkey engine mounted on a sled, used to yard distant logs and stack them within reach of the skidder; also loosely called a yarder. The cold decker is manned by a crew of seven: a rigger (high rigger), engineer, chaser, hooker (or, with a Clyde yarder, a rigging slinger), two chokermen, and a whistlepunk.

Cold-deck pile
In skidder logging, a pile of logs (as many as 4,000) yarded from a distance and stacked by a cold decker beneath a back spar for further transport to the landing by the skidder.

Cold-deck timber
Timber beyond a skidder's 1,100-foot reach, yarded into a back spar by a cold decker.

Cold-deck tree *See* Back spar.

Cribbing
A framework of logs, square timber, or railway ties used for supporting bridges, portable skidders, schoolhouses, and camp houses.

Crosscut saw
A steel saw for cutting across the grain of wood; it is long and flat, with two types of sharp teeth (cutters and rakers). *See also* Bucking saw; Falling saw.

Cruiser *See* Timber cruiser.

Crummy
Initially, an old boxcar or flatcar (railway crummy) with benches and perhaps a stove added, used as a log-train caboose to carry loggers to and from work; later, a crew bus (bus crummy).

Deadhead
Waterlogged timber that sinks out of the boom and can present a hazard to boats.

Dogs
U-shaped spikes used to fasten logs together when logging using skids and when yarding cedar poles from the forest.

Donkey engine (Donkey)
A steam-, gas-, or diesel-powered engine equipped with drums and an upright boiler; the engine is mounted on a heavy log sled and used for power in yarding logs. Also called steam pot or loading pot. *See also* Cherry picker; Cold decker; Rigging-up goat.

Double-bitted axe
A hand axe with two cutting blades, used for chopping undercuts, firewood, guyline stump notches, etc.; also called double-bladed axe.

Dump (Log dump)
A place where logs are literally dumped from the end of the railway into a lake or the ocean. Logs are generally pushed or rolled from a trestle built out over the water.

Engineer
(1) Any man operating a steam engine (thus used for yarding and loading levermen as well as locomotive engineers); (2) The boss of a survey crew, usually a civil engineer.

Eye splice
To make a secure loop of cable by bending the cable back and splicing it into itself. *See also* Long splice; Splice.

Fairlead
An enclosed steel roller that guides the line and permits it to spool properly on the drums of the donkey engine or Caterpillar tractor.

Faller
A woods worker who cuts down (fells) trees, usually by chopping and sawing with a partner.

Falling gang
Two fallers and one or two buckers responsible for falling and bucking trees, under the direction of a bull bucker.

Falling saw
A flexible crosscut saw, generally between 7½ and 11½ feet in length, used by a pair of fallers for cutting down trees. *See also* Bucking saw.

False bunk
A heavy, curved wooden platform, usually made of Douglas fir, set at one or both ends of a flatcar to provide additional support when hauling boomsticks and timber sticks. *See also* Bunk.

Farm timber
Timber growing on farms or pre-emptions, often sold to a logging company.

Filer *See* Sawfiler.

Fireman
A man who keeps the steam up on a skidder, donkey engine, or locomotive by stoking it with firewood or monitoring the flow of crude oil.

First loader *See* Head loader.

Flat boom
A 576-foot raft of timber for sea towing, enclosed by a boomstick frame and held together with swifters and boom chains. *See also* Bag boom.

Flying corps *See* Rigging-up crew.

Gandy dancer
A member of the steel or section gang who pounds spikes into wooden ties to hold down steel rails; also called tie tamper.

Gin pole
A small pole (resembling a telephone pole) used as a spar in parbuckle loading or in raising a head spar.

Goat *See* Rigging-up goat.

Grade crew
The workers who prepare railway grades by felling trees and digging and blasting out rock, earth, and stumps.

Ground yarding
Any yarding system in which logs are pulled over skids or simply over the ground, as opposed to highlead yarding.

Guylines (Guys)
Strong lines running from the top and centre of a spar tree to surrounding stumps. Both raised and standing trees need guylines to hold the spar firmly upright and in place. "Top guys," about 110 feet up, support the top section of the spar tree; "buckle guys," about 80 feet up, support the middle section of the spar tree.

Gyppo
A small, minor, or fly-by-night logging outfit.

Handcar
A small, hand-pumped vehicle used by the section crew for patrolling the railway track and monitoring repairs. *See also* Speeder.

Hang-up
(1) A tree that refuses to fall because it is caught in the canopy of nearby trees; (2) A turn of logs caught behind a stump, root, snag, or other obstruction when yarding; (3) Fouled-up rigging; (4) Any kind of problem in a logging show.

Haulback line
A light line running from yarder or skidder to skyline; the haulback line sends the mainline and chokers out to the hookers and chokermen to pick up a new turn of logs. *See also* Line.

Hayrack boom
In skidder logging, a device for loading logs onto flatcars, consisting of two timbers (originally 82 feet long) fastened together and attached to the head spar horizontally above the skidder. The hayrack boom rotates around the spar to allow logs to be picked up and loaded.

Head hooker *See* Hooker.

Head loader (First loader)
A member of the skidder crew in charge of loading at the landing. He selects logs from the landing pile and directs their loading onto flatcars, while two second loaders "swing the tongs" beneath the hayrack boom. The head loader is third in command of the skidder crew, after the head rigger and second rigger. A head loader is also in charge of loading logs with a cherry picker.

Head rigger (Rigger)
A high rigger and senior member of the skidder crew. He is in charge of the rigging, of the smooth running of the highlead logging system, and of the skidder crew itself. He and his crew of second rigger and two, three, or four rigging men climb, limb, top, and rig up head spars and back spars, swing blocks into position, and mend broken or frayed line. As necessary, the head rigger and his crew work in conjunction with crews attached to rigging-up goats and cold deckers.

Head spar
A topped and limbed Douglas fir tree (generally a raised tree) to which lines are attached for yarding and loading logs in skidder logging. Secured by guylines, most head spars are about 120 to 140 feet high and at least 4 feet in diameter at the base. Also called home spar, home tree, skidder spar (or simply spar), or skidder tree.

Heel boom
A loading boom attached to the sled of a cherry picker and fitted with a single line and tong. The engineer uses the heel boom to swivel, position, and place logs one at a time on a flatcar.

Highball
(1) To move quickly; (2) A fast and efficient logging operation.

Highlead logging
A logging method in which fully rigged spar trees—in combination with cold deckers and skidders—are used to lift one end of the logs during yarding for easier transport, and to load logs onto waiting flatcars; also called aerial, overhead, or skidder logging.

High rigger (Rigger)
The man on a rigging-up, cold decker, or skidder crew who is in charge of topping and rigging head spars and back spars. He and his crew of second rigger and rigging men climb, limb, top, and rig up spar trees. *See also* Head rigger.

Homeguard
"A man who works for the same company and never leaves his job" (Swanson, *Bunkhouse Ballads*); used here in a specific sense to refer to workers with strong family ties to the Comox Valley and often lifelong employment with the Comox Logging & Railway Company—a "family company." (Not to be confused with "Home Guard," the Canadian civil defence force in the Second World War, including the Pacific Coast Militia Rangers.)

Hooker (Hooktender, Head hooker)
(1) On a skidder, the boss of the front-end

yarding crew, directing the second hooker, chokermen, and whistlepunk. The hooker is in charge of the whole operation of hauling logs from where they are cut to the railway landing; he selects the logs to be yarded. (2) On a cold decker, the yarding boss in charge of the chokermen and whistlepunk. Also called hookerman.

Hooktender *See* Hooker.

Hostler
A man in charge of maintaining the steam plant at a logging camp at night, providing hot water for the cookhouse and bunkhouses and keeping the steam up on locies.

Humdirgen
A steam-powered log unloader at the log dump; it pushes logs off flatcars and into the water.

Jackpot
A snarl of criss-crossed logs, often created when logs are dumped on top of one another in shallow water.

Jagger
A sharp steel strand from a broken or worn line. Chokermen wear heavy gloves to avoid painful hand lacerations from jaggers.

Landing
(1) Any level ground to which logs are yarded for sorting or loading; (2) A cleared area near the skidder (skidder landing) where logs are piled (in the landing pile) before being loaded on flatcars.

Landing (loading) crew *See under* Skidder crew.

Leverman, loading
The skidder engineer who "swings the boom"—who loads logs onto flatcars with tongs and hayrack boom; also called loading engineer.

Leverman, yarding or skidder
The engineer who yards logs from every corner of the setting to the skidder landing; also called yarding engineer.

Lidgerwood
A brand of skidder engine made in Washington State, used by Comox Logging from 1911 until 1946.

Line
(1) The "wire rope" used extensively in rigging trees and in yarding and loading logs; only the two-inch-thick skyline was called a cable (smaller sizes were called line). *See also* Guylines; Haulback line; Mainline; Pass line; Skyline; Strawline. (2) A blazed survey line or boundary. (3) A stretch of logging railway; *see also* Branch line; Main line; Spur line.

Loader
A member of the skidder or cherry-picker crew who loads logs from the landing onto flatcars. Normally, on a skidder, a crew of loading leverman, head loader, and two second loaders can keep pace with the logs brought in by the yarding leverman.

Loading leverman *See* Leverman, loading.

Locie
A logging locomotive (engine). Logging locies are of two types: heavy rod engines (Baldwins) used for main-line hauling, and lighter but versatile geared engines (Heislers, Shays, Climaxes) used on steep grades and for the switch work of hauling flatcars in and out of skidder settings.

Log boom *See* Bag boom; Flat boom.

Log dump *See* Dump.

Logging
Often used casually to mean the entire timber extraction and transport process; strictly speaking, only the yarding and loading of logs.

Logging crew
All workers directly involved in yarding and loading logs (workers in the rigging-up, skidder, and cold-decker crews).

Log stamp
A large hammer imprinted with a logging company's insignia, used for stamping both ends of a log at the skidder landing.

Long corner
A distant area adjoining a setting and inaccessible to skidders, where logs are retrieved by cold deckers.

Long splice
To join two cables by weaving together the separated cable strands at the end of each cable. *See also* Eye splice; Splice.

L-stick
A device for calculating board footage; also called scale stick or gyp-stick.

Main line
The main stretch of a logging railway.

Mainline
The main yarding line on a skidder which, coupled to the haulback line, brings turns of choked logs in from the woods. The mainline, dragging its turn

of logs, is wound into the skidder by the yarding leverman.

Marion shovel
An American-made brand of electric shovel used for building railway grades. *See also* Mat; Mat-hooker.

Marlinspike
A steel spike some 16 inches long, tapering to a flat point; used in splicing line.

Mat
(1) A portable set of logs fastened together, upon which a Marion shovel rests when making grades; (2) A set of logs placed beneath a raised spar tree on swampy or unstable ground.

Mat-hooker
A man who hooks and unhooks the Marion shovel mat and oils the shovel; also called mat-man or swamper.

Outside tug
A powerful deep-sea tug used for towing log booms from the sheltered log storage area to the mill (in this case, across the Strait of Georgia).

Parbuckle
A line slung around a log, causing it to roll onto a skid road or flatcar when one end of the line is pulled by a donkey.

Pass line
In rigging a spar tree, a light line (sometimes the strawline) from the skidder used for taking the high rigger up and down the spar tree, pulling blocks and lines up to him, and hanging and adjusting the rigging on the tree.

Peavey
A six-foot tool with a metal spike and hinged cant hook at one end, used for rolling and positioning logs on land, adjusting swifters on log booms, and breaking up obstinate jackpots at log dumps.

Peeler
(1) A high-grade Douglas fir log that is free of large knots, cracks, and rot, and that is literally peeled into high-quality plywood; (2) An employee of a pole company who, armed with a wide-bladed axe, removes the bark from small cedar trees.

Pike pole
A pole about 16 feet long, with a sharp point and hook, used by boom men to sort and stow floating logs.

Pocket
A corral-like enclosure, fenced by pilings, in which log booms are made up and stored; also called rafting pocket.

Powderman
A man who blows up stumps from rights-of-way, blows up roots from blowdowns, and blasts bark off snags for safety.

Quarter
A quarter of a setting, or about 15 to 20 acres.

Raised tree
A spar tree (almost always a head spar) that is brought into a setting by rail and raised in the absence of a suitable standing tree. A raised tree enables riggers to avoid the swelled butt of standing trees and to rig the head spar always at trackside. Raised trees are often reused and hauled by rail to the next setting.

Rig (Rig up)
To set up a spar tree for logging by raising a tree, or using a standing one, and attaching to it guylines, blocks, skyline, and associated lines and gear.

Rigger
Generally refers to a high rigger or head rigger but may also mean a back rigger or second rigger. Riggers work on rigging-up, cold-decker, and skidder crews, an indication of their importance at every juncture in highlead logging.

Rigging
(1) The network of overhead lines, blocks, fittings, chokers, and other gear (also called tackle or jewellery) used in rigging trees and yarding and loading logs in highlead logging systems; (2) Working on jobs involving rigging.

Rigging slinger
A position on a Clyde cold decker analogous to that of hooker on a skidder; the rigging slinger chooses the logs to be yarded to the landing, helps carry and untangle rigging, and directs the chokermen and whistlepunk.

Rigging-up crew
The woods crew, led by a high rigger, that procures raised trees and puts the rigging on head spars in advance of skidder crews; the rigging-up crew also moves the skidder itself to new settings. Also called bull gang, flying corps, rigging-ahead crew, or rig-up gang.

Rigging-up goat
A donkey engine used by the rigging-up crew in raising and rigging head spars.

Right-of-way
A graded path through the forest, 60 or 80 feet wide, containing a logging railway.

Road
In highlead logging, the path or avenue along which logs are yarded between back spar and skidder landing.

Roundhouse
A building used for repairing and maintaining locomotives.

Sand house
A storage shed at a logging camp for sand used to give locomotives traction.

Sawfiler (Filer)
A man who sharpens crosscut saws with files.

Scaler
A man who scales bucked logs, i.e., assesses their quality and measures them to estimate board footage, using L-stick or calipers and steel tape. *See also* Water scaler.

Second hooker
The skidder crewman second in command to the hooker.

Second loader
(1) One of usually two skidder crewmen who, using tongs, transfer logs from the landing to the flatcar; (2) One of usually two cherry-picker crew-men who, using tongs, hook and load logs at trackside.

Second rigger
An assistant to a high rigger in topping and rigging spar trees. On a skidder crew, a second rigger is second in command after the head rigger.

Section
(1) A stretch of logging railway; (2) A 72-foot length of a log boom.

Section gang or crew
The men in charge of maintaining the ties, steel, and grade on a section of railway.

Setting
The 50-acre to 80-acre area logged from a single central head spar: the work site of falling, rigging-up, yarding, and loading crews.

Show
Any logging operation or outfit ("summer show," "skidder show").

Side
A complete log production unit within a company's territory, with its own settings, spur lines, and crews for falling, bucking, rigging-up, yarding, and loading.

Side-push (Side-rod)
(1) Boss of a side or a logging camp; (2) An assistant foreman.

Siwash
Chinook jargon for a line that has been diverted by getting bent around a stump, tree, or log. *See also* Bight.

Skid
(1) A partially sunk, debarked, and greased log upon which logs are dragged by horses or oxen (*see also* Skidding; Skid road); (2) A hewn Douglas fir timber attached to the bottom of a donkey engine to enable the machine to move through the bush and over felled timber (*see also* Sled).

Skidder
The powerhouse of highlead logging. A glorified donkey engine mounted on a flatcar and parked at the base of a fully rigged head spar, the skidder yards logs along a skyline with one engine and loads them with another engine by means of a hayrack boom. Also called flying machine, cable-way skidder, or overhead rig. *See also* Lidgerwood.

Skidder crew
A crew of about 21 men responsible for transporting logs from where they are cut to the flatcar at the skidder landing. Divided into two groups: a *landing (loading) crew*, consisting of yarding leverman, loading leverman, fireman, head loader, two second loaders, chaser, and bunkmaker; and the *yarding crew*, further divided into front-end and back rigging crews. The front-end crew consists of a hooker, second hooker, whistlepunk, and three or four chokermen. The back rigging crew consists of head rigger (who is boss of the entire skidder crew), second rigger, and two, three, or four rigging men who rig the back spars and repair and maintain lines, blocks, and rigging.

Skidder landing *See* Landing.

Skidder leverman *See* Leverman, yarding or skidder.

Skidder logging *See* Highlead logging.

Skidder setting *See* Setting.

Skidder spar or tree *See* Head spar.

Skidder spur
A small spur line at the landing where the skidder is parked.

Skidding
The process of yarding logs from where they are felled to where they are loaded. In ground yarding this takes place on a skid road; in highlead yarding, on a "road."

Skid greaser
A junior member of a horse or oxen logging crew whose job is to grease the skid road with oil or fat to facilitate movement of logs.

Skid road
In the days of horse and oxen logging, a road formed of skids along which logs were yarded.

Skyline
A stationary cable, about two inches in diameter, stretched between head spar and back spar some 80 feet above the ground. Logs are yarded along the skyline into the skidder landing by means of choker lines suspended from a travelling carriage and assisted by the moving mainline and haulback line.

Slash
(1) The debris resulting from logging; (2) A logged-off area strewn with such debris.

Slash fire or burn
A controlled fire set in logging slash, usually in fall, to remove debris, reduce risk of forest fire, and promote growth of Douglas fir.

Sled (Sleigh)
Two hewn Douglas fir skids bolted beneath a donkey engine for ease of transport.

Sleighmaker (Sledmaker)
A man who makes sleds for donkey engines.

Snag
A standing dead or dying tree, often lacking limbs and top, left standing by logging crews because it has no value, but constituting a safety and fire hazard.

Snipe
To bevel the end of a log so it will ride smoothly over a skid road during ground yarding.

Spar (Spar tree)
A rigged tree. *See also* Back spar; Head spar.

Sparkcatcher
(1) A cap resembling a chef's hat that is placed on top of locomotive, donkey, and skidder smokestacks to reduce the chance of fire (also called spark arrestor); (2) A boy or elderly man hired by a logging company to douse sparks produced by locomotives, donkeys, or skidders.

Speeder
A four-wheeled, gas-engined vehicle running on railway tracks; it replaced the handcar. The speeder transports logging and camp personnel and the section gang.

Splice
To join together two cables, or ends of the same cable, by unravelling and tying the individual strands. *See also* Eye splice; Long splice.

Spot
To position a flatcar in the ideal place for loading or unloading.

Springboard
A Douglas fir plank, five or six feet long and two inches thick, inserted into a notch in a tree as high as 12 feet from the ground. A faller stands on a springboard to avoid the swelled butt of the tree and large amounts of pitch, and also to gain a foothold on sloping ground. Also called chopping board or falling board.

Spur line
A minor line on a railway, usually leading off a branch line and into an individual skidder setting; also refers to the very short line upon which a skidder sits at a setting.

Stakey
Having money to spend; used to describe a logger's urge to quit and collect his paycheque, get out of camp, and go to town to spend his money, often with binge consequences.

Standing tree
A tree selected as a spar (generally a back spar) in its original position. Sometimes two standing trees are blazed in case one turns out to be punky or otherwise unsuitable as a back spar. *See also* Raised tree.

Steam pot *See* Donkey engine.

Steel gang or crew
The men who tamp ties, lay down steel rails, and "pull steel" (remove rails) on a logging railway.

Straw boss
An assistant foreman.

Strawline
A light line used when rigging up trees and later when moving or adjusting blocks and lines.

Stump ranch
A subsistence farm on logged-over land, often bought from a logging company.

Superintendent (Super)
The boss of a logging company, in charge of all shows, sides, foremen, and employees, and in charge of all aspects of log procurement, production, and transport.

Surveyors (Survey crew)
The civil engineer and his crew, who survey areas to be logged, lay out railways and logging camps, and generally plan a logging operation; used by loggers interchangeably with "engineers."

Swamper
An apprentice or assistant to a teamster on a horse or oxen logging show; later, an assistant on a steam shovel (*see* Mat-hooker) or Caterpillar.

Swedish fiddle
A crosscut falling or bucking saw.

Swifter
A bored boomstick between 66 and 72 feet long, placed across the top of a log boom for stability and linked to the boom with boom chains.

Tailhold stump
A stump located near a head spar and used to anchor a rigging line, such as the haulback. Sometimes a series of such stumps was required for extra stability.

Timber cruiser (Cruiser)
A man who initially surveys a forest stand and assesses the timber therein. After his field work he submits a map showing the contours and physical features of the land as well as the volume and type of timber.

Timberman
A man who prepares the timber required to maintain flatcars, sleds, donkeys, skidders, and other equipment.

Timber stick
An unusually long and high-quality log destined for structural or bridge timber; a timber stick usually requires two or three flatcars equipped with false bunks for transport.

Timekeeper
A logging camp official who keeps track of men's working hours and wages.

Tongs
A two-armed grasping device attached by line from hayrack or heel boom and used for loading logs.

Top (a tree)
In skidder logging, to remove the heavily branched top section of a spar tree in preparation for rigging.

Turn
A load of logs yarded to the landing in one trip; also called trip.

Undercut
A notch chopped in a tree trunk to direct the tree's fall when it is sawed from the other side.

Water scaler
A scaler at the booming grounds.

Wee McGregor
A portable bucking saw powered by a one-cylinder gas engine; precursor to the chainsaw.

Weight log
A log suspended from one end of a hayrack boom as a counterbalance to the boom's load.

Whistlepunk
A signalman on a skidder or cold-decker crew who relays to the engineer the hooker's signal that logs have been secured with chokers and may be hauled away. Whistlepunks are usually boys beginning their careers in the woods.

Widow-maker
A loose branch or piece of bark that may fall on a logger below.

Yarder
See Cold decker.

Yarding
The process of moving logs, whether using a skidder, cold-decker, A-frame, Cat, horse team, or other means.

Yarding crew *See under* Skidder crew.

Yarding leverman *See* Leverman, yarding or skidder.

METRIC CONVERSION CHART

1 inch = 2.5 centimetres
1 foot = 30.5 centimetres
1 mile = 1.6 kilometres
1 mile/hour = 1.6 kilometres/hour
1 square foot = 929 square centimetres
1 acre = 0.4 hectares
1 square mile = 2.6 square kilometres (259 hectares)
1 pound = 0.5 kilograms (454 grams)
1 short ton = 907 kilograms

bibliography

Interviews conducted by the author

Agar, Shawn, December 1997.

Aitken, Ken, 24 June 2009.

Anderson, Arne, 28 December 1997, 16 February 2003.

Anderson, Walter, 7 February 1999, 29 January 2001, 18 December 2001.

Annand, Art and Betty (*née* White), 7 January 1998, 8 June 2002.

Bailey, Dick and Eva (*née* Manson), 21 November 1998, 20 December 1998, 7 June 2001.

Bayly, Brian, 2 October 1996.

Bebenik, Wayne, 2 November 2008.

Beech, Art and Hazel, 5 and 13 November 2003.

Bell, Florence, 11 May 1998.

Berkeley, Bob, 5 December 2002.

Berkeley, Eddy, 25 November 1996.

Berkeley, Jean (*née* Crockett), 19 November 2002, 5 December 2002, 17 February 2003, 2 August 2006.

Biss, Frank and Ruth (*née* Christenson), 16 November 1998, 16 March 2004.

Blackburn, Fred and Mavis (*née* Harrison), 30 October 1996, 10 December 1996, 15 December 1996, 3 March 2004, 28 July 2004, 5 October 2006.

Blackburn, Gordon, 30 October 1996, 8 June 2002.

Bowen, Ron, 16 October 1999.

Brand, Karen (*née* Holm), 5 December 2000.

Brown, Bob, 29 November 2001.

Brown, Evelyn (*née* Vogel), 20 October 2001.

Buchanan, Melda (*née* McCulloch), 22 January 1996, 1 May 1999, 17 March 2000.

Burchell, Len, 14 June 2006.

Cameron, Dona (*née* Cliffe), 15 May 2004.

Cameron, Karl, 6 December 2002.

Cardriver, Keith (*né* Pirie), 27 May 2007.

Carter, Valerie (*née* Cessford), 28 January 2006.

Carwithen, Malcolm, October 1996.

Churchill, Hi and Mavis (*née* Janes), 7 September 1998, 27 March 1999, 28 March 1999, 13 May 1999, 6 June 1999, 24 June 1999.

Churchill, Wally, 4 January 2002, 9 October 2002.

Clark, Noëlle, 10 December 2002, 4 March 2004, 22 November 2005.

Cliffe, Bob and Lil (*née* Willis), 6 February 1998, 18 April 1998, 2 February 1999.

Cliffe, Inez (*née* Ledgerwood), 11 October 1996.

Cliffe, Lorna, 9 December 1998, 31 March 1999.

Clifford, Fred and Olga (*née* Blaseckie), 5 December 2002, 6 December 2002, 14 January 2003, 10 December 2003.

Currie, Phyllis (*née* Radford), 5 September 1998.

Dalby, Ralph, 5 September 1998, 8 December 1998, 28 August 1999.

Davis, Beatty and Penny (*née* Bailey), 10 November 1996, 13 November 1996, 23 November 1999, 4 December 2001.

Davis, Dave, 23 February 2001.

Davis, Ray, 18 March 2002, 4 December 2002.

Davison, Frank and Kay (*née* Blakely), 5 October 1996, 7 October 1996, 22 October 1999, 23 February 2000.

Delcourt, Doreen (*née* Holmstrom), 15 December 2002, 21 January 2003.

Denny, Roy, 22 February 1997.

Downey, Dick and Sheila, 10 October 1996, 13 October 1996, 29 October 1996, 20 November 1996, 1 May 2001, 2 February 2002, 6 April 2002, 23 February 2004, 8 November 2007.

Eastman, Terry and Nola (*née* Chalmers), 5 March 2002.

Edwards, Donna (*née* Piercy), 6 December 2002.

Edwards, Walt, 19 January 1996, 26 February 1998, 22 October 1999, 29 February 2000, 26 November 2002, 17 February 2003.

Ell, Sven, 8 December 1998, 15 October 1999, 17 October 1999, 11 January 2000, 27 March 2000, 27 April 2001, 16 August 2006.

Ellison, Marga (*née* Tame), 14 September 2002.

Ellison, Ted, 14 January 2004.

Emblem, Reg and Marg (*née* Dunn), 9 December 2002, 11 January 2003.

Emblem, Richard, 30 September 2002.

Erickson, Harry and Margaret, 6 December 1998, 8 December 1998, 21 March 1999,

30 March 1999, 24 April 1999, 17 December 2002, 21 January 2003, 22 May 2003, 23 November 2003, 20 February 2004, 23 February 2004, 31 March 2004.

Forsman, Elis, 19 December 2002.

Forsman, Michael, 27 May 2000.

Fossum, Jack, 20 November 1997.

Frame, Jody, 27 March 2001, 30 August 2001.

Gee, Leslie (*née* Parkin), 23 January 2001.

Grant, Gerry and Jean (*née* Taber), 3 March 2004.

Grant, Mack, 5 November 1995.

Gray, Jackie (*née* Gullett), 20 January 1996, 29 December 1997, 8 June 2002.

Gray, Len, 3 February 2001, 19 February 2004, 3 March 2004.

Green, Alvin, 8 March 2004, 12 March 2004, 13 June 2006.

Greig, Jim, 10 July 2002.

Grieve, Loggie, 23 February 2001.

Griffiths, Bus, 14 December 1997.

Haas, Donnie, 12 June 1999, 17 June 1999.

Hagg, Roy, 28 May 2007.

Halme, Arlene, 23 September 2002.

Halme, David, 23 September 2002.

Hanham, Stan and Betty, 1 October 1995, 3 June 1996, 5 May 2002, 8 May 2002.

Hansen, Gale, 6 December 2002.

Harding, Bob and Rene (Irene), 29 January 1996, 21 May 1996.

Harmston, Ralph and Thelma, 23 January 1996.

Harris, Bert and Ethel, 31 March 2002, 5 December 2002, 10 February 2004.

Hatfield, John and Phyllis (*née* Janes), 7 October 1996, 15 October 1996.

Hemstad, Betty Anne (*née* Wicklund), 29 October 2003.

Herbert, Madge (*née* Steere-Clark), 30 November 2003, 5 December 2003, 8 March 2004, 3 February 2007, 19 January 2008, 8 March 2008.

Heron, Myrtle (*née* McKenzie), 24 February 1996, 25 February 1996, February 1998, 9 August 1998, 23 August 1998, 25 November 1998, 27 December 1998, 6 February 1999, 13 February 2001, 29 January 2006, 3 September 2006.

Hodgins, Bevin, 20 May 2007.

Hodgins, Dale, 20 May 2007.

Hodgins, Stan, 2 October 1995, 26 January 1996.

Holmes, John and Jackie (*née* Harding), 19 January 1996, 21 January 1996, June 1996, 30 September 1996, 1 October 1996, 31 October 1996, 1 November 1996, 14 December 2005, 12 August 2006.

Holmes, Muriel, 4 March 2001, 28 February 2002, 5 March 2002, 25 September 2002, 26 September 2002, 7 September 2005.

Hunt, Bob, 1 October 1996, 26 November 1996.

Janes, Davey, 5 November 1995, 2 June 1996, 6 June 1999, 4 February 2001, 5 May 2002.

Jonsson, Gunnar, 11 October 1996, 13 October 1996.

Keller, Eleanor (*née* Richter), May 1996.

Kennedy, Sheila (*née* Haggarty), 18 December 1997.

King, Bruce, 9 November 1996, 30 January 1999.

Knappett, Al, 25 February 2003, 15 March 2003, 11 May 2003, 2 September 2006.

Knight, Ray and Mae, 18 June 2002.

Kreutziger, Gerald, 3 March 2002.

Kreutziger, Oscar, 19 April 1998, 29 June 1998.

Kulla, Bruce, 6 August 2002.

Lehtonen, Larry, 31 January 1996, 3 June 1996.

Lindorff, Robin, 20 May 2007.

Loukes, Jo (*née* Jennens), 5 December 2002.

Lyttback, Bill, 14 November 1996, 12 May 2002.

McAllister, Bob, 8 June 2002.

McIver, Don and Joan, 8 October 1996, 7 November 1996, 14 February 2000, 23 February 2000, 29 January 2006, 2 November 2008.

McLoughlin, Jake, 9 May 2003, 2 July 2005, 25 February 2006.

McMonnies, Angus, 20 April 2002, 22 May 2002.

McPhee, Bob, 16 December 1997.

McQuillan, Darrell and Patricia (*née* Black), 31 January 1996, 12 June 1996, 5 March 2000.

McQuinn, Bob, 16 September 2002.

McQuinn, Dick and Diana (*née* Robertson), 25 July 2000, 3 November 2000, 10 November 2000, 10 January 2001, 16 September 2002, 5 November 2003.

McQuinn, Howard, 26 September 1995.

McTaggart, Myrna (*née* Cliffe), 23 November 1997.

Maki, Roy, 25 September 2001.

Manly, Jim, 19 November 2000.

Marriott, Barbara (*née* Duncan), 20 December 1994, 17 January 1996, 4 October 1996, 13 October 1996, 17 November 1996, 17 February 2000, 23 February 2000, 10 March 2002.

Martin, Neil and Eleanor (*née* Burritt), 13 January 1996, 19 January 1996, 2 February 1996, 4 February 1996, 19 November 1996, 30 December 1997, 20 February 1998, 2 June 1999, 23 February 2000, 7 February 2001.

Masters, Ruth, 10 March 1994, 17 November 1996, 18 November 1996, 28 November 1996, 1 December 1996, 17 June 1999, 22 January 2000.

Mathers, Laurie, 23 May 2003, 8 June 2003.

Menzies, Audrey (*née* Grieve), 3 June 1996.

Michel, Chris, 28 December 2005.

Moncrief, Bill "Bronco" and Kay, 6 December 2002.

Morrison, Malcolm, 10 December 2003.

Muckle, Jim, 29 January 1996.

Muir, Bob and Cathy, 19 July 2006, 25 January 2007, 27 October 2008, 3 November 2008.

Murphitt, Peter, 10 January 2006.

Murtsell, Eddy, 6 February 1999, 23 February 2000.

Newson, Margaret, 21 January 2003, 31 March 2003.

Niemi, Arfie, 8 June 2002.

Norman, Al, 25 November 2002.

Norman, Rob, December 2003, 9 July 2006.

Oven, Tony and Norma (*née* Sheppard), 4 November 2002, 8 December 2002.

Owen, Victor, January 2001, 9 October 2001, 30 October 2005.

Parkin, Fred, 8 January 2004, 27 February 2004, 29 March 2004, 2 April 2004, 19 March 2007.

Passmann, Emmy, 16 February 2004, 17 February 2004.

Pedersen, Buck, 30 March 2002, 6 May 2002, 19 June 2002, 30 January 2007.

Pelto, Mauno, 11 November 1996.

Pentland, Waring, 11 October 2003.

Pettersen, Oscar and Eleanor (*née* McQuillan), 2 March 2004.

Pidcock, Herbert, 24 October 1998, 25 October 1998, 23 March 1999, 29 March 1999, 20 June 1999.

Piikkila, Erik, 12 April 2002.

Poje, Maria (*née* Adelberger), 19 September 2002.

Poje, Ray, 23 October 2002.

Rainer, Jim and Dorene (*née* Brinham), 25 February 2001, 11 October 2003.

Schaff, Lynn (*née* Chalmers), March 2001, 31 May 2002.

Schiefke, Ron, 2 December 2006.

Shilton, Pat (*née* Burgess), 28 December 1997, 16 February 2003.

Smith, Diane, 9 April 2006.

Smith, Margaret (*née* Dann), 23 September 2001.

Solmie, John, 16 September 2006.

Spencer, Hope, 16 November 1997, 4 December 2000.

Stelfox, Bob, 14 June 2006.

Stewart, Alan, 30 November 2000.

Stockand, Ray, 13 October 1996, 29 June 1998, 6 September 1998.

Storr, Don, 8 December 2001.

Stroulger, Charlie and Trudy, December 2000, 2 January 2001.

Surgenor, Tyke, 14 December 1997.

Taillefer, Marcel, 15 December 2002, 15 February 2003, 16 February 2003.

Tait, Joanna (*née* Smith), March 2007.

Taylor, Oscar and Laura (*née* Padula), 7 July 2006.

Telosky, Sam and Doreen (*née* Hahn), 31 December 1997, 18 March 1998, 4 December 1998, 19 November 2002, 7 April 2003, 15 January 2004, 31 January 2004, 28 April 2004, 14 October 2005, 28 January 2006, 23 February 2006, 5 September 2006, 28 October 2006.

Thran, Dan, 8 December 2001.

Tobacco, Chuna, 23 February 1996.

Trotter, Harry, 19 January 1996.

Tukham, Jack and Dorothy (*née* Duncan), 9 November 1996, 22 May 2003, 21 February 2004.

Turner, Tony and Mary, 9 November 1996, 20 April 1998, 22 June 1998, 29 February 2000.

Ulrich, Madelin, 26 April 2001, 23 September 2001, 4 March 2002.

Verchère, Bill, 5 March 2002.

Verchère, Sarah (*née* McNeil), 5 March 2002.

Vogel, Harry, 20 October 2001.

Vroom, Isabel (*née* Flinton), 5 November 2008.

Walker, Arnie, 25 February 2006.

Walker, Doris (*née* Marvin), 8 December 1997.

Ware, Jack Jr., 24 February 2001, 24 June 2001, 30 October 2001, 3 March 2002.

Weaver, Robert, 9 November 1996.

Westwood, Don and Marie (*née* Davis), 5 March 2002, 12 March 2002.

White, Archie, 15 May 2003.

White, Elwood, 11 May 2000, 24 February 2002, November 2006.

Whitehead, Warrick, 25 November 2008.

Whyte, Bill, 15 April 2003.

Whyte, Bryson and Terry (*née* Taber), 7 November 2002, 15 April 2003, 10 December 2003.

Williams, Patricia (*née* Holmes), 23 June 1997, 30 July 1998, 22 January 2000, 16 September 2006, 18 September 2006.

Wilson, Bill and Betty (*née* Lowe), 10 November 1996, 18 April 1998, 17 November 1998, 27 May 2002.

Wilson, Walter, 28 November 1996.

Woods, Bill and Rita, 4 October 1996, 7 September 1998, 8 January 2000, 11 January 2000, 26 September 2000, 18 November 2001, 1 December 2001, 7 December 2001, 7 April 2002, 8 June 2002.

Woods, Bud (Sidney Richard), 23 February 2001.

Yates, George Jr. and Amelia (*née* Williams), 8 October 1996, 30 November 1996, 11 December 1996, 3 March 2004, 14 June 2006, 8 November 2007.

Correspondence with the author

Anderson, Walter, 17 February 2000.

Annand, Betty (*née* White), 21 March 2001, 4 November 2002 (both e-mail).

Barnett, Ruth (*née* Pidcock), 19 April 1995.

Barr, Jennifer, 17 March 2008 (e-mail).

Berkeley, Gary, 30 October 2002, 14 November 2002, 18 November 2002, 19 January 2003 (all e-mail).

Berkeley, Jean (*née* Crockett), 25 April 2003.

Biss, Frank, 10 November 2003, 19 November 2003, 20 November 2003 (all e-mail).

Blackburn, Fred, 25 January 2000, December 2005, October 2006.

Blackburn, Gordon, 5 March 2003.

Brown, Bob, 29 November 2001, 16 April 2002.

Cessford, Ab, November 2005, 5 January 2006.

Cessford, Wayne, 23 November 2002 (e-mail).

Clark, Noëlle, 6 December 2002 (e-mail), 17 December 2002 (letter), 26 December 2002 (e-mail), 3 February 2004 (letter).

Davison, Frank, December 2002, April 2003, 2 May 2003, 2 June 2003, 22 January 2004, 4 March 2004, July 2006, April 2007.

Ell, Sven, 11 January 2000.

Forsman, Michael, 11 June 2000 (e-mail), 12 November 2002 (e-mail), 7 January 2003 (letter), 16 February 2003 (letter), 20 October 2008 (e-mail).

Frame, Jody, 15 June 2002 (letter), 13 May 2003 (e-mail).

Gall, George, 28 March 2003.

Grange, Bill, 28 October 2005 (e-mail).

Gray, Jackie (*née* Gullett), 16 March 1998.

Grieve, Loggie, 16 January 2004.

Halme, Arlene, 3 October 2002, 19 November 2002, 26 November 2002 (all e-mail).

Hatfield, John, 3 February 2001.

Helm, Elmer, 31 October 2002, 15 December 2002.

Helm, Eugene, 31 December 2001.

Hemstad, Betty Anne (*née* Wicklund), 30 May 2003, 4 November 2003.

Heron, Myrtle (*née* McKenzie), 25 November 2005, 14 September 2006.

Hindmarch, Nancy (*née* Lyttback), 12 October 2006, 26 October 2006.

Hodgins, Jack, 27 January 2002, 22 October 2002, 6 December 2005 (all e-mail).

Holgate, Marguerite (*née* Rennie), 4 January 2000, 3 February 2001, November 2001, 9 October 2002.

Hunter, Jinny, 17 November 2008 (e-mail).

Irwin, Gaileen (*née* Pottage), 30 November 2008 (e-mail).

Irwin, Rich, 30 November 2008 (e-mail).

James, Rick, 1 October 2005, February 2007, March 2007.

Jennings, Bob, 16 December 1999, 3 January 2001, 8 June 2001, 30 June 2001, 17 April 2002.

Knappett, Al, 28 January 2004.

McLoughlin, Shirley, 7 January 2006.

McPhee, Nan (*née* Grant), 22 December 1997, 29 October 2002.

McQuillan, Darrell, 20 January 2004.

McQuinn, Dick, 25 July 2001, 14 November 2002 (both e-mail).

Malbon, Marie (*née* Miner), 28 November 2003, 3 December 2003 (both e-mail).

Marsden, Tom and Marilyn, October 2005.

Martin, Neil, 16 January 1998, 29 January 1998, 8 December 2000, 21 November 2001, 25 March 2003.

Masters, Ruth, 19 April 2007.

Miller, John, 19 May 2000 (e-mail).

Moncrief, Bill "Bronco," 31 August 2002, 28 October 2002.

Muir, Bob and Cathy, 22 April 2004, 24 November 2005.

Musgrave, Philip, 25 January 2002, 27 June 2002 (both e-mail).

Niemi, Arfie, December 2002, 3 March 2003, 28 April 2003.

Nordin, Charlie and Harry, May 1999.

Oven, Norma (*née* Sheppard), 4 November 2002, 1 December 2002.

Parminter, John, 20 December 2000, 7 July 2001, 21 July 2001, 12 September 2003 (all e-mail).

Pedersen, Buck, 3 April 2007.

Pentland, Waring, 27 July 2003, 13 December 2003.

Pettersen, Eleanor (*née* McQuillan), 20 December 2005.

Piikkila, Erik, 20 December 2000, 26 April 2002, 29 June 2007, 8 October 2008 (all e-mail).

Pritchard, Allan, 28 June 2000, 2 November 2002, 15 January 2003, 21 July 2006.

Rainer, Jim, 22 October 2007 (e-mail).

Rogers, Tom and Joy (*née* Robertson), 10 April 2002, 9 May 2002.

Schaff, Lynn (*née* Chalmers), 8 July 2002.

Shackleton, Leona (*née* Spencer), 29 March 2003.

Sheasgreen, Joe, 16 October 2002.

Sjuberg, Ernie, 30 April 1999.

Spearing, Bob, 22 April 2002, 5 June 2002.

Storr, Don, 7 January 2004 (e-mail).

Taylor, Jeanette, 2 September 2004.

Telosky, Sam and Doreen (*née* Hahn), 12 March 2001, January 2004, April 2004.

Trofymow, Tony, 1 October 2007.

Tukham, Dorothy (*née* Duncan), 8 March 2004.

Twamley, Gloria (*née* Berkeley), 19 November 2002, 21 November 2002 (both e-mail).

Ulrich, Madelin, 15 September 2001.

Verchère, Bill, 1 March 2002 (e-mail).

Vogel, Harry, 1 November 2001, 15 April 2002.

Ware, Jack Jr., 8 March 2001, 28 March 2001, 27 June 2001, 30 August 2001, 11 September 2001, 23 October 2001, 30 October 2001.

Williams, Patricia (*née* Holmes), October 1996, 11 October 1996, 1 November 1996, 24 November 1996, 6 December 1996, 30 January 2000, 20 March 2001.

Winger, Robert, 14 November 2002.

Woods, Bud (Sidney Richard), February 2006.

Yates, George Jr., March 2004.

Other unpublished sources

Anderson, Walter. Photographs. Private collection of family.

Anon. "Headquarters, The Old Townsite." [*c.* 1960.] TS. Private collection of Len Burchell.

Baikie, Wallace. "Seven Years with the Comox Logging Company." Undated TS. Museum at Campbell River.

Barnett, Ruth (*née* Pidcock). "Edwardian Memoirs: Mrs. Janetta Dundas." Undated TS. Private collection of Rick James.

Berkeley, Bob. Photographs. Private collection.

Berkeley, Eddy. Papers and photographs. Private collection of Gloria Twamley (*née* Berkeley).

Berkeley, Jack and Jean (*née* Crockett). Interview by Don Berkeley. 14 April 1995. VHS videotape. Private collection of family.

Berkeley, Jean. Photographs. Private collection of Gloria Twamley (*née* Berkeley).

Biss, Frank. Photographs. Private collection.

Blackburn, Fred. Papers and photographs. Private collection.

Brown, Nancy (*née* Thomson). Photographs. Private collection.

Bullock, Roger. Photographs. Courtenay & District Museum & Archives.

Burchell, Len. Photographs. Private collection.

Cameron, Karl. Photographs. Private collection.

"Canadian Collieries (Dunsmuir), Limited. Wellington Colliery Railway Company and Comox Logging and Railway Company. Lease dated 1 April 1927." Buckham Collection, British Columbia Archives.

Cardriver, D.K. [Keith]. "Timberdown: Adventures in the Americas." Winter 2004/05. TS. Private collection.

Cartwright, Helena. "An Outline of Logging in the Comox Valley." In Isenor, "The Comox Valley."

Churchill, Wally. Papers. Private collection.

Clark, Noëlle, and Roger Emery. Papers and photographs. Private collection.

Cliffe, Inez (*née* Ledgerwood). Photographs. Private collection.

Cliffe, Joe. "Data on Steam Locomotives." 22 September 1980. TS. Courtenay & District Museum & Archives.

———. "Evolution of Logging, or One Man's Life in Logging on the West Coast." November 1976. TS. Courtenay & District Museum & Archives.

———. "The Filberg I Knew." [*c*. 1980.] Courtenay & District Museum & Archives.

———. Interview by Evelyn Henderson. 31 July 1980. Transcript. Courtenay & District Museum & Archives.

Cliffe, Lorna. Papers and photographs. Private collection; also Courtenay & District Museum & Archives.

Clifford, Fred and Olga (*née* Blaseckie). Photographs. Private collection.

Comox Logging & Railway Company. Company records. [*c*. 1910–55.] TimberWest office, Courtenay.

———. Company records. [*c*. 1910–55.] Courtenay & District Museum & Archives.

———. Employment and employee registration records. [*c*. 1918–55.] TimberWest head office, Vancouver; to be donated to Courtenay & District Museum & Archives.

Crown Zellerbach Canada. "How Our Trees Serve Us" [publicity kit]. [*c*. 1966]. Private collection of the author.

Currie, Phyllis (*née* Radford). Papers and photographs. Private collection of family.

Davis, Ray. Photographs. Private collection.

Davison, Frank. Papers and photographs. Private collection.

Downey, Dick. Photographs. Private collection.

Duncan, Eric. Newspaper cuttings and scrapbook. [*c*. 1927–36.] Private collection of family.

Eastman, Nola (*née* Chalmers). Papers and photographs. Private collection.

Edwards, Donna (*née* Piercy). Photographs. Private collection.

Erickson, Harry. Photographs. Private collection.

Filberg, Bob. Interview by C.D. Orchard. 14 June 1960. Transcript. University of British Columbia Special Collections.

———. Letter to Jack McQuinn. 17 December 1933. Private collection of Dick McQuinn.

———. Photographs. [*c*. 1938–1943.] Filberg Heritage Lodge & Park Association, Comox.

Filberg Heritage Lodge & Park Association. Notes and photographs relating to the Comox Logging & Railway Company. [*c*. 1920–55.] Filberg Heritage Lodge & Park Association, Comox.

Forsman, Michael. Photographs. Private collection.

Greenwood, Gwen (*née* Walker). Photographs. Private collection.

Halme, Arlene. Papers. Private collection.

Hanham, Stan. Photographs. Private collection of family.

Hansen, Cora (*née* Cliffe). "Comox Schools." Undated TS. Private collection of Gale Hansen.

Harding, Rene. "The Story of Logging in the Comox Valley." In Isenor, "The Comox Valley."

Harris, Bert. Photographs. Private collection.

Herbert, Madge (*née* Steere-Clark). Papers and photographs. Private collection.

Heron, Myrtle (*née* McKenzie). Papers and photographs. Private collection.

Hilton, Arthur. "The Story of the Comox Logging Company" [as told to Rene Harding in 1954]. In Isenor, "The Comox Valley."

Hind, Babs. Photographs. Private collection.

Hindmarch, Nancy (*née* Lyttback). Photographs. Private collection.

Holgate, Marguerite (*née* Rennie). "Miss Helena Cartwright (Nellie) 1900–1993." 1996. TS. Private collection.

Holmes, Chris. "Comox Lake Camp" [guest book and scrapbook for Comox Logging & Railway Company]. 1933–44. Courtenay & District Museum & Archives.

Holmes, John. Papers and photographs. Private collection.

Holmes, Muriel. Papers and photographs. Private collection; also Comox Archives & Museum Society.

Isenor, D.E., ed. "The Comox Valley: Its Pioneers. Parts 1–8." 1962. TS. Buckham Collection, British Columbia Archives.

Jennings, Bob. Papers and photographs. Private collection.

Keller, Eleanor (*née* Richter). Photographs. Private collection.

Knight, Ray. Papers and photographs. Private collection of family.

Laing, Hamilton Mack. "The Slashing." In book MS entitled "Rural Felicity." 1922. H.M. Laing Papers, British Columbia Archives.

McKelvie, B.A. "The 'S' Sign. A Biographical Sketch of Eustace Smith, Logger, Timber Cruiser and Forest Engineer." Undated TS. Courtenay & District Museum & Archives.

McKenzie, Wilma. Photographs. Comox Archives & Museum Society.

McQuinn, Dick. Papers and photographs. Private collection.

Marsden, Tom. Photographs. Private collection.

Masters, Ruth. "Lest We Forget" [scrapbook of Comox Valley men and the Second World War]. Compiled 1978. Courtenay & District Museum & Archives.

Mathers, Laurie. Photographs. Private collection.

Mooney, Trudy. Photographs. Private collection.

Muir, Bob. Papers and photographs. Private collection.

Niemi, Arfie. Photographs. Private collection.

Nordin, Charlie Jr. Papers and photographs. Private collection.

Norman, Al. Photographs. Private collection of Rob Norman.

Norman, Rob. Photographs. Private collection.

Oven, Norma (*née* Sheppard). Papers and photographs. Private collection.

Owen, Victor. Photographs. Private collection.

Parkin, Fred. Photographs. Private collection.

Parkin, Norman. Interview notes made by Martin Hagarty. [*c.* 1984.] Filberg Heritage Lodge & Park Association, Comox.

Pedersen, Buck. Photographs. Private collection of Dorn Pedersen.

Pelto, Babs (*née* Burnett). Photographs (including the collection of Joe Cliffe). Courtenay & District Museum & Archives.

Pidcock, Reg. Interview notes made by Martin Hagarty. [*c.* 1984.] Filberg Heritage Lodge & Park Association, Comox.

Poje, Ray. Papers and photographs. Private collection.

Rogers, Fred. Papers. Private collection.

Ross, Brian. Photographs. Private collection.

Schiefke, Ron. Papers. Private collection.

Shackleton, Leona (*née* Spencer). Photographs. Private collection.

Shilton, Pat (*née* Burgess). Photographs. Private collection.

Smith, Diane. Papers. Private collection.

Smith, Margaret (*née* Dann). Papers. Private collection.

Steere-Clark, Major Harry. "Biographical Sketch. Major Harry Steere-Clark. Safety Director, British Columbia Loggers' Association." [*c.* 1927.] TS. Private collection of Noëlle Clark.

Stelfox, Bob. Photographs. Private collection.

Stewart, Florence M.J. (*née* Cliffe). "The History of the Cliffe Family in the Comox Valley (1862–1979)." Undated TS. Private collection of Inez Cliffe.

Stewart, John. "Early Days at Fraser Mills, B.C. From 1889 to 1912." Undated TS. Private collection of Ron Schiefke.

Stroulger, Charlie. Photographs. Private collection of Gertrude Stroulger.

Taillefer, Marcel. Photographs. Private collection.

Tait, Joanna (*née* Smith). Papers. Private collection.

Telosky, Sam. Photographs. Private collection.

Tukham, Dorothy (*née* Duncan). Papers and photographs. Private collection.

Verchère, Bill. Photographs. Private collection.

Ware, Jack Jr. Papers and photographs. Private collection of family.

White, Elwood. Photographs. Private collection.

Whyte, Bryson. Papers and photographs. Private collection.

Williams, Patricia (*née* Holmes). Photographs. Private collection.

Yates, George Jr. Papers and photographs. Private collection.

Selected Newspapers and Periodicals

Circle F Newsletter (Canadian Western Lumber Company), 1955–56

Comox Argus, 1919–55

Comox District Free Press, 1960–90

Crown Zellerbach News, 1957–81

The Timberman, 1925–46, 1961

West Coast Lumberman, 1925–46

Published Sources

Annand, Betty. *Voices from Bevan*. Pitt Meadows, BC: Rising Sun Publications, 2002.

Baikie, Harper. *A Boy and His Axe*. Campbell River, BC: Harper Baikie, 1991.

Baikie, Wallace. *Rolling with the Times*. Campbell River, BC: Kask Graphics, 1985.

———. *Strathcona: A History of British Columbia's First Provincial Park*. With Rosemary Phillips. Campbell River, BC: Ptarmigan Press, 1986.

Baptie, Sue. *First Growth: The Story of British Columbia Forest Products Limited*. Vancouver: British Columbia Forest Products, 1975.

Bosher, J.F. "Vancouver Island in the Empire." *Journal of Imperial and Commonwealth History* 33, no. 3 (September 2005): 349–68.

Bowen, Lynne. *Boss Whistle: The Coal Miners of Vancouver Island*

Remember. Lantzville, BC: Oolichan Books, 1982.

Carney, Pat. "Bob Filberg—A Smiling, Swearing Logger." *Vancouver Sun*, 3 November 1967.

Cliffe, Joe. "Early Methods of Logging." In Isenor and others, *Land of Plenty*, 184–87.

Conway, Steve. *Logging Practices: Principles of Timber Harvesting Systems.* San Francisco: Miller Freeman Publications, 1976.

Coulson, Barry. *The Logger's Digest.* Vol. 1, *From Horses to Helicopters.* Victoria: Orca Books, 1992.

Cummings, Doug, Gary Oliver, and Keith Anderson. "Comox Logging and Railway Company: Railway History." *The Steamchest* 5, no. 6 (August 1962): 1–11.

Drushka, Ken. *Working in the Woods: A History of Logging on the West Coast.* Madeira Park, BC: Harbour Publishing, 1992.

Duncan, Eric. *From Shetland to Vancouver Island: Recollections of Seventy-eight Years.* Edinburgh: Oliver and Boyd, 1934.

———. "Winter Notes from Comox Valley." *Vancouver Province*, February 1932 [from Duncan's scrapbook, private collection of family].

Egan, Brian. *The Ecology of the Coastal Douglas-Fir Zone.* Victoria: Ministry of Forests, 1995.

———. *The Ecology of the Coastal Western Hemlock Zone.* Victoria: Ministry of Forests, 1995.

Englebert, Renny. "Monarchs of the Forest Fall to Men of Brawn and Skill." *Vancouver Sun Supplement*, 24 May 1947.

———. "Sixty-one Years in the Forests of British Columbia: The Story of Eustace Smith, Timber Cruiser." *Forest and Mill*, 13 April 1948, 3.

Farley, A.L. *Atlas of British Columbia. People, Environment, and Resource Use.* Vancouver: UBC Press, 1979.

Filberg, R.J. "In My Time: Filberg Recalls His Early Days in Logging on Coast." *The Truck Logger*, January 1968. Reprinted in W. Baikie, *Rolling with the Times*, 146–49.

Fleming, R.B. *The Railway King of Canada: Sir William Mackenzie, 1849–1923.* Vancouver: UBC Press, 1991.

Fossum, Jack. *Mancatcher: An Immigrant's Story of Logging, Policing and Pioneering in the Canadian West.* Comox, BC: Lindsay Press, 1990.

Garner, Joe. *Never Chop Your Rope.* Nanaimo, BC: Cinnabar Press, 1988.

Glover-Geidt, Janette. *The Friendly Port: A History of Union Bay, 1880–1960.* Campbell River, BC: Kask Graphics, 1990.

Gold, Wilmer. *Logging as It Was: A Pictorial History of Logging on Vancouver Island.* Victoria: Morriss Printing, 1985.

———. "Logging the Way It Was." *Times Colonist* (Victoria) *Islander*, 7 October 1979.

Gough, John. *The Story of British Columbia.* Toronto: J.M. Dent & Sons, 1952.

Gould, Ed. *Logging: British Columbia's Logging History.* Saanichton, BC: Hancock House, 1975.

Grainger, Martin Allerdale. *Woodsmen of the West.* Victoria: Horsdal & Schubart, 1994 [originally published 1908].

Great Central Book Committee. *When the Whistle Blew: The Great Central Story, 1925–1952.* Duncan, BC: Firgrove Publishing, 2002.

Green, Mervyn T. *British Columbia Industrial Locomotives.* Richmond, BC: M.T. Green, 1986.

Griffiths, Bus. *Now You're Logging.* Madeira Park, BC: Harbour Publishing, 1978.

Haig-Brown, Roderick. *Timber: A Novel of Pacific Coast Loggers.* New York: William Morrow, 1942.

Harding, Rene. "History of Lumbering in the Comox Valley." *Comox Argus*, 9 February 1955.

Harris, Cole. "Industry and the Good Life around Idaho Peak." Chap. 7 in *The Resettlement of British Columbia: Essays on Colonialism and Geographical Change.* Vancouver: UBC Press, 1997.

———. "The Pattern of Early Canada." *The Canadian Geographer* 31, no. 4 (1987): 290–98.

———. "The Struggle with Distance." Chap. 6 in *The Resettlement of British Columbia: Essays on Colonialism and Geographical Change.* Vancouver: UBC Press, 1997.

Hodgins, Reta Blakely, ed. *Merville and Its Early Settlers 1919–1985.* Campbell River, BC: Kask Graphics, 1985.

Hughes, Ben. *History of the Comox Valley 1862–1945.* Nanaimo, BC: Evergreen Press, [*c.* 1962].

Isenor, D[ick] E., Margaret McGill, and Donna Watson,

eds. *"For Our Children": A History of Comox Valley Schools.* Courtenay, BC: School Board District #71, [*c.* 1980].

Isenor, D.E., W.N. McInnis, E.G. Stephens, and D.E. Watson, eds. *Land of Plenty: A History of the Comox District.* Campbell River, BC: Ptarmigan Press, 1987.

Isenor, D.E., E.G. Stephens, and D.E. Watson, eds. *Edge of Discovery: A History of the Campbell River District.* Campbell River, BC: Ptarmigan Press, 1989.

———, eds. *One Hundred Spirited Years: A History of Cumberland, 1888–1988.* Campbell River, BC: Ptarmigan Press, 1988.

James, Rick. "Logging Train." *Times Colonist* (Victoria) *Islander,* 20 July 1997.

Johnstone, Bill. *Coal Dust in My Blood: The Autobiography of a Coal Miner.* Lantzville, BC: Oolichan Books, 2000.

——— [as William Johnstone]. "The E&N Land Grant." *Times Colonist* (Victoria) *Islander,* 23 March 1980.

Keenleyside, Hugh. *Memoirs of Hugh L. Keenleyside.* Vol. 1, *Hammer the Golden Day.* Toronto: McClelland & Stewart, 1981.

Kipling, Rudyard. "Mountains and the Pacific." In *Letters of Travel 1892 to 1913.* London: Macmillan, 1920.

Knight, Vera. *Air Raid Precautions for Canada.* Toronto: Macmillan, 1942.

Lembcke, Jerry, and William M. Tattam. *One Union in Wood: A Political History of the International Woodworkers of America.* Madeira Park, BC: Harbour Publishing, 1984.

Lyons, Don. "Tops in the Woods" [based in part on interviews with Mauno Pelto]. *Nanaimo Times,* 30 May 1985.

McCann, L.D., ed. *Heartland and Hinterland: A Geography of Canada.* 2nd ed. Scarborough, ON: Prentice Hall Canada, 1987.

McEwan, Ken. *International Woodworkers of America AFL – CIO – CLC: Local 1-80 Duncan, BC.* Duncan, BC: Duncan Print-Craft, 1982.

McGillivray, Brett. *Geography of British Columbia: People and Landscapes in Transition.* Vancouver: UBC Press, 2005.

MacKay, Donald. *Empire of Wood: The MacMillan Bloedel Story.* Vancouver: Douglas & McIntyre, 1983.

———. *The Lumberjacks.* Toronto: McGraw-Hill Ryerson, 1978.

McKenzie, Rose, and Jean Feely. Untitled contribution. In Isenor and others, *"For Our Children,"* 161–62.

Mackie, Richard Somerset. "'The Big Gyppo': Comox Logging at Ladysmith, 1936–1984." In *Ladysmith 100 Years: 1904–2004,* edited by Marina Sacht, 25–29. Ladysmith, BC: Ladysmith Centennial Committee, 2004.

———. "The Colonization of Vancouver Island, 1849–1858." *BC Studies* 96 (Winter 1992/93): 3–40.

———. "Cougars, Colonists, and the Rural Settlement of Vancouver Island." In *Beyond City Limits: Rural History in British Columbia,* edited by Ruth Sandwell, 120–42. Vancouver: UBC Press, 1999.

———. *Hamilton Mack Laing: Hunter-Naturalist.* Victoria: Sono Nis Press, 1985.

———. *Island Timber: A Social History of the Comox Logging Company, Vancouver Island.* Victoria: Sono Nis Press, 2000.

———. *The Wilderness Profound: Victorian Life on the Gulf of Georgia.* Victoria: Sono Nis Press, 1995.

McLeod, Vikie, and others. *Comox Golf Club's "First Fifty."* Courtenay, BC: Cartwright Printing, 1984.

McNight, George. *Sawlogs on Steel Rails: A Story of the 45 Years of Railway Operations in the Logging Camps of the Port Alberni Area.* Port Alberni, BC: Port Alberni Seniors' History Committee, 1995.

Mayse, Arthur. Untitled column. *Times Colonist* (Victoria), 29 September 1985.

Mayse, Susan. *Ginger: The Life and Death of Albert Goodwin.* Madeira Park, BC: Harbour Publishing. 1990.

Meade, Don. "Roaming B.C. with a Camera: W.F. Montgomery Has Half-Century Record." *Vancouver Sun,* 5 February 1955, magazine section.

Murray, Peter, ed. *Riding the Skyline* [a collection of articles by Martin Grainger]. Victoria: Horsdal & Schubart, 1994.

Neufeld, Andrew, and Andrew Parnaby. *The IWA in Canada: The Life and Times of an Industrial Union.* Vancouver: IWA Canada / New Star Books, 2000.

Nikiforuk, Ted, and others. *Coquitlam 100 Years: Reflections of the Past.* Coquitlam, BC: District of Coquitlam, 1990.

Ormsby, Margaret A. *British Columbia: A History.* Toronto: Macmillan, 1958.

Pierre, Joseph H. *When Timber Stood Tall.* Seattle: Superior Publishing, 1979.

Pojar, Jim, Karel Klinka, and Dennis A. Demarchi. "Coastal Western Hemlock Zone." In *Ecosystems of British Columbia*, edited by Del Meidinger and Jim Pojar, 95–111. Victoria: Ministry of Forests, 1991.

Pritchard, Allan. "The Royal Navy and the Comox Settlement." *BC History* 40, no. 2 (2007): 20–27.

Rajala, Richard A. *The Legacy and the Challenge: A Century of the Forest Industry at Cowichan Lake.* Victoria: Beach Holme Publishers, 1993.

———. *Up-Coast: Forests and Industry on British Columbia's North Coast, 1870–2005.* Victoria: Royal British Columbia Museum, 2006.

Reksten, Terry. *The Dunsmuir Saga.* Vancouver: Douglas & McIntyre, 1991.

Rushton, Gerald. *Echoes of the Whistle: An Illustrated History of the Union Steamship Company.* Vancouver: Douglas & McIntyre, 1980.

Schulz, Henry. *A New Frontier: The Canadian Chronicles of Henry Schulz.* Campbell River, BC: Ptarmigan Press, 1984.

Sirois, James. *Afloat in Time.* Surrey, BC: Hancock House, 1999.

Skelton, Robin. *Landmarks.* Victoria: Sono Nis Press, 1979.

Smith, Eustace. *British Columbia Forests: Rambling Notes and General Remarks Regarding British Columbia Forests.* Vancouver: G.A. Roedde, 1944.

Sproule, Gwyn. *Whyte's Bay Days. The Memoirs of Mavis Wall.* Courtenay, BC: Percolator Press, 2007.

Stubbs, Dorothy I. *"All About Us": A History of the City of Courtenay.* Courtenay, BC: Glacier Press, 1975.

Swanson, Robert E. *Bunkhouse Ballads.* Toronto: Thomas Allen, 1945.

———. *Rhymes of a Western Logger: A Book of Verse.* Vancouver: Lumberman Printing Company, 1942.

———. *Whistle Punks and Widow-Makers: Tales of the B.C. Woods.* Madeira Park, BC: Harbour Publishing, 1993.

Taylor, G.W. *Timber: History of the Forest Industry in B.C.* Vancouver: J.J. Douglas, 1975.

Taylor, Jeanette. *River City: A History of Campbell River and the Discovery Islands.* Madeira Park, BC: Harbour Publishing, 1999.

Turner, Robert D. *Logging by Rail: The British Columbia Story.* Victoria: Sono Nis Press, 1990.

———. *Vancouver Island Railroads.* 2nd ed. Victoria: Sono Nis Press, 1997.

Victoria Heritage Foundation. *This Old House: Victoria's Heritage Neighbourhoods.* Vol. 1, *Fernwood and Victoria West.* Victoria: Victoria Heritage Foundation for the City of Victoria, 2008.

Ware, Reuben. "Land and a New Life: British Columbia's Land Settlement Board, 1918–1925." *British Columbia Historical News*, Spring 1982, 13–16.

Weatherby, Hugh. "Comox's Old Counting Dog Still Barks up to Five." *Forest and Mill*, 14 May 1948, 2.

———. "Safety Is His Business." *Vancouver Sun Supplement*, July 1948 [from "Logging" file of clippings, Courtenay & District Museum & Archives].

Wellwood, Vida. "The Smith Saga: Progress in Lumbering over the Years" [based on notes compiled by the late Eustace Smith]. *Comox District Free Press*, 16 August 1967.

Whitford, H.N., and R.D. Craig. *Forests of British Columbia.* Ottawa: Commission of Conservation, 1918.

Workers' Compensation Board of British Columbia. *Yarding and Loading Handbook.* Vancouver: Workers' Compensation Board of British Columbia, 1980.

Wynn, Graeme. *Timber Colony: A Historical Geography of Early Nineteenth Century New Brunswick.* Toronto: University of Toronto Press, 1981.

index